MW00490004

The History of The Standard Oil Company

In Two Volumes

Vol. II

IDA M. TARBELL

INTRODUCTION BY DANNY SCHECHTER

NEW YORK

The History of The Standard Oil Company, Vol. II (in two volumes) was originally published in 1904.

For information, address:
P.O. Box 416, Old Chelsea Station
New York, NY 10011

or visit our website at:
www.cosimobooks.com

Ordering Information:
Cosimo publications are available at online bookstores. They may also be purchased for educational, business or promotional use:
- *Bulk orders:* special discounts are available on bulk orders for reading groups, organizations, businesses, and others. For details contact Cosimo Special Sales at the address above or at info@cosimobooks.com.
- *Custom-label orders:* we can prepare selected books with your cover or logo of choice. For more information, please contact Cosimo at info@cosimobooks.com.

Cover Design by www.popshopstudio.com

ISBN: 978-1-60520-763-6

There was no characteristic of Mr. Rockefeller and his great corporation which from the beginning had been more exasperating to the oil world than the secrecy with which operations were conducted. The plan of the South Improvement Company had only been revealed to those who signed an agreement to keep secret all transactions they might have with it. The purchase in 1874 and 1875 by the Standard Oil Company of Lockhart, Frew and Company of Pittsburgh, of Warden, Frew and Company of Philadelphia, and of Charles Pratt and Company of New York was so thoroughly concealed that Mr. Rockefeller, five years after it occurred, dared make an affidavit that it had never occurred!

—from Chapter Fourteen: "The Breaking Up of the Trust"

JOHN D. ROCKEFELLER

A sketch from life by George Varian, made in Cleveland, October, 1903

CONTENTS

CHAPTER NINE

THE FIGHT FOR THE SEABOARD PIPE-LINE

CHAPTER TEN

CUTTING TO KILL

CHAPTER ELEVEN

THE WAR ON THE REBATE

CHAPTER TWELVE

THE BUFFALO CASE

CHAPTER THIRTEEN

THE STANDARD OIL COMPANY AND POLITICS

CHAPTER FOURTEEN

THE BREAKING UP OF THE TRUST

CONTENTS

CHAPTER FIFTEEN

A MODERN WAR FOR INDEPENDENCE

CHAPTER SIXTEEN

THE PRICE OF OIL

CHAPTER SEVENTEEN

THE LEGITIMATE GREATNESS OF THE STANDARD OIL COMPANY

CHAPTER EIGHTEEN

CONCLUSION

LIST OF ILLUSTRATIONS

LIST OF ILLUSTRATIONS

LIST OF ILLUSTRATIONS

THE HISTORY OF
THE STANDARD OIL COMPANY

CHAPTER NINE

THE FIGHT FOR THE SEABOARD PIPE-LINE

PROJECT FOR SEABOARD PIPE-LINE PUSHED BY INDEPENDENTS—TIDEWATER PIPE COMPANY FORMED—OIL PUMPED OVER MOUNTAINS FOR THE FIRST TIME—INDEPENDENT REFINERS READY TO UNITE WITH TIDEWATER BECAUSE IT PROMISES TO FREE THEM FROM RAILROADS—THE STANDARD FACE TO FACE WITH A NEW PROBLEM—DAY OF THE RAILROADS OVER AS LONG DISTANCE TRANSPORTERS OF OIL—NATIONAL TRANSIT COMPANY FORMED—WAR ON THE TIDEWATER BEGUN—PLAN TO WRECK ITS CREDIT AND BUY IT IN—ROCKEFELLER BUYS A THIRD OF THE TIDEWATER'S STOCK—THE STANDARD AND TIDEWATER BECOME ALLIES—NATIONAL TRANSIT COMPANY NOW CONTROLS ALL PIPE-LINES—AGREEMENT ENTERED INTO WITH PENNSYLVANIA RAILROAD TO DIVIDE THE BUSINESS OF TRANSPORTING OIL.

THE project for a seaboard pipe-line to be built by the producers and to be kept independent of Standard capital and direction had been pushed with amazing energy. Early in the fall of 1878 General Haupt reported that his right of way was complete from the Allegheny River to Baltimore; contracts were let for the telegraph line and preparation begun to lay the pipe. Before much actual work had been done it became clear to the company that it was not from the Butler oil field but from that of Bradford that a seaboard pipe-line should run; that the former field was showing signs of exhaustion, while the latter was evidently going to yield abundantly. With a promptness which would have done credit to Mr. Rockefeller himself, Messrs. Benson, Hopkins and McKelvy changed their plan.

The new idea was to lay a six-inch line from Rixford, in the Bradford field, to Williamsport, on the Reading Railroad, a distance of 109 miles. The Reading, not having had so far any oil freight, was happy to enter into a contract with them to run oil to both Philadelphia and New York until they could get through to the seaboard themselves. In November, 1878, a limited partnership, called the Tidewater Pipe Company, was organised with a capital of $625,000 to carry out the scheme. Many of the best known producers of the Oil Regions took stock in the company, the largest stockholders being A. A. Sumner and B. D. Benson.*

The first work was to get a right of way. The company went at the work with secrecy and despatch. Its first move was to buy from the Equitable Pipe Line, the second independent effort to which, as we have seen, the Producers' Union lent its support in 1878, a short line it had built, and a portion of a right of way eastward which Colonel Potts had been quietly trying to secure. This was a good start, and the chief engineer, B. F. Warren, pushed his way forward to Williamsport near the line which Colonel Potts had projected. The Standard, intent on stopping them, and indeed on putting an end to all future ventures of this sort, set out at once to get what was called a "dead line" across the state. This was an exclusive right for pipe-line purposes from the northern to the southern boundary of Pennsylvania. As there was no free pipe-line bill in those days, this "dead line," if it had been complete, would have been an effectual barrier to the Tidewater. Much money was spent in this sordid business, but they never succeeded in completing a line. The Tidewater, after a little delay, found a gap not far from where it wanted to cross, and soon had pushed itself through to Williamsport. With the actual laying of the pipe there was no interference which proved serious, though the railroads frequently held back

* See Appendix, Number 37. Articles of incorporation of the Tidewater Pipe Line.

ALANSON A. SUMNER

Prominent supporter of the Tide-water Pipe Company, still active in its counsels.

HENRY HARLEY

President of the Pennsylvania Transportation Company. Projector of the first seaboard pipe line.

SAMUEL VAN SYCKEL

The first successful pipe line for gathering and transporting oil was completed by Mr. Van Syckel in 1865.

GENERAL HERMAN HAUPT

Civil Engineer for the first and second pipe lines projected to the seaboard.

shipments of supplies. At Williamsport, where the pipe crossed under the railroad, it was torn out once. The Tidewater had no trouble in this case in getting an injunction which prevented further lawlessness.

By the end of May the company was ready for operation. The plant which they had constructed proposed to transport 10,000 barrels of oil a day over a distance of 109 miles. The apparatus for doing this consisted simply of tanks, pumps and pipes. At Coryville, on the edge of the Bradford field, two iron tanks, each holding 25,000 barrels of oil, were connected with an enormous pump of a new pattern devised by the Holly Company especially for this work. This pump, which was driven by an engine of seventy horse-power, was expected to force the oil through a six-inch pipe to a second station twenty-eight miles away and about 700 feet higher. Here a second pump took up the oil again, driving it to the summit of the Alleghanies, a few miles east. From this point the oil ran by gravitation to Williamsport.

It was announced that the pumps would be started on the morning of May 28. The experiment was watched with keenest interest. Up to that time oil had never been pumped over thirty miles, and no great elevation had been overcome. Here was a line 109 miles long, running over a mountain nearly 2,600 feet high. It was freely bet in the Oil Regions that the Tidewater would get nothing but a drizzle for its pains. However, oil men, Standard men, representatives of the Pennsylvania Railroad, newspaper men and natives gathered in numbers at the stations, and indeed all along the route, to watch the result.

The pump at station one was started by B. D. Benson, the president of the company. There were present with him several members of the concern, and to-day these men speak with emotion of the moment when Mr. Benson opened the valve to admit the oil to the pump. Would the great venture, on which they had staked all, be a success? Without a hitch the oil flowed in a full stream into the pipe and began its long journey over the mountains. It travelled about as fast as a man could walk and, as the pipe lay on the ground, the head of the stream could be located by the sound. Patrolmen followed the pipe the entire length watching for leaks. There was now and then a delay from the stopping of the pumps; but the cause was trivial enough, never anything worse than chips under the valves or clogging in the pipe by stones and bits of wood which the workmen had carelessly left in when joining the pipe. When the oil reached the second station there was general rejoicing; nevertheless, the steepest incline, the summit of the Alleghanies, had yet to be overcome. The oil went up to the top of the mountain without difficulty, and on June 4, the seventh day after Mr. Benson opened the valve at Station One, oil flowed into the big receiving tank beyond Williamsport. A new era had come in the oil business. Oil could be pumped over the mountains. It was only a matter of time when the Tidewater would pump to New York.

Once at the seaboard, the Tidewater had a large and sure outlet for its oil in the group of independent refiners left at the mercy of the Standard in the fall of 1877 by the downfall of the Empire Line. These refiners had most of them run the entire gamut of experiences forced on the trade by the railroads and the Standard. Take, for instance, the experience of Ayres, Lombard and Company, related by Josiah Lombard in 1879 in the Pennsylvania suits. They had gone into the business in 1869 in West Sixty-sixth street. At the beginning they had shipped principally over the Erie, sometimes as high as 50,000 barrels a month; but when that road came into the hands of Fisk and Gould those gentlemen began to try to build up a refining business in New York for their own friends. Edward Stokes was at that time hand in glove with Fisk; he had in the Oil Regions an able friend, Henry Harley. Harley bought and shipped the oil over the Erie; special rates were given him, and the Stokes refinery soon began to flourish at the expense of the former shippers of the Erie. Mr. Lombard finding, as he says, that there was no possibility of doing business with that road under the Fisk and Gould management, went over to the New York Central. Here he furnished his own cars. Ayres, Lombard and Company owned 100 cars on the Central in 1872, worth about $35,000, and in these they shipped the bulk of their oil. The South Improvement Company manœuvres in the spring of 1872 completely stopped their shipping over that road and in 1872 they sold their cars. Mr. Lombard said in his testimony: "We sold them (the cars) because the Standard Oil Company were getting the ascendency so much over the New York roads that we could not get a rate of freight from the lower districts and the Parker district, where the bulk of the oil was produced at that time, that would enable us to compete with them in the New York market, so there was no use in owning the cars."

Driven off the Erie and Central, the firm made a running arrangement with Mr. Rockefeller for a year; the Standard bought the cars and agreed to furnish Ayres, Lombard and Company crude oil for a certain price at a certain time, and take the refined oil from them at a fixed price. This contract was made probably under the Refiners' Association which Mr. Rockefeller succeeded in effecting in August, 1872, after the failure of the South Improvement Company, which association, as we have already seen, took in fully four-fifths of the refining interests of the country. The contract continued, Mr. Lombard said in testimony, for a year or more, and was then terminated by notice from the Standard Oil Company. Soon after the termination of the contract with the Standard, which was either late in 1873 or early in 1874 (Mr. Lombard was not able to decide this when he was under examination), the firm began shipping over the Pennsylvania road. They bought part of their oil at this time from Adnah Neyhart. Now, sometime in 1875, as we have seen, Mr. Neyhart began to feel the Standard pressure and his business was sold to the Standard. Again Ayres, Lombard and Company found a large part of their supply of oil cut off. For about a year they shipped over the Pennsylvania. It was not long, however, before the concern found that even on the Pennsylvania they were under a disadvantage, that road having made in 1875 discriminating contracts with the Standard. Again the firm changed, buying its oil from J. A. Bostwick and Company of New York. Now Bostwick was the Standard Oil buyer, one of the original South Improvement Company, and a stockholder in the Standard Oil Company. Mr. Lombard swore that he had not been taking oil of Bostwick for more than a year before the Standard began to draw its lines around him, as he put it, and again the question arose how were they to get oil for their refinery. There seemed no way but to try to make a contract with the Pennsylvania Company. On the 18th of May, 1877, he went to Philadelphia and saw Colonel Potts, who told him he would be glad to have his shipments on the Pennsylvania. Accordingly a contract was made for a year, the company guaranteeing them as low a rate as anybody else had. But this contract of Mr. Lombard was destined to end as speedily and as disastrously as all of those he had been making for over five years, for in the fall of the year the Empire Line was sold to the Standard, and in the spring of 1878, when Mr. Lombard's contract ran out, the Pennsylvania refused to renew it on the terms they gave the Standard. Mr. Lombard gave a very interesting account of the interview he and his fellow refiners of New York had with Mr. Cassatt in reference to this matter:

"In March, 1878, I think it was by appointment, we had an interview with Mr. Cassatt, third vice-president of the Pennsylvania Railroad. There were present Mr. Bush, Mr. Gregory, Mr. Burke, Mr. Ohlen, and myself, besides Mr. Cassatt. It was held in Mr. Bush's office, 123 Pearl street, New York. We sought that interview for the purpose of finding out what our position would be on the Pennsylvania Railroad after the termination of our contract with the Empire Line, which they had assumed. We had quite a plain talk on the subject. We began by telling Mr. Cassatt something that he already knew—that we for the past year had been probably the largest shippers over the Pennsylvania Railroad that they had had; largest shippers of petroleum. He acknowledged it, and we asked him if we should, after the first of May, be on the same footing and have as low a rate of freight as anybody else, which was guaranteed by contract up to that time. He said no, we would not. We asked him why not. Well, he said, it would not be satisfactory to the Standard Oil Company. I then put the question to him what difference it made to the Pennsylvania Railroad Company whether it was satisfactory to the Standard Oil Company or not. He said that the Standard Oil Company was the only party which could keep peace between the trunk lines. I said, It seems to me you have the matter very much in your own hands; there are but four of you; if you agree upon a certain rate of freight the oil is to come forward at, I see no use of the intervention of a third party or a fifth party in this case. He said, I cannot trust—or rather, he said, They are the only people that can keep harmony. Then we had a little discussion about the rates. He said that they had been bringing oil for the past year at a very low rate. I told him I understood it was a little over seventy cents an average on crude petroleum. He denied it, and said it was not. Then when we were talking about the subject of rates, he said of course the rates on petroleum were very profitable, and said we could find out the rate at which they could bring petroleum, if they were compelled to, by looking up their annual report, and seeing the cost a ton per mile, which was something like five or six mills per ton per mile, and which if we figured that it would be a very profitable business. We told him we did not object to him making a good profit at any time; all we wished was to have as low a rate of freight as anybody else had, which we could not get.

"He said we had better make an arrangement with the Standard and we would all of us make money, and that they had a very large business and proposed to make money, and the discrimination would be so light against us that we would hardly notice it, and we formed the idea from what he said. We asked him whether the discrimination against us would be larger if the rate of freight were high than it would if the rate of freight were low. He said, yes, it would be, but he said the discrimination would be very small. We tried to find out by asking what it would be, but did not succeed. He then said if we would unite with the Standard we would do better and everything would be peaceable and harmonious, and he would use his efforts to promote such a union if we wished it. We told him we did not wish to unite with the Standard; we dealt on freight matters with the Pennsylvania Railroad, not with the Standard Oil Company.

"There was another interview at which Mr. Bush, Mr. Ohlen, Mr. Cassatt, and myself were the only parties as I remember it; it was held in Pennsylvania, at the office of the Pennsylvania Railroad Company, in the last part of May or early part of June; it was at the time of what we called the squeeze in cars. Previous to that time we had had all the cars we wanted without any difficulty; at that time and when we were wanting just about the same kind of cars we had previously been wanting, and business was running on very easily, we found we were unable to get anything like the amounts we had before; instead of getting for the firm I represented from twelve to fifteen cars a day, we were getting only one or two—utterly insufficient for the business. We came over to see Mr. Cassatt about it—Mr. Bush, Mr. Ohlen, and myself. He said he knew there was trouble; that the other side, the Standard Oil Company, had some five hundred cars full here at Philadelphia and Baltimore; that he had not discovered it until recently, but that he would have it remedied. They had been holding them here full. I asked him why, if he knew of the cars being detained, he kept giving them cars. He said he did not know exactly how that was. I told him if these cars were shipped here and held, it seemed to me they ought to stop giving cars to parties holding them. He said the matter would be remedied soon. We asked him how soon. He could not tell exactly. I said, 'Can't you stop giving them cars?' He said he would remedy the matter, we should have all the cars we needed; and it was at that time that he made the remark to which Mr. Bush testified, when we had some little general conversation, that if we built a pipe-line he would buy it up for old iron in sixty days. I think I remarked that the Conduit Pipe brought a good price for old iron, in a laughing way. The interview was pleasant enough. Then early in July—I think it was the last part of June or early part of July—Mr. Ohlen, Mr. Bush, Mr. Wilson, Mr. King, Mr. Gregory, and myself came to Philadelphia and met Colonel Scott, president of the Pennsylvania Railroad, Mr. Cassatt, and Mr. Brundred at the office of the Pennsylvania road, with the same trouble, the same two troubles as of old, a scarcity of cars and a discrimination in freight. As to scarcity of cars, they claimed that we were getting our allotment. We told them we knew nothing about an allotment, that previous to the first of May we had sufficient cars for our business; since that time we got scarcely any; that if they had not sufficient cars to do the business with we would put on cars. Mr. Scott said they would not allow that, they had bought out one line and did not propose to have another; we then demanded cars for the business, making again the offer to put on cars if they could not furnish them, with the same result. He said they had already fought one fight in our behalf which cost them a million and a half of dollars. We told them not at all in our behalf, we had nothing to do with it; we were simply shippers over the road and did not participate in the matter at all; it was a matter of their own. He seemed to be a little sore about that. When he made the remark which has been given in evidence before, he said there would be no peace or profit in the business until we made some arrangement with the

Standard Oil Company; he would be very glad to have such an arrangement made, and would do all in his power to accomplish it. We told him we did not wish any arrangement with the Standard Oil Company; we had been dealing for years with the Pennsylvania Railroad Company, and we wished to deal with them now on all transportation and freight matters. I think there was nothing further in that interview.

"He asked why we did not apply to the other roads for transportation. We told him we had. He said, with what results? That the Central Road had no cars of their own. He said that was a very flimsy pretext. I said that the Erie road cars were controlled by the Standard Oil Company, and the Central cars were controlled by the Standard Oil Company. That in fact the whole transportation of the oil country seemed to be controlled by the Standard Oil Company, and the New York Central, and the Erie, and the Pennsylvania Central, and the Baltimore and Ohio, they controlled the whole thing, and there was no chance, and in addition to that we had been shippers and customers of the Pennsylvania road for years."

Naturally enough, men who had been through such experiences as these of Mr. Lombard were glad to unite with the Tidewater, which promised to free them from the railroads and their chief competition, and they promised to take all their supply from the line.

The success of the Tidewater experiment brought Mr. Rockefeller face to face with a new situation. Just how serious this situation was is shown by the difference in the cost of transporting a barrel of oil to the seaboard by rail and transporting it by pipe. According to the calculation of Mr. Gowen, the president of the Reading Railroad, the cost by rail was at that time from thirty-five to forty-five cents. The open rate was from $1.25 to $1.40, and the Standard Oil Company probably paid about eighty-five cents, when the roads were not protecting it from "injury by competition." Now, according to General Haupt's calculation in 1876, oil could be carried in pipes from the Oil Regions to the seaboard for 16 2-3 cents a barrel. General Haupt calculated the average difference in cost of the two systems to be twenty-three cents, enough to pay twenty-eight per cent. dividends on the cost of a line even if the railway put their freights down to cost.

This little calculation is enough to show that the day of the railroads as long-distance transporters of crude oil was over; that the pipe-lines were bound to replace them. Now, Mr. Rockefeller had by ten years of effort made the roads his servant; would he be able to control the new carrier? A man of lesser intellect might not have foreseen the inevitableness of the new situation; a man of lesser courage would not have sprung to meet it. Mr. Rockefeller, however, is like all great generals: he never fails to foresee where the battle is to be fought; he never fails to get the choice of positions. He wasted no time now in deciding what should be done. He proposed not merely to control future long-distance oil transportation; he proposed to own it outright.

Hardly had the news of the success of the Tidewater's experiment reached the Standard before this truly Napoleonic decision was being carried out. Mr. Rockefeller had secured a right of way from the Bradford field to Bayonne, New Jersey, and was laying a seaboard pipe-line of his own. At the same time he set out to acquire a right of way to Philadelphia, and soon a line to that point was under construction. Even before these seaboard lines were ready, pipes had been laid from the Oil Regions to the Standard's inland refining points—Cleveland, Buffalo and Pittsburg. With the completion of this system Mr. Rockefeller would be independent of the railroads as far as the transportation of crude oil was concerned. It was, of course, a new department in his business, and, to manage it, a new company was organised in April, 1881—the National Transit Company—with a capital of five million dollars, and a charter of historical interest, for it was a mate of the charter of the ill-fated South Improvement Company, granted by the same Legislature and giving the same omnibus privileges—the right in fact to do any kind of business, except banking, in any part of the world. The South Improvement Company charter, as we have seen, was repealed. The charter

BYRON D. BENSON

The first president of the Tidewater Pipe Company.

DAVID K. MC KELVY

The successor of Mr. Benson as president of the Tidewater.

MAJOR ROBERT E. HOPKINS

Treasurer of the Tidewater from its organization until his death in 1901.

SAMUEL Q. BROWN

The present president of the Tidewater, successor to Mr. McKelvy.

which the National Transit Company now bought seems to have gone into hiding when the character of its mate was disclosed and so had been forgotten. How it came to be unearthed by the Standard or what they paid for it, the writer does not know. However, as H. H. Rogers aptly told the Industrial Commission in 1899, when he was asked if a considerable sum was not given for it: "I should suppose every good thing had to be paid for; I should say a man owning a charter of that kind would sell it at the best price he could get."

And while Mr. Rockefeller was making this lavish expenditure of money and energy to meet the situation created by the bold development of the Tidewater, what was his attitude toward that company? One would suppose that Mr. Rockefeller, of all men, would be the first to acknowledge the service the Tidewater had rendered the oil business; that in this case he would have felt an obligation to make an exception to his claim that the oil business was his; that he would have allowed the new company to live. But Mr. Rockefeller's commercial vision is too keen for that; that would *not* be business. The Tidewater had been built to feed a few independent refineries in New York. If these refineries operated outside of him, they might disturb his system; that is, they might increase the output of refined and so lower its price. The Tidewater must not be allowed to live, then. But how could it be put out of commission? It had money to operate. There were plenty of oil producers glad to give it their product, because it was independent. The Reading Railroad had gone heart and soul into its fight—it had refiners pledged to take its oil, and these refiners had markets of their own at home and abroad. What was he going to do about it? There were several ways to accomplish his end; in two of them, at least, Mr. Rockefeller excelled from long practice. The first was to get out of the way the refineries which the Tidewater expected to feed, and this was undertaken at once. The refiners were approached usually by members of the Standard Oil Company as private individuals, and terms of purchase or lease so generous made to them that they could not afford to decline. At the same time they were assured confidentially that the Tidewater scheme was a pure chimera, that they understood the pipe-line business better than anybody else and they knew oil could not be pumped over the mountains. All but one firm yielded to the pressure. Ayres and Lombard stood by the Tidewater, but soon after their refusal to sell they were condemned as a public nuisance and obliged to move their works! The Tidewater met the situation by beginning to build refineries of its own—one at Bayonne, New Jersey, and another near Philadelphia—in the meantime storing the oil it had expected to sell.

Having done his best to cut off his rival's outlet, Mr. Rockefeller called upon the railroads to carry out that article of their contract with him which bound them to protect him from "injury by competition." What was done was told a few months later to the Committee on Commerce in the House of Representatives by Franklin B. Gowen, the president of the Reading Railroad. According to Mr. Gowen the Tidewater and Reading were no sooner ready to run oil than a meeting of the trunk lines was held at Saratoga, at which the representatives of the Standard Oil Company were present, and on that day the through rate on oil was reduced to twenty cents per barrel to the Standard Oil Company. "It was subsequently reduced to fifteen cents," Mr. Gowen told the Committee, "and I believe, though I do not certainly know, to ten cents per barrel in cars of the Standard Oil Company; . . . and I am told that at the meeting at Saratoga a time was fixed by the Standard Oil Company within which they promised to secure the control of the pipe-line—provided the trunk lines would make the rate for carrying oil so low that all concerned in transportation would lose money.

"I know this, that only three or four months ago we were told—I do not mean myself, but the gentlemen who directly represented the pipe-line which leads to our road—that if they would agree to give all their oil to the Standard Oil Company to be refined, we could carry 10,000 barrels a day, and the rates would be advanced by the trunk lines. But, to use the language of those making the offer, 'we' (meaning the Standard Oil Company) 'will never permit the trunk lines to advance the rate on oil until your pipe-line gives us all its product to refine,' and the prophesy of four months ago has become the history of to-day." Mr. Flagler differs with Mr. Gowen in his explanation of this cut in rates. Mr. Flagler contends that the Standard Oil Company really opposed it, but that the railroads insisted on it. Mr. Flagler's testimony is interesting reading in connection with all that we know about the Tidewater Company. It will be found in the appendix.*

This was the Tidewater's first year's experience. The second and third were not unlike it. But the company lived and expanded. It bought and built refineries, it sent its president to Europe to open markets, it extended its pipe-line still nearer to the seaboard, and it did this by a series of amazingly plucky and adroit financial moves—borrowing money, speculating in oil, exchanging credit, chasing checks from bank to bank, "hustling," in short, as few men ever did to keep a business alive. And every move had to be made with caution, for the Standard's eye was always on them, its hand always outstretched. Samuel Q. Brown, the present president of the organisation, when on the witness stand in December, 1882, said that so much did the Tidewater fear espionage that they were accustomed to keep their oil transactions as a private and not a general account, in order that they might not be reported to the Standard; that even matters which they believed they

* See Appendix, Number 38. Testimony of Henry M. Flagler in regard to the Tidewater contest.

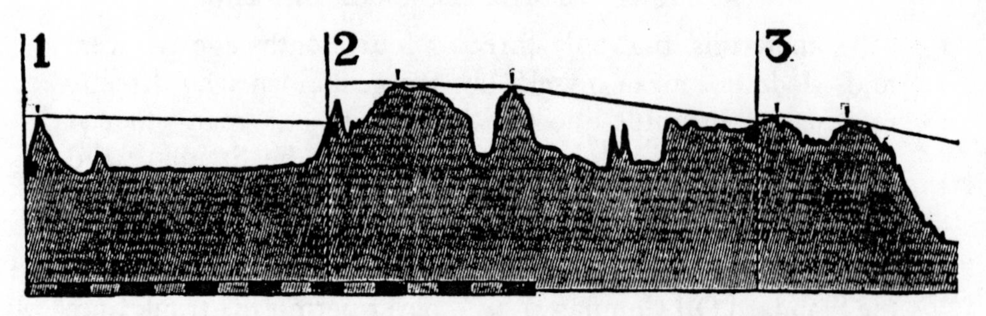

Scale—3 miles to each division. CONDENSED PROFILE OF TIDEWATER PIPE LINE

The pipe followed the jagged line representing surface of the ground. The numbers above the Station 1 lifted the oil over 600 feet. From here it flowed by gravitation until the gradient line—the oil to the next high point, the crest of the Alleghanies. As the gradient line shows, the oil now would the speed of the flow.

were keeping in an absolutely private way frequently leaked out, to the injury of the business.

By January, 1882, the Tidewater was in such a satisfactory condition that it decided to negotiate a loan of $2,000,000 to carry out plans for enlargement. The First National Bank of New York, after a thorough examination of the business, agreed to take the bonds at ninety cents on the dollar, but trouble began as soon as the probable success of the bond issue was known. The officials of the First National Bank were called upon by stockholders of the Tidewater, men holding nearly a third of the company's stock, and assured that the company was insolvent, and that it would be unsafe for the bank to take the loan. The First National declined to be influenced by the information, on the ground that the disgruntled stockholders had sold themselves to the Standard Oil Company, and were trying to discredit the Tidewater, so that the Standard might buy it in. It had been planned to place some of these bonds in Europe, and Franklin B. Gowen was sent over for that purpose. Mr. Brown said on the witness stand, a few months later, that as soon as Mr. Gowen started from this side it was cabled to Europe that he was going over to place bonds which were not sound; that the stockholders

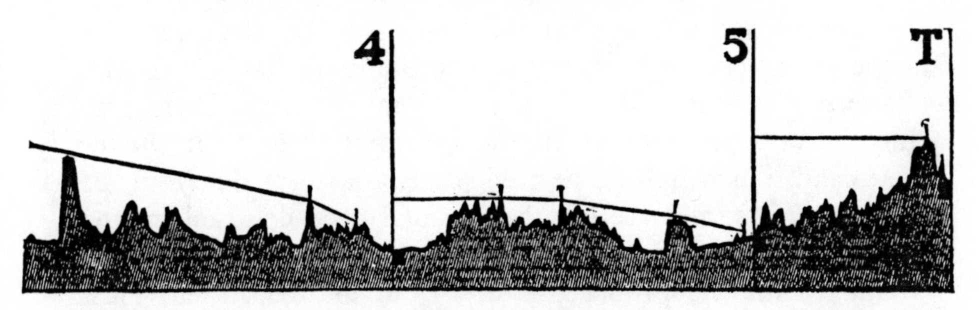

BETWEEN RIXFORD AND TAMANEND, PENNSYLVANIA

surface line show the location of the pumping stations from which the oil was forced. The pump at sloping straight line above the surface line—touched the ground. A new station, No. 2, then lifted the flow to Station 4, making many steep ascents without further pumping. Station 3 was added to increase

were all of them wealthy men, and if the bonds had been good property they would have taken them themselves. Mr. Brown declared this report was spread so generally on the other side that it interfered seriously with Mr. Gowen's attempt to place the loan.

These manœuvres failing to ruin the Tidewater's credit, a more serious attack was made in the fall of 1882, by the filing of a long bill of complaint against the management of the company, followed by an appeal that a receiver be appointed and the business wound up. The appeal came from E. G. Patterson, a stockholder of the Tidewater, and a man who, up to this time, had been one of the most intelligent opponents of the Standard in the Oil Regions. Mr. Patterson was one of the few who had realised, from the first development of Mr. Rockefeller's pretensions, that it was a question of transportation, and that, if the railroads could be forced by courts and legislatures to do their duty, the coal-oil business would not belong to Mr. Rockefeller. He had been one of the strongest factors in the great suits compromised in 1880, and his disgust at the outcome had been so great that he had washed his hands of the Producers' Union. Later he had been engaged by the state of Pennsylvania to collect evidence on which to support a claim against the Standard Oil Company for some $3,000,000 of back taxes. The Standard had made Mr. Patterson's services unnecessary by coming forward and giving the attorney-general all the information as to its financial condition which he desired. Exasperated at the result of all his efforts, and feeling that he had been deserted by the public he had tried to serve, Mr. Patterson sent word to the Standard that he proposed still further to attack them (just how he never explained) unless they would give him, not to attack, as much as there was in the contract from the state.* They seem to have thought it worth while to buy peace, and agreed to give Mr. Patterson some $20,000 in all, and secure him a position for a term of years. The first payment was made at the end of April, 1882, and $5,000 of the money received Mr. Patterson paid to the Tidewater for stock he had taken at its organisation. No sooner was the stock in his hands than he began the preparation of the bill of complaint above referred to, and in December the case was heard.

The Oil Regions watched it with keenest interest. That Mr. Patterson had made some settlement with the Standard was generally known, and the charge was freely circulated that they had bribed him to bring this suit in hopes of blasting the credit of the Tidewater and getting its stock for a song. The testimony brought out in the trial did not bear out this popular notion. The case was rather more complicated. That the suit was backed by the Standard, one would have to be very naïve to doubt, but they were using other and stronger parties than Mr. Patterson, and that was a faction of the company known as the "Taylor-Satterfield crowd." These men, controlling some $200,000 worth of Tidewater stock, had been professing themselves dissatisfied with the management of the business for some months, though always

* Court of Common Pleas, Crawford County, Pennsylvania. Patterson *vs.* Tidewater Pipe Company, Limited. Testimony of E. G. Patterson, December, 1882.

refusing to sell their holdings at an advanced price. It was generally believed in the Oil Regions that their "dissatisfaction" was fictitious, that they were in reality in league with the Standard in an attempt to create a panic in Tidewater stock, a belief which was strengthened when it was learned that a big oil company, which the gentlemen controlled, the Union, had been sold about that time to the Standard Oil Trust for something like $500,000 in its stock. The first manœuvre of the Taylor-Satterfield faction had been the attempt to dissuade the First National Bank from taking the Tidewater loan referred to above. Failing in this, they seem to have imbued Mr. Patterson thoroughly with their pretended dissatisfaction and to have persuaded him to bring the suit. For some reason which is not clear they failed properly to support him in the suit, and when it came off they practically deserted him. The Tidewater had no trouble in proving that the complaints of insolvency and mismanagement were without foundation, and Judge Pierson Church, of Meadville, before whom the case was argued, refused to appoint the receiver, intimating strongly that, in his judgment, the case was an attempt to levy a species of blackmail, in which it must not be expected that his court would co-operate. Judge Church's decision was given on January 15. Two days later a sensation came in Tidewater affairs, which quite knocked the Patterson suit out of the public mind; it was nothing less than a bold attempt by the Taylor party, or, as it was now known, "the Standard party," to seize the reins of government. It was a very cleverly planned coup.

The yearly meeting for the election of officers in the company was fixed for a certain Wednesday in January. By verbal agreement it had been postponed, in 1882, to some time in February, the controller, D. B. Stewart, a member of the Taylor faction, representing that he could not have his statement ready earlier. No notices were sent out to this effect, although this should have been done. Taylor and his party, taking advantage of this fact perfectly well known to them, appeared at the Tidewater offices on January 17, and although one of the Benson faction, as the majority was known from the name of the company's president, was present with sufficient proxies to vote nearly two-thirds of the stock, they overruled him and elected themselves to the control. They also elected to the Board of Managers, Franklin B. Gowen, the president of the Reading, and James R. Keene, the famous speculator, both large holders of Tidewater bonds. They followed their election immediately by sending out notices to the banks with which the company did business not to honour checks drawn by the Benson party, and to the post-office to deliver mail to no one but themselves.

The announcement caused a terrible commotion in oil circles. Both Mr. Keene and Mr. Gowen refused to recognise the new board, Mr. Gowen telegraphing in answer to the notification of his election:

JOHN SATTERFIELD,
Titusville.

At quarter of three o'clock to-day I received a despatch signed with your name as manager and chairman, stating that a meeting of the Board of Managers would be held at noon to-day. While the notice itself is sufficient to render invalid any action you may have attempted at such meeting as has been held, even if you had power to act at all, I deny your right to call any meeting or act in any manner as an officer of the company, and will hold you and all your associates responsible at law for the occurrences of yesterday, and for your subsequent action thereunder.

(Signed) F. B. GOWEN.

The Benson party took immediate action, applying for an injunction restraining the new board from taking possession of the books and offices. This was granted and a date for a hearing appointed. Up to the hearing the old board did business behind barricaded doors! The case was heard in Meadville before Judge Pierson Church—the same who had heard the Patterson case. As it was a case to be decided on purely technical matters—the rules governing elections—no sensation was looked for, but one came immediately. It was a long affidavit from James R. Keene, even more notorious then than now—there were fewer of his kind—for deals and corners and devious stock tricks, declaring that both the Patterson case and this attempt to obtain control were dictated by the "malicious ingenuity" of the Standard for the purpose of destroying the Tidewater and getting hold of its property:

"From my first connection with the company," said Mr. Keene, "it has been hampered and embarrassed in its business by the unscrupulous competition of the Standard Oil Company. When it first began to transport and deliver oil at tidewater, the refineries which purchased and refined oil were one after another bought up by the Standard Oil Company or driven out of business by vexatious and oppressive annoyances. The most private details of our business have been communicated to the officers of the Standard Oil Company, and they have, by every means in their power, interfered with our affairs. By the arrangement which they were able to make with the railroads leading from the Oil Regions, other than the Philadelphia and Reading Railroad Company and the Central Railroad of New Jersey, the Standard Oil Company have been able to obtain a control of the business of transporting and refining oil, with the exception of that part of the business which has been carried on by the Tidewater Pipe Company and their refineries, to which it had made deliveries. Repeated efforts have been made by parties in their interest to secure the control of the Tidewater Pipe Company, and if they could succeed, the monopoly thereby secured would add many million dollars a year to their profit."

Mr. Keene's putting of the case was undoubtedly correct, but pious horror of commercial brigandage, coming from "Jim" Keene, was useful only to give joy to a cynical world, unencumbered by the possession of stock in either concern. The Keene sensation was followed by a second, an affidavit from John D. Archbold, of the Standard Oil Company, denying that his company had any interest in the present suit, but adding that for some time the officers of the Tidewater had been seeking an alliance with the Standard:

"Byron D. Benson and David McKelvy have at various times for the past years met me at their own instance, and have proposed to combine the business of the Tidewater Pipe Company with that of the Standard Oil Company, desiring the Standard Oil Company to agree on a division of the business of transporting and refining oil, and to agree with the Tidewater Pipe Company in fixing the rate of transporting oil and the price of refined oils. These proposals were renewed to me by B. D. Benson during the summer of 1882, he coming to my office at his own instance and urging, by various arguments, such an arrangement. These proposals, in whatever shape made, have always been declined. This deponent has also had many interviews with James R. Keene, and always at his request, upon the same subject, in which interviews said Keene has earnestly urged such a combination and has used many arguments in favour of the advantage which would result from such a combination. These proposals have always been declined."

Naturally they were declined—the Standard was not seeking an alliance, it was seeking ownership of the Tidewater; and it expected so to discredit the company that it could buy in its stock for a song. Mr. Archbold's affidavit cooled popular sympathy for the hunted concern no little, however. A suggestion of any kind of a compromise with the Standard was looked upon as rank disloyalty by the Oil Regions, free competition in rates and in prices being, they contended, the only hope of the country. Mr. Archbold's affidavit must have something in it, everybody thought, though it might be, as Mr. Benson immediately swore, "grossly inaccurate."

Such was the character of the charges and countercharges in this purely technical case. The judge took little notice of them in his decision, but, after an exhaustive discussion of the points involved in the election, decided it was illegal and continued the injunction he had granted against the new board. Judge Church's decision aroused general exultation in the Oil Regions—as any failure of the Standard to get what it wanted was bound to do, and with good reason. The Tidewater's growth in the face of the Standard's constant interference with its business was proof that independent pipelines and independent refineries could be built up if men had sufficient brains and courage and patience. What one set of men had done, another could do. Their hope of restoring freedom of competition to the oil business was still further brightened in June by the news that the Legislature of Pennsylvania had passed a free pipe-line bill—the measure that they had been urging for twelve years without avail. With a sturdy example of independence, like the Tidewater, before them, and the right of eminent domain for pipes, the future of competition in oil seemed to be up to the oil men themselves.

But the Oil Regions have always been prone to jump at conclusions. They were forgetting Mr. Rockefeller's record when they concluded that he was through with the Tidewater. Because he had failed in his old South Improvement Company trick, that is, failed to create a panic among Tidewater stockholders, and so get their property at panic prices, was no reason at all to suppose he had abandoned the chase. There still remained a legitimate method of getting into the company, and, as a last resort, Mr. Rockefeller accepted it. He bought the minority stock of the concern, held by the Taylor party. Up to this time Mr. Rockefeller had appeared in Tidewater affairs as a destroyer. He now appeared in a rôle in which he is quite as able—as a pacifier, and his extraordinary persuasiveness was never exercised to better effect. "We own $200,000 worth of your stock," he could tell the people he had been fighting. "If you will consent to confine yourselves to a fixed percentage of our joint business, and will sustain pipage rates and the price of refined oil, we will let you alone. Let us dwell together in peace."

The Tidewater, tired of the fight, accepted. And so these men—to whom the oil business owes one of its most remarkable developments, who, in face of the most powerful and unscrupulous opposition, had in four years built up a business worth five and one-half millions of dollars—signed contracts in October, 1883, fixing the relative amount of business they were henceforth to do as 11½ per cent. of the aggregate, the Standard having 88½ per cent. The two simply became allies.

The agreement between them was the same in effect as all Mr. Rockefeller's running agreements—it limited and kept up prices.* Any benefit the oil business might have reaped from natural and decent competition between the two was of course ended by the alliance. For all practical purposes the two were one. In the phrase of the region, the Tidewater had "gone over to the Standard," and there it has always remained. The contract was made for fifteen years, but since its expiration it has been lived up to honourably by both parties without other than a verbal understanding. For, note this: Mr. Rockefeller always keeps his word. Indeed, in studying his career, one is frequently reminded of Tom Sawyer's great resolution—never to sully piracy by dishonesty!

The Tidewater has prospered within the boundary Mr. Rockefeller drew for it, as those who have accepted submissively his boundaries have never failed to do. Mr. Rockefeller is right when he says, as he does so often, that all who come with him prosper. That the company would have succeeded in becoming eventually a formidable rival of the Standard, and in controlling much more than eleven per cent. of the business, no one can doubt who knew Mr. Benson, Major Hopkins, Mr. McKelvy, and their colleagues. They were business men of the first order, as their tremendous work from 1878 to 1883 shows.

Once more the good of the oil business was secure, and Mr. Rockefeller at once proceeded to arrange his great house in the new order made necessary by the introduction of the seaboard pipe-line. The entire transportation department of the business had to be reorganised. When the seaboard pipe-line became a factor in the oil business, in 1879, the Standard Oil

* See Appendix, Number 39 A. Agreement between Standard and Tidewater refineries.

See Appendix, Number 39 B. Agreement between Standard and Tidewater Pipe Lines.

Company owned practically the entire system of oil-gathering pipe-lines—that is, the lines carrying oil from the wells to the storing or shipping points. These lines were organised under the name of the United Pipe Lines, and the organisation was magnificent in both extent and in character of service rendered. Never, indeed, has the ability of the men Mr. Rockefeller gathered into his machine shone to better advantage than in the building up and management of the pipe-line business. At the end of 1883, when the alliance was made with the Tidewater, the United Pipe Lines were taking from the wells of Pennsylvania fully a million and a half barrels of oil a month. Their pipes, of an aggregate length of 3,000 miles, connected with thousands of wells scattered all over the wide Oil Regions.

Whenever the oil men opened a new field, no matter how remote from those already developed, the United Pipe Lines immediately went there to care for the oil. In more than one case, in these years of rapid and excessive development of oil territory, the pipe-line company invested great sums in preparing to take care of oil fields whose yield never paid the cost of the pipe laid. Thus, in 1882, there was a tremendous excitement over the opening of the Cherry Grove field. The Standard spent $2,000,000 getting ready to take care of a great outpouring of oil—which came, but did not stay. In 1882 Cherry Grove produced 2,345,400 barrels; in 1883, 755,512! It cost the company forty-six cents a barrel to take care of the production of one short-lived group of wells in this field, on which they never realised more than twenty cents pipage.

The Standard not only gathered this oil; it stored it, to wait its owner's demand. At this date it controlled 40,000,000 barrels of iron tankage, in which it stored the enormous stocks, over 35,000,000 barrels, which had accumulated in the five previous years. When the oil passed to the pipe-line, the owner received his money for it at once, if he wished, or the line "carried" it. When a producer had 1,000 barrels in the line, he received a pipe-line certificate for it. In December of 1883 the United Pipe Lines had issued certificates for nearly all of the 35,000,000 barrels of stocks above ground. The oil men thus had a bank for their oil, a bank recognised generally as sound as any in the United States.

Such were the returns from the pipe-line for its services that no business ever justified more fully the extraordinary outlays of money and energy which it had taken to perfect it. For each barrel of oil the United Pipe Lines gathered, they received, when it was taken from the lines, twenty cents. The service cost them perhaps two cents after installation, though in these years, when they were obliged to carry some 30,000,000 barrels, they had constantly $6,000,000 on their books on which they did not at once realise. They could afford to let this sum stand because of the storage charge. For every 1,000 barrels carried in their tanks they received $6.25 each fifteen days—$152 a year. Now, tankage did not cost over $250 per 1,000 barrels, so that the storage more than paid its cost in two years. There were often great losses by fire, but these were paid by the owners of the oil—a pro rata assessment being made. There was a deterioration in quantity and quality of oil from holding, but this again was paid by the owners in a shrinkage charge of three per cent., deducted from the quantity of oil when run. Thus on every side the pipe-line business was guarded. So long as it could keep out competition and hold up its prices, there was no better paying business in the United States than piping oil.

As we have seen, Mr. Rockefeller began to add long-distance pipe-lines to his business as soon as the Tidewater demonstrated their feasibility, and before the time the Tidewater was brought into harmony he had a complete system to the seaboard and to his inland refinery points, organised under the name of the National Transit Company. The United Pipe

Lines and the National Transit Company were really one business, the former consisting of local lines and the other of trunk lines, and to make the organisation more compact the former was transferred to the latter on April 1, 1884. The paid-up capital of the concern at this date was $31,000,000. Just as Mr. Rockefeller claimed, in 1878, that he was "prepared to enter into a contract to refine all the petroleum that could be sold in the markets of the world," so now he could announce that he was prepared to gather, store and transport all the crude petroleum not only that the markets of the world demanded, but that the producers took from the ground. As things now stood the only remaining point where he could possibly be affected by competition was the railroads. A new relation to the railroads was created by the new development. Mr. Rockefeller was not only independent of them, he was their competitor, for, like them, he was a common carrier obliged to transport what was offered. His open rate to New York was forty-five cents, to Philadelphia forty, though the actual service probably did not cost over ten cents. By the alliance with the Tidewater any danger of competition from a pipe-line, which could of course afford to cut the price, was shut off. The railroads might possibly, however, lower the prices a little and still make a profit. It was very necessary that the price be kept up in order that too much encouragement should not be given to outside refiners. The only group which threatened to grow to large proportions, at this time, was in the Oil Regions, a group which was the direct outgrowth of the compromise of 1880. As will be remembered, the agreement with the Pennsylvania Railroad made then stipulated that all rates should be open, and that if a rebate was given to one shipper another could have it on demand. After the compromise the Pennsylvania had undertaken again to stimulate the growth of independent refineries, and several plants had been built in Titusville and Oil City. Having removed the New York group from competition by the alliance with the Tidewater it was Mr. Rockefeller's business to make it as hard as possible for the independents in the Oil Regions to do business, and to do this he must make a contract with the Pennsylvania.

Moreover, when Mr. Rockefeller entered New Jersey with his seaboard pipe-line, he had been obliged to cross the Pennsylvania Railroad. He could not do so without the consent of the company, there being no free pipe-line in the country. He accordingly had been obliged to make a traffic arrangement with them to get his pipe through. A new arrangement was now necessary in order to prevent competition, and in August, 1884, a contract was signed, for "considerations mutually interchanged," by which the National Transit Company agreed to give to the Pennsylvania Railroad twenty-six per cent. of "all petroleum brought to the Atlantic seaboard by all existing carriers, whether rail or pipe, now engaged in transporting such property, or which may hereafter engage in such transportation in conjunction with the Transit Company's pipes." At the same time that the Transit Company agreed to give the railroad this amount of oil, it also signed an agreement to carry this oil for the railroad on a sliding scale. When the open rate of the pipe-line was forty cents to Philadelphia the railroad was to pay the company eight cents—with each five cents difference, up or down, in the open rate, there was to be one cent difference to the railroad, the Transit never to receive less than six or more than ten cents.* Suppose, for example, that the entire seaboard shipment of oil in the month ending December 20, 1884, had been 1,000,000 barrels. 260,000 barrels belonged to the Pennsylvania. If the Transit Company ran all the railroad's percentage it would get eight cents a barrel for the service, $20,800,

* See Appendix, Number 40. Two agreements of even date, August 22, 1884, between the Pennsylvania Railroad Company and the National Transit Company.

and it would pay the railroad $104,000 less $20,800, or $83,200. The pipe-line probably never ran the whole amount. More or less refined oil—naphtha, benzine, and other petroleum products—would necessarily go by rail. Large sums were paid monthly by the National Transit, however, to the railroad. Mr. Rockefeller seems to have been paying the Pennsylvania Railroad this money not to compete with him as an oil carrier. It would be difficult to find in our variegated commercial history a more beautiful example of the beneficence of combination—to those in the deal!

With the removal of danger of any competition by the Pennsylvania Railroad, the transportation department of the Standard Oil Trust seems to have been as nearly a perfect machine, both in efficiency and in its monopolistic power, as ever has been devised. It was more perfect, indeed, than the refining end of the trust, for independent refiners did exist, and since 1880 they had been showing increasing vigour, whereas there seemed now no opportunity for an independent pipe-line ever again to develop. Who, with the Tidewater's story in mind, would be bold enough to attempt to reach the sea? For the time being, then, the Standard Oil Company had things all its own way. It collected with its ally, the Tidewater, practically the entire output of a great raw product. It manufactured fully ninety per cent. of this product, and aimed to manufacture 100 per cent. It was a common carrier, and so obliged to deliver oil to rival refineries if they called for it, but these refineries paid forty or forty-five cents for a service which cost the Standard Oil Trust not over one-fourth of the sum.

Mr. Rockefeller had every reason to be satisfied with oil transportation in 1884, but there was a part of the oil business which was not so completely in his grasp. The markets of the country were still open. There the few independent refiners who had escaped strangulation were free to barter as they could. But the right to make all the oil in the world, which Mr. Rockefeller claimed, carried with it the right to sell all the oil the world consumed. The independent was therefore a poacher in the market and must be driven out.

CHAPTER TEN

CUTTING TO KILL

ROCKEFELLER NOW PLANS TO ORGANISE OIL MARKETING AS HE HAD ALREADY ORGANISED OIL TRANSPORTING AND REFINING—WONDERFULLY EFFICIENT AND ECONOMICAL SYSTEM INSTALLED—CURIOUS PRACTICES INTRODUCED—REPORTS OF COMPETITORS' BUSINESS SECURED FROM RAILWAY AGENTS—COMPETITORS' CLERKS SOMETIMES SECURED AS ALLIES—IN MANY INSTANCES FULL RECORDS OF ALL OIL SHIPPED ARE GIVEN STANDARD BY RAILWAY AND STEAMSHIP COMPANIES—THIS INFORMATION IS USED BY STANDARD TO FIGHT COMPETITORS—COMPETITORS DRIVEN OUT BY UNDERSELLING—EVIDENCE FROM ALL OVER THE COUNTRY—PRETENDED INDEPENDENT OIL COMPANIES STARTED BY THE STANDARD—STANDARD'S EXPLANATION OF THESE PRACTICES IS NOT SATISFACTORY—PUBLIC DERIVES NO BENEFIT FROM TEMPORARY LOWERING OF PRICES—PRICES MADE ABNORMALLY HIGH WHEN COMPETITION IS DESTROYED.

TO know every detail of the oil trade, to be able to reach at any moment its remotest point, to control even its weakest factor—this was John D. Rockefeller's ideal of doing business. It seemed to be an intellectual necessity for him to be able to direct the course of any particular gallon of oil from the moment it gushed from the earth until it went into the lamp of a housewife. There must be nothing—*nothing* in his great machine he did not know to be working right. It was to complete this ideal, to satisfy this necessity, that he undertook, late in the seventies, to organise the oil markets of the world, as he had already organised oil refining and oil transporting. Mr. Rockefeller was driven to this new task of organisation not only by his own curious intellect; he was driven to it by that thing so abhorrent to his mind—competition. If, as he claimed, the oil business belonged to him, and if, as he had announced, he was prepared to refine all the oil that men would consume, it followed as a corollary that the markets of the world belonged to him. In spite of his bold pretensions and his perfect organisation, a few obstinate oil refiners still lived and persisted in doing business. They were a fly in his ointment—a stick in his wonderful wheel. He must get them out; otherwise the Great Purpose would be unrealised. And so, while engaged in organising the world's markets, he incidentally carried on a campaign against those who dared intrude there.

When Mr. Rockefeller began to gather the oil markets into his hands he had a task whose field was literally the world, for already, in 1871, the year before he first appeared as an important factor in the oil trade, refined oil was going into every civilised country of the globe. Of the five and a half million barrels of crude oil produced that year, the world used five millions, over three and a half of which went to foreign lands. This was the market which had been built up in the first ten years of business by the men who had developed the oil territory and invented the processes of refining and transporting, and this was the market, still further developed, of course, that Mr. Rockefeller inherited when he succeeded in corralling the refining and transporting of oil. It was this market he proceeded to organise.

The process of organisation seems to have been natural and highly intelligent. The entire country was buying refined oil for illumination. Many refiners had their own agents out looking for markets; others sold to wholesale dealers, or jobbers, who placed trade with local dealers, usually grocers. Mr.

JOHN D. ROCKEFELLER IN 1880

FROM A PHOTOGRAPH BY SARONY

Rockefeller's business was to replace independent agents and jobbers by his own employees. The United States was mapped out and agents appointed over these great divisions. Thus, a certain portion of the Southwest—including Kansas, Missouri, Arkansas and Texas—the Waters-Pierce Oil Company, of St. Louis, Missouri, had charge of; a portion of the South —including Kentucky, Tennessee and Mississippi—Chess, Carley and Company, of Louisville, Kentucky, had charge of. These companies in turn divided their territory into sections, and put the subdivisions in the charge of local agents. These local agents had stations where oil was received and stored, and from which they and their salesmen carried on their campaigns. This system, inaugurated in the seventies, has been developed until now the Standard Oil Company of each state has its own marketing department, whose territory is divided and watched over in the above fashion. The entire oil-buying territory of the country is thus covered by local agents reporting to division headquarters. These report in turn to the head of the state marketing department, and his reports go to the general marketing headquarters in New York.

To those who know anything of the way in which Mr. Rockefeller does business, it will go without saying that this marketing department was conducted from the start with the greatest efficiency and economy. Its aim was to make every local station as nearly perfect in its service as it could be. The buyer must receive his oil promptly, in good condition, and of the grade he desired. If a customer complained, the case received prompt attention and the cause was found and corrected. He did not only receive oil; he could have proper lamps and wicks and burners, and directions about using them.

The local stations from which the dealer is served to-day are models of their kind, and one can easily believe they have always been so. Oil, even refined, is a difficult thing to handle without much disagreeable odour and stain, but the local stations of the Standard Oil Company, like its refineries, are kept orderly and clean by a rigid system of inspection. Every two or three months an inspector goes through each station and reports to headquarters on a multitude of details —whether barrels are properly bunged, filled, stencilled, painted, glued; whether tank wagons, buckets, faucets, pipes, are leaking; whether the glue trough is clean, the ground around the tanks dry, the locks in good condition; the horses properly cared for; the weeds cut in the yard. The time the agent gets around in the morning and the time he takes for lunch are reported. The prices he pays for feed for his horses, for coal, for repairs, are noted. In fact, the condition of every local station, at any given period, can be accurately known at marketing headquarters, if desired. All of this tends, of course, to the greatest economy and efficiency in the local agents.

But the Standard Oil agents were not sent into a territory back in the seventies simply to sell all the oil they could by efficient service and aggressive pushing; they were sent there to sell all the oil that was bought. "The coal-oil business belongs to us," was Mr. Rockefeller's motto, and from the beginning of his campaign in the markets his agents accepted and acted on that principle. If a dealer bought but a barrel of oil a year, it must be from Mr. Rockefeller. This ambition made it necessary that the agents have accurate knowledge of all outside transactions in oil, however small, made in their field. How was this possible? The South Improvement scheme provided perfectly for this, for it bound the railroad to send daily to the principal office of the company reports of all oil shipped, the name of shipper, the quantity and kind of oil, the name of consignee, with the destination and the cost of freight.* Having such knowledge as this, an agent could

* The Eighth Section of Article Second of this contract, defining the duties of the railroads reads: "To make manifests or way-bills of all petroleum or its products

immediately locate each shipment of the independent refiner, and take the proper steps to secure the trade. But the South Improvement scheme never went into operation. It remained only as a beautiful ideal, to be worked out as time and opportunity permitted. The exact process by which this was done it is impossible to trace. The work was delicate and involved operations of which it was wise for the operator to say nothing. It is only certain that little by little a secret bureau for securing information was built up until it is a fact that information concerning the business of his competitors, almost as full as that which Mr. Rockefeller hoped to get when he signed the South Improvement Company contracts, is his to-day. Probably the best way to get an idea of how Mr. Rockefeller built up this department, as well as others of his marketing bureau, is to examine it as it stands to-day. First, then, as to the methods of securing information which are in operation.

Naturally and properly the local agents of the Standard Oil Company are watchful of the condition of competition in their districts, and naturally and properly they report what they learn. "We ask our salesmen and our agents to keep their eyes open and keep us informed of the situation in their respective fields," a Standard agent told the Industrial Commission in 1898. "We ask our agents, as they visit the trade, to make reports to us of whom the different parties are buying; principally to know whether our agents are attending to their business or not. If they are letting too much business get away from them, it looks as if they were not attending to their

transported over any portion of the railroads of the party of the second part or its connections, which manifests shall state the name of the consignor, the place of shipment, the kind and actual quantity of the article shipped, the name of the consignee, and the place of destination, with the rate and gross amount of freight and charges, and to send daily to the principal office of the party of the first part duplicates of all such manifests or way-bills."—Proceedings in Relation to Trusts, House of Representatives, 1888. Report Number 3,112, page 360.

business. They get it from what they see as they go around selling goods." But there is no such generality about this part of the agent's or salesman's business as this statement would lead one to believe. As a matter of fact it is a thoroughly scientific operation. The gentleman who made the above statement, for instance, sends his local agents a blank like the following to be made out each month:

EXHIBIT "B—R."*

MONTHLY REPORT.

USE COPYING PENCIL.
40830

Town————————————*Date*————————*1897.*

DEALER	ADDRESS	Estimated Sales per month of		Brand or Kind of Goods	Price	If by Tank Wagon mark "T"	BUY FROM	
		R. Oil	Gaso.				Name	Point of Shipment

————————————*Salesman or Agent.*

The local agent gets the information to fill out such a report in various ways. He questions the dealers closely. He watches the railway freight stations. He interviews everybody in any way connected with the handling of oil in his territory. All of which may be proper enough. When, in the early eighties, Howard Page, of the Standard Oil Company, was in charge of the Standard shipping department in Kentucky, his agents visited the depots once a day to see what oil

* Record of pleadings and testimony in Standard Oil Trust quo warranto cases in the Supreme Court of Ohio, 1899, page 681.

arrived there from independent shippers. A record of these shipments was made and reported monthly to Mr. Page. He was able to tell the Interstate Commerce Commission, in 1887, almost exactly what his rivals had been shipping by rail and by river. Mr. Page claimed that his agents had no special privileges; that anybody's agents would have been allowed to examine the incoming cars, note the consignor, contents and consignee. It did not appear in the examination, however, that anybody but Mr. Page had sent agents to do such a thing. The Waters-Pierce Oil Company, of St. Louis, once paid one of its Texas agents this unique compliment: "We are glad to know you are on such good terms with the railroad people that Mr. Clem (an agent handling independent oil) gains nothing by marking his shipments by numbers instead of names." In the same letter the writer said: "Would be glad to have you advise us when Clem's first two tanks have been emptied and returned, also the second two to which you refer as having been in the yard nine and sixteen days, that we may know how long they have been held in Dallas. The movement of tank cars enters into the cost of oil, so it is necessary to have this information that we may know what we are competing with." *

The superior receiving the filled blanks carefully follows them by letters of instructions and inquiries, himself keeping track of each dealer, however insignificant, in the local agent's territory, and when one out of line has been brought in, never failing to compliment his subordinate. But however diligent the agent may be in keeping his eyes open, however he may be stirred to activity by the prodding and compliments of his superiors, it is of course out of the question that he get anything like the full information the South Improvement scheme insured. What he is able to do is supplemented by a system which compares very favourably with that famous

* Trust Investigation of Ohio Senate, 1898, page 370.

scheme and which undoubtedly was suggested by it. For many years independent refiners have declared that the details of their shipments were leaking regularly from their own employees or from clerks in freight offices. At every investigation made these declarations have been repeated and occasional proof has been offered; for instance, a Cleveland refiner, John Teagle, testified in 1888 to the Congressional Committee that one day in 1883 his bookkeeper came to him and told him that he had been approached by a brother of the secretary of the Standard Oil Company at Cleveland, who had asked him if he did not wish to make some money. The bookkeeper asked how, and after some talk he was informed that it would be by his giving information concerning the business of his firm to the Standard. The bookkeeper seems to have been a wary fellow, for he dismissed his interlocutor without arousing suspicion and then took the case to Mr. Teagle, who asked him to make some kind of an arrangement in order to find out just what information the Standard wanted. The man did this. For twenty-five dollars down and a small sum per year he was to make a transcript of Mr. Teagle's daily shipments with net price received for the same; he was to tell what the cost of manufacturing in the refinery was; the amount of gasoline and naphtha made and the net price received for them; what was done with the tar; and what percentage of different grades of oil was made; also how much oil was exported. This information was to be mailed regularly to Box 164 of the Cleveland post-office. Mr. Teagle, who at that moment was hot on the tracks of the Standard in the courts, got an affidavit from the bookkeeper. This he took with the money which the clerk had received to the secretary of the Standard Oil Company and charged him with bribery. At first the gentleman denied having any knowledge of the matter, but he finally confessed and even took back the money. Mr. Teagle then gave the whole story to the newspapers. where it of course made much noise.

Several gentlemen testified before the recent Industrial Commission to the belief that their business was under the constant espionage of the Standard Oil Company. Theodore Westgate, an oil refiner of Titusville, told the Commission that all of his shipments were watched. The inference from his testimony was that the Standard Oil Company received reports direct from the freight houses. Lewis Emery, Jr., of Bradford, a lifelong contestant of the Standard, declared that he knew his business was followed now in the same way as it was in 1872 under the South Improvement Company contract. He gave one or two instances from his own business experience to justify his statements, and he added that he could give many others if necessary. Mr. Gall, of Montreal, Canada, declared that these same methods were in operation in Canada. "When our tank-cars come in," Mr. Gall told the Commission, "the Standard Oil Company have a habit of sending their men, opening a tank-car, and taking a sample out to see what it contains." Mr. Gall declared that he knew this a long time before he was able to get proof of it. He declared that they knew the number of cars that he shipped and the place to which they went, and that it was their habit to send salesmen after every shipment. Mrs. G. C. Butts, a daughter of George Rice, an independent refiner of Marietta, Ohio, told the Ohio Senate Committee which investigated trusts in 1898 that a railroad agent of their town had notified them that he had been approached by a Standard representative who asked him for a full report of all independent shipments, to whom and where going. The agent refused, but, said Mrs. Butts: "We found out later that someone was giving them this information and that it was being given right from our own works. . . . A party writing us from the Waters-Pierce office wrote that we had no idea of the network of detectives, generally railroad agents, that his company kept, and that everything that we or our agents said or did was reported back to the managers through a regular network of detectives who were agents of the railroads and oil company as well."

But while the proofs the independents have offered of their charges show that such leaks have occurred at intervals all over the country, they do not show anything like a regular system of collecting information through this channel. From the evidence one would be justified in believing that the cases were rare, occurring only when a not over-nice Standard manager got into hot competition with a rival and prevailed upon a freight agent to give him information to help in his fight. In 1903, however, the writer came into possession of a large mass of documents of unquestionable authenticity, bearing out all and more than the independents charge. They show that the Standard Oil Company receives regularly to-day, at least from the railroads and steamship lines represented in these papers, information of *all* oil shipped. A study of these papers shows beyond question that somebody having access to the books of the freight offices records regularly each oil shipment passing the office—the names of consignor and consignee, the addresses of each, and the quantity and kind of oil are given in each case. This record is made out usually on a sheet of blank paper, though occasionally the recorder has been indiscreet enough to use the railroad company's stationery. The reports are evidently intended not to be signed, though there are cases in the documents where the name of the sender has been signed and erased; in one case a printed head bearing the name of the freight agent had been used. The name had been cut out, but so carelessly that it was easy to identify him. These reports had evidently been sent to the office of the Standard Oil Company, where they had received a careful examination, and the information they contained had been classified. Wherever the shipment entered was from one of the distributing stations of the Standard Oil Company, a line was drawn through it, or it was checked off in some way. In every other case in the mass of reports there was written, opposite the name of the consignee, the name of a person *known* to be a Standard agent or salesman in the territory where the shipment had gone.

Now what is this for? Copies of letters and telegrams accompanying the reports show that as soon as a particular report had reached Standard headquarters and it was known that a carload, or even a barrel, of independent oil was on its way to a dealer, the Standard agent whose name was written after the shipment on the record had been notified. "If you can stop car going to X, authorise rebate to Z (name of dealer) of three-quarters cent per gallon," one of the telegrams reads. There is plenty of evidence to show how an agent receiving such information "stops" the oil. He *persuades* the dealer to countermand the order. George Rice, when before the House Committee on Manufactures in 1888, presented a number of telegrams as samples of his experience in having orders countermanded in Texas. Four of these were sent on the same day from different dealers in the same town, San Angelo. Mr. Rice investigated the cause, and, by letters from the various firms, learned that the Standard agent had been around "threatening the trade that if they bought of me they would not sell them any more," as he put it.

Mrs. Butts in her testimony in 1898 said that her firm had a customer in New Orleans to whom they had been selling from 500 to 1,000 barrels a month, and that the Standard representative made a contract with him to pay him $10,000 a year for five years to stop handling the independent oil and take Standard oil! Mrs. Butts offered as evidence of a similar transaction in Texas the following letter:

"LOCKHART, TEXAS, November 30, 1894.

"Mr. Keenan, who is with the Waters-Pierce people at Galveston, has made us several visits and made us propositions of all kinds to get us out of the business. Among others, he offered to pay us a monthly salary if we would quit selling oil and let them have full control of the trade, and insisted that we name a figure that we would take and get out of the business, and also threatened that if we did not accept his proposition they would cut prices below what oil cost us and force us out of business. We asked him the question, should we accept his proposition, would they continue to sell oil as cheap as we were then selling it, and he stated most positively that they would advance the price at once should they succeed in destroying competition.

"J. S. LEWIS AND COMPANY."

In the Ohio Investigation of 1898 John Teagle, of Cleveland, being upon his oath, said that his firm had had great difficulty in getting goods accepted because the Standard agents would persuade the dealers to cancel the orders. "They would have their local man, or some other man, call upon the trade and use their influence and talk lower prices, or make a lower retail price, or something to convince them that they'd better not take our oil, and, I suppose, to buy theirs." Mr. Teagle presented the following letter, signed by a Standard representative, explaining such a countermand:

"JOHN FOWLER, "DES MOINES, IOWA, January 14, 1891.
Hampton, Iowa.

"*Dear Sir:*—Our Marshalltown manager, Mr. Ruth, has explained the circumstances regarding the purchase and subsequent countermand of a car of oil from our competitors. He desires to have us express to you our promise that we will stand all expense provided there should be any trouble growing out of the countermand of this car. We cheerfully promise to do this; we have the best legal advice which can be obtained in Iowa, bearing on the points in this case. An order can be countermanded either before or after the goods have been shipped, and, in fact, can be countermanded even if the goods have already arrived and are at the depot. A firm is absolutely obliged to accept a countermand. The fact that the order has been signed does not make any difference. We want you to absolutely refuse, under any circumstances, to accept the car of oil. We are standing back of you in this matter, and will protect you in every way, and would kindly ask you to keep this letter strictly confidential. . . .

"Yours truly, E. P. PRATT."

Peter Shull, of the Independent Oil Company of Mansfield, Ohio, testified before the same committee to experiences similar to those of Mr. Teagle.

"If I put a man on the road to sell goods for me," said Mr. Shull, "and he takes orders to the amount of 200 to 300 barrels a week, before I am able to ship these goods possibly, the Standard Oil Company has gone there and compelled those people to countermand those orders under a threat that, if they don't countermand them, they will put the price of oil down to such a price that they cannot afford to handle the goods."

In support of his assertion Mr. Shull offered letters from firms he has been dealing with. The following citations show the character of them:

"TIFFIN, OHIO, February 1, 1898.

"INDEPENDENT OIL COMPANY,
Mansfield, Ohio.

"*Dear Sirs:*—The Standard Oil Company, after your man was here, had the cheek to come in and ask how many barrels of oil we bought and so forth, then asked us to countermand the order, saying it would be for our best; we understand they have put their oil in our next door and offer it at six cents per gallon, at retail. Shall we turn tail or show them fight? If so, will you help us out any? . . .

"Yours truly,

"TALBOTT AND SON."

"TIFFIN, OHIO, January 24, 1898.

"INDEPENDENT OIL COMPANY.

"*Dear Sirs:* . . . I am sorry to say that a Standard Oil man from your city followed that oil car and oil to my place, and told me that he would not let me make a dollar on that oil, and was dogging me around for two days to buy that oil, and made all kinds of threats and talked to my people of the house while I was out, and persuaded me to sell, and I was in a stew what I should do, but I yielded and I have been very sorry for it since. I thought I would hate to see the bottom knocked out of the prices, but that is why I did it—the only reason. The oil was all right. I now see the mistake, and that is of getting a carload—two carloads coming in here inside of a week is more than the other company will stand. . . .

"Yours truly,

"H. A. EIRICK."

In case the agent cannot persuade the dealer to countermand his order, more strenuous measures are applied. The letters quoted above hint at what they will be. Many letters have been presented by witnesses under oath in various investigations showing that Standard Oil agents in all parts of the country have found it necessary for the last twenty-five years to act at times as these letters threaten. One of the most aggressive of these campaigns waged at the beginning of this war of exterminating independent dealers was by the Standard marketing agent at Louisville, Kentucky—Chess, Carley and Company. This concern claimed a large section of the South as its territory. George Rice, of Marietta, Ohio, had been in this field for eight or ten years, having many regular customers. It became Chess, Carley and Company's business to secure these customers and to prevent his getting others. Mr. Rice was handicapped to begin with by railroad discrimination. He was never able to secure the rates of his big rival on any of the Southern roads. In 1888 the Interstate Commerce Commission examined his complaints against eight different Southern and Western roads, and found that no one of them treated him with "relative justice." Railroad discriminations were not sufficient to drive him out of the Southwest, however, and a war of prices was begun. According to the letters Mr. Rice himself has presented he certainly in some cases began the cutting, as he could well afford to do. For instance, Chess, Carley and Company were selling water-white oil in September, 1880, in Clarksville, Tennessee, at twenty-one cents a gallon delivered in carloads—export oil was selling in barrels in New York at that date at 10⅝ cents a gallon. Rice's agent offered at eighteen cents. The dealer to whom he made the offer, Armstrong by name, wished to accept, but as he had been buying of Chess, Carley and Company, went first to see them about the matter. He came back "scared almost out of his boots," wrote the agent to Rice.

"Carley told him he would break him up if he bought oil of anyone else; that the Standard Company had authorised him to spend $10,000 to break up any concern that bought oil from anyone else; that he (Carley) would put all his drummers in the field to hunt up Armstrong's customers and sell his customers groceries at five per cent. below Armstrong's prices, and turn all Armstrong's trade over to Moore, Bremaker and Company, and settle with Moore, Bremaker and Company for their losses in helping to break Armstrong up, every thirty days.

"That if Armstrong sent any other oil to Clarksville, Tennessee, he (Carley) would put the price of oil so low in Clarksville as to make the party lose heavily, and that they (the Standard) would break up anyone that would sell him (Armstrong) oil, and that he (Carley) had told Stege and Reiling the same thing. Did you ever? What do you think of that?"

Very soon after this, Chess, Carley and Company took in hand a Nashville firm, Wilkinson and Company, which was

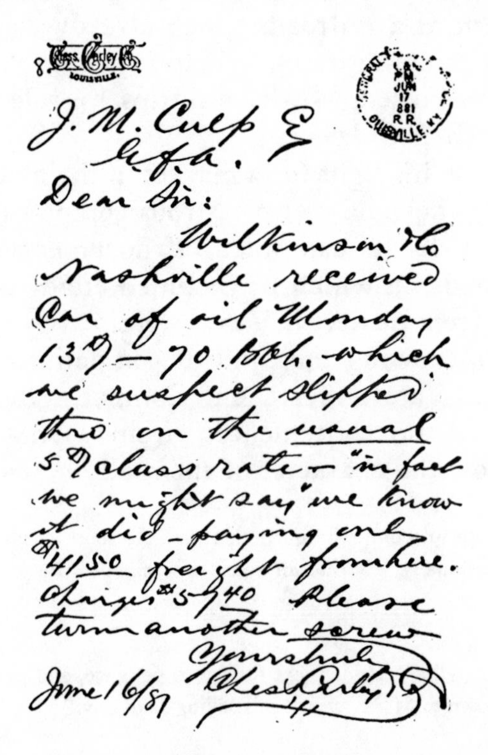

Chess Carley Co.
Louisville.

J. M. Culp Esq
G. F. A.
Dear Sir:
Wilkinson & Co
Nashville received
Car of oil Monday
1397 — 70 Bbls. which
we suspect slipped
thro on the usual
5th class rate — "in fact
we might say" we know
it did — paying only
$41.50 freight from here.
Charges $57.40 Please
turn another screw
Yours truly
Chess Carley Co
June 16/81

buying of Rice. "It is with great reluctance," they wrote, "that we undertake serious competition with any one, *and certainly this competition will not be confined to coal-oil or any one article, and will not be limited to any one year.* We always stand ready to make reasonable arrangements with any one who chooses to appear in our line of business, and it will be unlike anything we have done heretofore if we permit any one to force us into an arrangement which is not reasonable. Any loss, however great, is better to us than a record of this kind." And four days later they wrote: "If you continue to bring on the oil, it will simply force us to cut down our price, and no other course is left to us but the one we have intimated." Wilkinson and Company seem to have stuck to Rice's oil, for, sixteen months later, we find Chess, Carley and Company calling on the agent of a railroad, which already was giving the Standard discriminating rates, to help in the fight.

The screw was turned, Mr. Rice affirms, his rate being raised fifty per cent. in five days.

Rice carried on his fight for a market in the most aggressive way, and everywhere he met disastrous competition. In 1892 he published a large pamphlet of documents illustrating Standard methods, in which he included citations from some seventy letters from dealers in Texas, received by him between 1881 and 1889, showing the kind of competition his oil met there from the Waters-Pierce Oil Company, the Standard's Texas agents. A dozen sentences, from as many different towns, will show the character of them all:

"I have had wonderful competition on this car. As soon as my car arrived the Waters-Pierce Oil Company, who has an agent here, slapped the price down to $1.80 per case 110."

". . . Oil was selling at this point for $2.50 per case, and as soon as your car arrived it was put down to $1.50, which it is selling at to-day."

"The Waters-Pierce Oil Company reduced their prices on Brilliant oil from $2.60 to $1.50 per case and is waging a fierce war."

"Waters-Pierce Oil Company has our state by the throat and we would like to be extricated."

"I would like to handle your oil if I could be protected against the Waters-Pierce Oil Company. I am afraid if I would buy a car of oil from you this company would put the oil way below what I pay and make me lose big money. I can handle your oil in large quantities if you would protect me against them."

"The Waters-Pierce Oil Company has cut the stuffing out of coal-oil and have been ever since I got in my last car. They put the price to the merchants at $1.80 per case."

"We have your quotations on oil. While they are much lower than what we pay, yet unless a carload could be engaged it would pay no firm to try and handle, as Waters-Pierce Oil Company would cut below cost on same."

"The day your oil arrived here, their agent went to all my customers and offered their Eupion oil at ten cents per gallon in barrels and $1.50 per case, and lower grades in proportion, and told them if they did not refuse to take the oil he would not sell them any more at any price, and that he was going to run me out of the business, and then they would be at his mercy."

"Now we think Waters-Pierce Oil Company have been getting too high a price for their oil. They are able and do furnish almost this entire state with oil. They cut prices to such an extent when any other oil is offered in this state that they force the parties handling the oil to abandon the trade."

"Trace and hurry up car of oil shipped by you. We learn it is possible that your oil is side-tracked on the line, that Waters-Pierce might get in their work."

"If we were to buy a car or more, the Waters-Pierce Oil Company would manage to sell a little cheaper than we could, and continue doing so until they busted me up."

"In regard to oil, we are about out now, and Waters-Pierce have put their oil up again and quote us at the old price."

"Jobbers say when they take hold of another oil they are at once boycotted by Waters-Pierce Oil Company, who not only refuse to sell them, but put oil below what they pay for it, and thus knock them out of the oil trade, unless they sell at a loss."

"If I find that I can handle your oil in Texas without being run out and losing money by this infernal corporation, the Waters-Pierce Oil Company, I want to arrange with you to handle it extensively. I received verbal notice this morning from their agent that they would make it hot for me when my oil got here."

Mr. Rice claims, in his preface to the collection of letters here quoted from, that he has hundreds of similar ones from different states in the Union, and the writer asked to examine them. The package of documents submitted in reply to this request was made up literally of hundreds of letters. They came from twelve different states, and show everywhere the same competitive method—cutting to kill. One thing very noticeable in these letters is the indignation of the dealers at the Standard methods of securing trade. They resent threats. They complain that the Standard agents "nose" about their premises, that they ask impudent questions, and that they generally make the trade disgusting and humiliating. In Mississippi, in the eighties, the indignation of the small dealers against Chess, Carley and Company was so strong that they formed associations binding themselves not to deal with them.

These same tactics have been kept up in the Southwest ever since. A letter, dated April 28, 1891, from the vice-president of the Waters-Pierce Oil Company, A. M. Finlay, to his agent at Dallas, Texas, says bluntly: "We want to make the prices at Dallas and in the neighbourhood on Brilliant and water-white oil, that will prevent Clem (an independent dealer) from doing any business." And Mr. Finlay adds: "Hope you will make it a point to be present at the next meeting of the city council, to-morrow night, and do everything possible to prevent granting a permit to build within the city limits, unless building similar to ours is constructed, for it would not be fair to us to allow someone else to put up constructions for the storage of oil, when they had compelled us to put up such an expensive building as we have." *

Mr. Rice is not the only independent oil dealer who has produced similar testimony. Mr. Teagle and Mr. Shull, in Ohio, have furnished considerable. "The reason we quit taking your oil is this," wrote a Kansas dealer to Scofield, Shurmer and Teagle, in 1896: "The Standard Oil Company notified us that if we continued handling your oil they would cut the oil to ten cents retail, and that we could not afford to do, and for that reason we are forced to take their oil or do business for nothing or at a loss." "The Standard agent has repeatedly told me that if I continued buying oil and gasoline from your wagon," wrote an Ohio dealer to the same firm in 1897, "they would have it retailed here for less than I could buy. I paid no attention to him, but yesterday their agent was here and asked me decidedly if I would continue buying oil and gasoline from your wagon. I told him I would do so; then he went and made arrangements with the dealers that handle their oil and gasoline to retail it for seven cents."

Mr. Shull summed up his testimony before the same committee to which Mr. Teagle gave the above, by declaring: "You take $10,000 and go into the business and I will guarantee you won't be in business ninety days. Their motto is that anybody going into the oil business in opposition to them they will make life a burden to him. That is about as near as you can get to it."

Considerable testimony of the same sort of practices was offered in the recent "hearing before the Industrial Commission," most of it general in character. The most significant special case was offered by Mr. Westgate, the treasurer of the American Oil Works, an independent refinery of Titusville, Pennsylvania.

The American Oil Works, it seems, were in 1894 shipping oil called "Sunlight" in barrels to South Bend, Washington.

* Trust Investigation of Ohio Senate, 1898, page 370.

This was in the territory of the Standard agents at Portland, Oregon, one of whom wrote to a South Bend dealer when he heard of the intrusion: "We will state for your information that never a drop of oil has reached South Bend of better quality than what we have always shipped into that territory. They can name it 'Sunlight,' 'Moonlight,' or 'Starlight,' it makes no difference. You can rest assured if another carload of 'Sunlight' arrives at your place, it will be sold very cheap. We do not purpose to allow another carload to come into that territory unless it comes and is put on the market at one-half its actual cost. You can convey this idea to the young man who imported the carload of 'Sunlight' oil."

When John D. Archbold, of the Standard Oil Company, had his attention called to this letter by Professor Jenks, of the Industrial Commission, Mr. Archbold characterised the letter as "a foolish statement by a foolish and unwise man" and promised to investigate it. Later he presented the commission with an explanation from the superior of the agent, who declared that the writer of the letter did not have any authority to say that oil would be sold on the basis mentioned. "The letter," he continued, "was intended to be written in a jocular manner to deny a claim that he was selling oil inferior in quality to that sold by others." It is hard for the mere outsider to catch the jocularity of the letter, and it must have been much more difficult for the dealer who received it to appreciate it.

Independent oil dealers of the present day complain bitterly of a rather novel way employed by the Standard for bringing into line dealers whose prejudices against buying from them are too strong to be overcome by the above methods. This is through what are called "bogus" oil companies. The obdurate dealer is approached by the agent of a new independent concern, call it the A B C Oil Company, for illustration. The agent seeks trade on the ground that he represents an independent concern and that he can sell at lower prices than the firm from which the dealer is buying. Gradually he works his way into the independent's trade. As a matter of fact, the new company is merely a Standard jobbing house which makes no oil, and which conceals its real identity under a misleading name. The mass of reports from railroad freight offices quoted from in this article corroborate this claim of the independents. The A B C Oil Company is mentioned again and again as shipping oil, and in the audited reports it is always checked off in the same fashion as the known Standard companies, and none of its shipments is referred to Standard agents. Independents all over the country tell of loss of markets through underselling by these "bogus" companies. The lower price which a supposedly independent concern gives to a dealer who will not, under any condition, buy of the Standard, need not demoralise the Standard trade in the vicinity if the concession is made with caution. After the trade is secure, that is, after the genuine independent is ousted, the masquerading concern always finds itself obliged to advance prices. When the true identity of such a company becomes known its usefulness naturally is impaired, and it withdraws from the field and a new one takes its place.

There is never a dealer in oil too small to have applied the above methods of competition. In recent years they have frequently been applied even to oil peddlers. In a good many towns of the country oil is sold from door to door by men whose whole stock in trade is their peddling wagons. Many of these oil peddlers build up a good trade. As a rule they sell Standard oil. Let one take independent oil, however, and the case is at once reported. His customers are located and at once approached by a Standard tank wagon man, who frequently, it is said, not only sells at a lower price than they have been paying, but even goes so far as to clean and fill the lamps! In these raids on peddlers of independent oil, refined oil has been sold in different cities at the doors of consumers at less than crude oil was bringing at the wells, and several cents per gallon less than it was selling to wholesale dealers in refined. It is claimed by independents that at the present time the "bogus" companies generally manage this matter of driving out peddlers, thus saving the Standard the unpopularity of the act and the dissatisfaction of the rise in price which, of course, follows as soon as the trade is secured.

The general explanation of these competitive methods which the Standard officials have offered, is that they originate with "over-zealous" employees and are disapproved of promptly if brought to the attention of the heads of the house. The cases seem rather too universal for such an explanation to be entirely satisfactory. Certainly the system of collecting information concerning competitive business is not practised by the exceptional "over-zealous" employee, but is a recognised department of the Standard Oil Company's business. In the mass of documents from which the reports of oil shipments referred to above were drawn, are certain papers showing that the system is nearly enough universal to call for elaborate and expensive bookkeeping at the headquarters of each Standard marketing division. For instance, on the next page is a fragment illustrating the page of a book kept at such a headquarters.

What does this show? Simply that every day the reports received from railroad freight agents are entered in records kept for the purpose; that there is on file at the Standard Oil headquarters a detailed list of the daily shipments which each independent refiner sends out, even to the initials and number on the car in which the shipment goes. From this remarkable record the same set of documents shows that at least two sets of reports are made up. One is a report of the annual volume of business being done by each particular independent refiner or wholesale jobber, the other of the business of each individual local dealer, so far as the detectives of the Standard have been able to locate it. For instance, among the documents is the report on a well-known oil jobbing house in one of the big cities of the country—reproduced on the next page.

Competing Oil Receipts, Georgeville **Territory**

Bemis Reports	Date Ship'd	Date Rec'd	SHIPPER	FROM	CONSIGNEE	DESTINATION	Refined Barrels	Naphtha Barrels	Lub'ting Barrels	CAR		REMARKS
										Initial	No.	
	May											
5/20	5/14	/	Penn. Mfg. Co.	Oil City	L. M. O. Co	Georgeville		93		L. M	7741	
5/22	5/15	16	Warren L. & G. Co	Struthers	A. Spence Co	"		76		P. L. M.	432	
5/7	5/2	7	Crystal Oil Co	Oil City	Kennebunk Oil Co	"	64		12	G. K.	1684	
5/12	5/6	9	Crew Levick Co	Struthers	X. Y. Z. Oil Co	"	112			L M	43	
"	"	/	Empire Oil Wks.	Reno	Gordon [illegible] Co	"	63			P L M	64328	
5/27	5/23	25	Warren L. & G. Co	Struthers	"	"		87		G K	63748	
6/7	6/1	4	Tiona Rfg Co	Clarendon	"	"	75			G. K	66042	
6/14	5/30	30	Titusville Oil Wks	Titusville	Lenox Martin Co	"		92		B & O.	37421	
"	5/22	22	Pittsburg Rfg Co	Coraopolis	Henry Whitehead	"		92		"	94	
6/11	27	5/1	Emery Rfg Co	Bradford	X Y Z Oil Co	"		122		G K	496	
6/14	5/26	26	Climax L. & G. Co	Titusville	"	"		126		P. L M	643	

The figures, dates, consignees and destination on the above are fictitious. The names of shippers were copied from the original in possession of the writer.

A comparison of this report with the firm's own accounts shows that the Standard came within a small per cent. of an accurate estimate of the X Y Z's business.

Another curious use made of these reports from the freight offices is forming a card catalogue of local dealers. (See form on page 55.) Oil is usually sold at retail by grocers. It is with them that the local agents deal. Now the daily reports from the freight offices show the oil they receive. The competition reports from local agents also give more or less information concerning their business. A card is made out for each of them, tabulating the date on which he received oil, the name and location of the dealer he got it from, the quality, and the price he sells at. In a space left for remarks on the card there is written in red ink any general information about

Statement showing Receipts and Deliveries for December

	Old Places	New Places	Car Load	Less Car Load	Barrels 1901 Coal Oil	1902 Coal Oil	1901 Gasoline	1902 Gasoline
Total Receipts of Competitor					3540	5070	1102	2214
Less shipments not in our District					420	1849	198	562
Net Shipments in our District					3120	3221	904	1652
Territory covered by Competitor — Oil	2140		2140	927	3067			
	2412	63	2475	742		3217		
				361			361	
Gaso			1051	411				1462
Balance					53	4	543	190
Not accounted for							211	87
					53	4	332	103

For Year	Old Places	New Places	Car Load	Less Car Load	1901 Coal Oil	1902 Coal Oil	1901 Gasoline	1902 Gasoline
Receipts					23787	26742	8764	13141
Less Shipments not in our territory					1410	1921	1262	2167
					22377	24821	7502	10974
Territory covered by competitor — Oil	8146	1179	9325	6127	15452			
	9691	487	10178	1129		11307		
				678			678	
Gaso.	729	11	623	849				2212
					6925	13514	6827	8762
Not accounted for						3122	196	2171
					6925	10392	6628	6591

The above is similar to the form compiled by the Standard Oil Company.

the dealer the agent may have picked up. Often there is an explanation of why the man does not buy Standard oil—not infrequently this explanation reads: "Is opposed to monopolies." It is impossible to say from documentary evidence how long such a card catalogue has been kept by the Standard; that it has been a practice for at least twenty-five years the following quotation from a letter written in 1903 by a prominent Standard official in the Southwest to one of his agents shows: "Where competition exists," says the official, "it has been our custom to keep a record of each merchant's daily

Name E. C. Link　　Town Georgeville

P. O.　　Sh. Pt.　　Salesman Merrill　　Nearest Tank Wagon Station　　Rate

Date Shipped	Year	Shipped by	From	No. Bbls.	Kind of Oil	Price	Remarks
5/14	1901	B & Co	Pasadena	2	Refd		Formerly dealt altogether with X.Y.Z Co. & it is hard to wean him away. Is somewhat prejudiced & Opposed to monopolies & won't buy of S.O. Co. Merrill
5/19		"	"	3	"		
5/26		"	"	3	"		
6/3		"	"	3	"		
6/10		X.Y.Z. Co	"	3	L.M.N		
6/20		"	"	3	"		
6/26		B. & Co	"	2	Refd		
7/4		X.Y.Z Co	"	4	L.M.N		
7/16		B & Co	"	3	Refd		

The names, figures, and locations on the above form are fictitious. The remarks are copied from cards in possession of the writer.

purchase of bulk oil; and I know of one town at least in the Southern Texas Division where that record has been kept, whether there was competition or not, for the past fifteen years." *

The inference from this system of "keeping the eyes open" is that the Standard Oil Company knows practically where

* Trust Investigation of Ohio Senate, 1898, page 371.

every barrel shipped by every independent dealer goes; and where every barrel bought by every corner-grocer from Maine to California comes from. The documents from which the writer draws the inference do not, to be sure, cover the entire country, but they do cover in detail many different states, and enough is known of the Standard's competitive methods in states outside this territory to justify one in believing that the system of gathering information is in use everywhere. That it is a perfect system is improbable. Bribery is not as dangerous business in this country as it deserves to be—of course nothing but a bribe would induce a clerk to give up such information as these daily reports contain—but, happily, such is the force of tradition that even those who have practised it for a long time shrink from discovery. It is one of those political and business practices which are only respectable when concealed. Naturally, then, the above system of gathering information must be handled with care, and can never have the same perfection as that Mr. Rockefeller expected when he signed the South Improvement Company charter.

The moral effect of this system on employees is even a more serious feature of the case than the injustice it works to competition. For a "consideration" railroad freight clerks give confidential information concerning freight going through their hands. It would certainly be quite as legitimate for post-office clerks to allow Mr. Rockefeller to read the private letters of his competitors, as it is that the clerks of a railroad give him data concerning their shipments. Everybody through whose hands such information passes is contaminated by the knowledge. To be a factor, though even so small a one, in such a transaction, blunts one's sense of right and fairness. The effect on the local Standard agent cannot but be demoralising. Prodded constantly by letters and telegrams from superiors to secure the countermand of independent oil, confronted by statements of the amount of sales which have gotten away from him, information he knows only too well to have been secured by underhand means, obliged to explain why he cannot get this or that trade away from a rival salesman, he sinks into habits of bullying and wheedling utterly inconsistent with self-respect. "Is there nothing you independents can do to prevent our people finding out who you sell to?" an independent dealer reports a hunted Standard agent asking him. "My life is made miserable by the pressure brought on to chase up your sales. I don't like such business. It isn't right, but what can I do?"

The system results every now and then, naturally enough, in flagrant cases of bribing employees of the independents themselves. Where the freight office does not yield the information, the rival's own office may, and certainly if it is legitimate to get it from one place it is from the other. It is not an unusual thing for independent refiners to discharge a man whom they have reason to believe gives confidential information to the Standard. An outrageous case of this, which occurred some ten years ago, is contained in an affidavit which has been recently put at the writer's disposition. It seems that in 1892 the Lewis Emery Oil Company, an independent selling concern in Philadelphia, employed a man by the name of Buckley. This man was discharged, and in September of that year he went into the employ of the leading Standard refinery of Philadelphia, a concern known as the Atlantic Refining Company. According to the affidavit made by this man Buckley, the managers of the Standard concern, some time in February, 1893, engaged him in conversation about affairs of his late employer. They said that if they could only find out the names of the persons to whom their rival sold, and for what prices, they could soon run him out of business! And they asked Buckley if he could not get the information for them. After some discussion, one of the Standard managers said: "What's the matter with the nigger?" alluding to a coloured boy in the employment of the Lewis Emery concern. Buckley told them that he would try him. "You can tell the nigger," said one of the men, "that he needn't be afraid, because if he loses his position there's a position here for him."

Buckley saw the negro and made a proposition to him. The boy agreed to furnish the information for a price. "Starting from February, 1893," says Mr. Buckley, "and lasting up to about August of the same year, this boy furnished me periodically with the daily shipments of the Lewis Emery concern, which I took and handed personally, sometimes to one and sometimes to the other manager. They took copies of them, and usually returned the originals." The negro also brought what is known as the price-book to Buckley, and a complete copy of this was made by the Standard managers. "In short," says Mr. Buckley in his affidavit, "I obtained from the negro all the inside facts concerning the Lewis Emery Oil Company's business, and I furnished them all to the Standard managers." In return for this information the negro lad was paid various sums, amounting in all to about ninety dollars. Buckley says that they were charged upon the Standard books to "Special Expenses." The transaction was ended by the discharge of the coloured boy by the Lewis Emery concern.

The dénouement of this case is tragic enough. The concern was finally driven out of business by these and similar tactics, so Mr. Emery and his partner both affirm. The negro was never taken into the Atlantic Refinery, and Buckley soon after lost his position, as he of course richly deserved to. A man who shows himself traitorous, lying, thieving, even for the "good of the oil business," is never kept long in the employment of the Standard Oil Company. It is notorious in the Oil Regions that the people who "sell" to the Standard are never given responsible positions. They may be shifted around to do "dirty work," as the Oil Regions phrase goes, but they are pariahs in the concern. Mr. Rockefeller knows as well as any man ever did the vital necessity of honesty in an organisation, and the Buckleys and negroes who bring him secret intelligence never get anything but money and contempt for their pains.

For the general public, absorbed chiefly in the question, "How does all this affect what we are paying for oil?" the chief point of interest in the marketing contests is that, after they were over, the price of oil has always gone back with a jerk to the point where it was when the cutting began, and not infrequently it has gone higher—the public pays. Several of the letters already quoted in this chapter show the immediate recoil of the market to higher prices with the removal of competition. A table was prepared in 1892 to show the effect of competition on the price of oil in various states of the Union. The results were startling. In California, oil which sold at non-competitive points at 26½ cents a gallon, at competitive points brought 17½ cents. In Denver, Colorado, there was an "Oil War" on in the spring of 1892, and the same oil which was selling at Montrose and Garrison at twenty-five cents a gallon, in Denver sold at seven cents. This competition finally killed opposition and Denver thereafter paid twenty-five cents. The profits on this price were certainly great enough to call for competition. The same oil which was sold in Colorado in the spring of 1892 at twenty-five cents, sold in New York for exportation at 6.10 cents. Of course the freight rates to Colorado were high, the open rate was said to be nine cents a gallon, but that it cost the Standard Oil Company nine cents a gallon to get its oil there, one would have to have documentary proof to believe, and, even if it did, there was still some ten cents profit on a gallon—five dollars on a barrel. In Kansas, at this time, the difference between the price at competitive and non-competitive points was seven cents; in Indiana six cents; in South Carolina four and one-half cents.*

In 1897 Scofield, Shurmer and Teagle, of Cleveland, prepared a circular showing the difference between prices at competitive and non-competitive points in Ohio, and sent it out to the trade. According to this circular the public paid from 25 to 33⅓ per cent. more where there was no competition. The fact that oil is cheaper where there is competition, and also that the public has to pay the cost of the expensive "Oil Wars" which have been carried on so constantly for the last twenty-five years all over the country, is coming to be recognised, especially in the Middle West of this country, by both dealers and communities. There is no question that the attempts of Standard agents to persuade or bully dealers into countermanding orders, or giving up an independent with whose oil they are satisfied, meet with much less general success than they once did. It even happens now and then that communities who have had experience with "Oil Wars" will stand by an independent dealer for months at a time, resisting even the temptation to have their lamps cleaned and filled at next to nothing.

Briefly put, then, the conclusion, from a careful examination of the testimony on Standard competitive methods, is this:

The marketing department of the Standard Oil Company is organised to cover the entire country, and aims to sell all the oil sold in each of its divisions. To forestall or meet competition it has organised an elaborate secret service for locating the quantity, quality, and selling price of independent shipments. Having located an order for independent oil with a dealer, it persuades him, if possible, to countermand the order. If this is impossible, it threatens "predatory competition," that is, to sell at cost or less, until the rival is worn out. If the dealer still is obstinate, it institutes an "Oil War." In late

* See Appendix, Number 41. Table showing prices of oil at competitive and non-competitive points in 1892.

years the cutting and the "Oil Wars" are often intrusted to so-called "bogus" companies, who retire when the real independent is put out of the way. In later years the Standard has been more cautious about beginning underselling than formerly, though if a rival offered oil at a less price than it had been getting—and generally even small refineries can contrive to sell below the non-competitive prices of the Standard—it does not hesitate to consider the lower price a declaration of war and to drop its prices and keep them down until the rival is out of the way. The price then goes back to the former figure or higher. John D. Archbold's testimony before the Industrial Commission in 1898 practically confirms the above conclusion. Mr. Archbold said that the Standard was in the habit of fighting vigorously to hold and advance its trade—even to the extent of holding prices down to cost until the rival gives way—though he declared it to be his opinion that the history of the company's transactions would show that the competitor forces the fight. Mr. Archbold told the commission that he personally believed it was not advisable to sell below cost for the sake of freezing out a smaller rival, save in "greatly aggravated cases," though he admitted the Standard sometimes did it. The trouble is that, accepting Mr. Rockefeller's foundation principle that the oil business belongs to him, any competition is "an aggravated case." All that is reassuring in the situation has come from the obstinate stand of individuals—the refiners who insisted on doing an independent business, on the theory that "this is a free country"; the grocers who resented the prying and bullying of Standard agents, and asserted their right to buy of whom they would; the rare, very rare, community that grasped the fact that oil sold below cost temporarily, meant later paying for the fight. These features of the business belong to the last decade and a half. At the period we have reached in this history—that is, the completion of the monopoly of the pipelines in 1884 and the end of competition in transporting oil—

there seemed to the independents no escape from Mr. Rockefeller in the market.

The sureness and promptness with which he located their shipments seemed uncanny to them. The ruthlessness and persistency with which he cut and continued to cut their prices drove them to despair. The character of the competition Mr. Rockefeller carried on in the markets, particularly of the South and Middle West of this country, at this time, aggravated daily the feeble refining element, and bred contempt far and wide among people who saw the cutting, and perhaps profited temporarily by it, but who had neither the power nor the courage to interfere. The knowledge of it fed greatly the bitterness in the Oil Regions. Part of the stock in conversation of every dissatisfied oil producer or ruined refiner became tales of disastrous conflicts in markets. They told of crippled men selling independent oil from a hand cart, whose trade had been wiped out by a Standard cart which followed him day by day, practically giving away oil. They told of grocers driven out of business by an attempt to stand by a refiner. They told endless tales, probably all exaggerated, perhaps some of them false, yet all of them believed, because of such facts as have been rehearsed above. There came to be a popular conviction that the "Standard would do anything." It was a condition which promised endless annoyance to Mr. Rockefeller and his colleagues. It meant popular mistrust, petty hostilities, misinterpretations, contempt, abuse. There were plenty of people even willing to deny Mr. Rockefeller ability. That the Standard was in a venture was enough in those people's minds to damn it. Anything the Standard wanted was wrong, anything they contested was right. A verdict for them demonstrated the corruption of the judge and jury; against them their righteousness. Mr. Rockefeller, indeed, was each year having more reason to realise monopoly building had its trials as wells as its profits.

CHAPTER ELEVEN

THE WAR ON THE REBATE

ROCKEFELLER'S SILENCE—BELIEF IN THE OIL REGIONS THAT COMBINED OPPOSITION TO HIM WAS USELESS—INDIVIDUAL OPPOSITION STILL CONSPICUOUS—THE STANDARD'S SUIT AGAINST SCOFIELD, SHURMER AND TEAGLE—SEEKS TO ENFORCE AN AGREEMENT WITH THAT FIRM TO LIMIT OUTPUT OF REFINED OIL—SCOFIELD, SHURMER AND TEAGLE ATTEMPT TO DO BUSINESS INDEPENDENTLY OF THE STANDARD AND ITS REBATES—FIND THEIR LOT HARD—THEY SUE THE LAKE SHORE AND MICHIGAN SOUTHERN RAILWAY FOR DISCRIMINATING AGAINST THEM—A FAMOUS CASE AND ONE THE RAILWAY LOSES—ANOTHER CASE IN THIS WAR OF INDIVIDUALS ON THE REBATE SHOWS THE STANDARD STILL TO BE TAKING DRAWBACKS—THE CASE OF GEORGE RICE AGAINST THE RECEIVER OF THE CINCINNATI AND MARIETTA RAILROAD.

THE apathy and inaction which naturally flow from a great defeat lay over the Oil Regions of Northwestern Pennsylvania long after the compromise with John D. Rockefeller in 1880, followed, as it was, by the combination with the Standard of the great independent seaboard pipe-line which had grown up under the oil men's encouragement and patronage. Years of war with a humiliating outcome had inspired the producers with the conviction that fighting was useless, that they were dealing with a power verging on the superhuman—a power carrying concealed weapons, fighting in the dark, and endowed with an altogether diabolic cleverness. Strange as the statement may appear, there is no disputing that by 1884 the Oil Regions as a whole looked on Mr. Rockefeller with superstitious awe. Their notion of him was very like that which the English common people had for Napoleon in the first part of the 19th century, which the peasants of Brittany have even to-day for the English—a dread power, cruel, omniscient, always ready to spring.

This attitude of mind, altogether abnormal in daring, impetuous, and self-confident men, as those of the Oil Regions were, was based on something more than the series of bold and admirably executed attacks which had made Mr. Rockefeller master of the oil business. The first reason for it was the atmosphere of mystery in which Mr. Rockefeller had succeeded in enveloping himself. He seems by nature to dislike the public eye. In his early years his home, his office, and the Baptist church were practically the only places which saw him. He did not frequent clubs, theatres, public meetings. When his manœuvres began to bring public criticism upon him, his dislike of the public eye seems to have increased. He took a residence in New York, but he was unknown there save to those who did business with him or were interested in his church and charities. His was perhaps the least familiar face in the Standard Oil Company. He never went to the Oil Regions, and the Oil Regions said he was afraid to come, which might or might not have been true. Certainly the Oil Regions never hesitated to express opinions about him calculated to make a discreet man keep his distance.

Even in Cleveland, his home for twenty-five years, Mr. Rockefeller was believed to conceal himself from his townsmen. It is certain that the operations of his great business were guarded with the most jealous care. The New York Sun sent an "experienced observer" to Cleveland in 1882 to write up the Standard concern. He speaks with amazement in his letters of the atmosphere of secrecy and mystery which he found enveloping everything connected with Mr. Rockefeller. You could not get an interview with him, the observer complained; even his home papers had ceased to go to the Standard offices to inquire about the truth of rumours which reached them from the outside. The hundreds of employees of the trust in the town were as silent as their master in all that concerned the business, and if one talked—well, he was not long an employee of Mr. Rockefeller. There was between the Standard Oil Company and the town and press of Cleveland none of the *camaraderie,* the mutual good-will and pride and confidence which usually characterise the relations between great businesses and their environment.

In Cleveland, as in the Oil Regions, Mr. Rockefeller's careful effort to cover up his intentions and his tracks had been at first met with jeers and blunt rebuffs, but he had finally succeeded in silencing and awing the people. It is worth noting that while all of the members of the Standard Oil Company followed Mr. Rockefeller's policy of saying nothing, there was no such popular dread of any other one of them. In the Oil Regions, for instance, there was a bitter hatred of the Standard Oil Company as an organisation, but for the most part the people liked the men who served it, and certainly had no awe of them, for these men circulated freely among their fellow-townsmen; they were active in all the pleasures and enterprises of the communities in which they lived; they were generous, able, cordial, and whatever the people said of the concern they served, they generally qualified it by expressing their personal likings for the men themselves.

A second reason for the popular dread of Mr. Rockefeller was that this man, whom nobody saw and who never talked, knew everything—even unexpected and trivial things—and those who saw the effect of this knowledge and did not see how he could obtain it, regarded him as little short of an omniscient being. There was really nothing in the least occult about Mr. Rockefeller's omniscience. He obtained part of his knowledge of other people's affairs by a most extensive and thoroughly organised system of news-gathering, such as any bright business man of wide sweep might properly employ. But he combined with this perfectly legitimate work the sordid methods of securing confidential information described in the last chapter. Certainly there is nothing of the transcendental in this kind of omniscience, and the feeling of supernaturalism which Mr. Rockefeller had inspired by 1884 has entirely evaporated since, as evidence of his methods has been circulated. The source was, however, long secret, and when again and again men who could hardly suppose their existence known to Mr. Rockefeller saw movements anticipated which they believed known only to themselves and their confidential agents, they began to dread him and to invest him with mysterious qualities. If Mr. Rockefeller had been as great a psychologist as he is business manipulator he would have realised that he was awakening a terrible popular dread, and he would have foreseen that one day, with the inevitable coming to light of his methods, there would spring up about his name a crop of scorn which would choke any crop of dollars and donations which the wealth of the earth could produce.

The effect of this dread was deplorable, for it intensified the feeling, now wide-spread in the Oil Regions, that it was useless to make further effort at a combined resistance. And yet these men, who were now lying too supine in Mr. Rockefeller's steel glove even to squirm, had laid the foundation of freedom in the oil business. It has taken thirty years to demonstrate the inestimable value of the efforts which in 1884 they regarded as futile—thirty years to build even a small structure on the foundation they had laid, though that much has been done.

The situation was saved at this critical time by individuals scattered through the oil world who were resolved to test the validity of Mr. Rockefeller's claim that the coal-oil business belonged to him. "We have a right to do an independent business," they said, "and we propose to do it." They began this effort by an attack on the weak spot in Mr. Rockefeller's armour. The twelve years just passed had taught them that the realisation of Mr. Rockefeller's great purpose had been made possible by his remarkable manipulation of the railroads. It was the rebate which had made the Standard Oil Trust, the rebate, amplified, systematised, glorified into a power never equalled before or since by any business of the country. The rebate had made the trust, and the rebate, in spite of ten years of combination, Petroleum Associations, Producers' Unions, resolutions, suits in equity, suits in quo warranto, appeals to Congress, legislative investigations—the rebate still was Mr. Rockefeller's most effective weapon. If they could wrest it from his hand they could do business. They had learned something else in this period—that the whole force of public opinion and the spirit of the law were against the rebate, and that the railroads, knowing this, feared exposure of discrimination, and could be made to settle rather than have their practices made public. Therefore, said these individuals, we propose to sue for rebates and collect charges until we make it so harassing and dangerous for the railroads that they will shut down on Mr. Rockefeller.

The most interesting and certainly the most influential of these private cases was that of Scofield, Shurmer and Teagle, of Cleveland, one of the firms which, in 1876, entered into a "joint adventure" with Mr. Rockefeller for limiting the output and so holding up prices.* The adventure had been most successful. The profits were enormous. Scofield, Shurmer and Teagle had made thirty-four cents a barrel out of their refinery the year before the "adventure." With the same methods of manufacture, and enjoying simply Mr. Rocke-

* See Chapter V, page 165.

feller's control of transportation rates and the enhanced prices caused by limiting output, they made $2.52 a barrel the first year after. This was the year of the Standard's first great coup in refined oil. The dividends on 88,000 barrels this year were $222,047, against $41,000 the year before. In four years Scofield, Shurmer and Teagle paid Mr. Rockefeller $315,-345 on his investment of $10,000—and rebates.

After four years the Standard began to complain that their partners in the adventure were refining too much oil—the first year the books showed they had exceeded their 85,000-barrel limitation by nearly 3,000, the second year by 2,000, the third by 15,000, the fourth by 5,000. Dissatisfied, the Standard demanded that the firm pay them the entire profit upon the excess refined; for, claimed Mr. Rockefeller, our monopoly is so perfect that we would have sold the excess if you had not broken the contract, consequently the profits belong to us. Scofield, Shurmer and Teagle paid half the profit on the excess, but refused more, and they persisted in exceeding their quota; then Mr. Rockefeller, controlling by this time the crude supply in Cleveland through ownership of the pipe-lines, shut down on their crude supply. If they would not obey the contract of their own will they could not do business. The firm seems not to have been frightened. "We are sorry that you refuse to furnish us crude oil as agreed," they wrote Mr. Rockefeller; "we do not regard the limitation of 85,000 barrels as binding upon us, and as we have a large number of orders for refined oil we must fill them, and if you refuse to furnish us crude oil on the same favourable terms as yourselves, we shall get it elsewhere as best we can and hold you responsible for its difference in cost."

Mr. Rockefeller's reply was a prayer for an injunction against the members of the firm, restraining them individually and collectively "from distilling at their said works at

WILLIAM C. SCOFIELD

Senior member of the firm of Scofield, Schurmer and Teagle, of Cleveland. Plaintiff in important suits against Lake Shore Railroad for freight discriminations.

DANIEL SCHURMER

Associate of Mr. Scofield and Mr. Teagle in the war on railroad rebates which the firm waged for nearly twenty years.

JOHN TEAGLE

Independent refiner of Cleveland, Ohio, prominent in struggle against freight discriminations by the railroads.

CHARLES B. MATTHEWS

Independent refiner of Buffalo. Plaintiff in "Buffalo case," where members of the Standard Oil Company were indicted for conspiracy.

Cleveland, Ohio, more than 85,000 barrels of crude petroleum of forty-two gallons each in every year, and also from distilling any more than 42,500 barrels of crude petroleum of forty-two gallons each, each and every six months, and also from distilling any more crude petroleum until the expiration of six months from and after July 20, 1880, and also from directly and indirectly engaging in or being concerned in any business connected with petroleum or any of its products except in connection with the plaintiff under their said agreement, and that on the final hearing of this case the said defendants may in like manner be restrained and enjoined from doing any of said acts until the expiration of said agreement, and for such other and further relief in the premises as equity can give." In this petition, really remarkable for its unconsciousness of what seems obvious—that the agreement was preposterous and void because confessedly in restraint of trade—the terms of the joint adventure are renewed in a way to illustrate admirably the sort of tactics with refiners which, at this time, was giving Mr. Rockefeller his extraordinary power over the price of oil.*

Scofield, Shurmer and Teagle did not hesitate to take up the gauntlet, and a remarkable defence they made. In their answer they declared the so-called agreement had at all times been "utterly void and of no effect as being by its terms in restraint of trade and against public policy." They declared that the Standard Oil Company had never kept the terms of the agreement, that it had intentionally withheld the benefits of the advantages it enjoyed in freight contracts, and that it now was pumping crude oil from the Oil Regions to Cleveland at a cost of about twelve cents a barrel and charging them (Scofield, Shurmer and Teagle) twenty cents. They denied that the Standard had sustained any damage through

* See Appendix, Number 42. Standard Oil Company's petition for relief and injunction.

them, but claimed that their business had been carried on at a large profit. "There is such a large margin between the price of crude oil and refined," declared the defendants, "that the manufacture and sale of refined oil is attended with large profit; it is impossible to supply the demand of the public for oil if the business and refineries of both plaintiff and defendant are carried on and run to their full capacities, and if the business of the defendants were stopped, as prayed for by the plaintiff, it would result in a still higher price for refined oil and the establishment of more perfect monopoly in the manufacture and sale of the same by plaintiff." To establish such a monopoly, the defendants went on to declare, had been the sole object of the Standard Oil Company in making this contract with them, and similar ones with other firms, to establish a monopoly and so maintain unnaturally high prices,* and certainly Scofield, Shurmer and Teagle knew whereof they swore, for they had shared in the spoils of the winter of 1876 and 1877, and at this very period, October, 1880, they were witnessing an attempt to repeat the coup.

The charge of monopoly Scofield, Shurmer and Teagle sustained by a remarkable array of affidavits—the most damaging set for the Standard Oil Company which had ever been brought together. It contained the affidavits of various individuals who had been in the refining business in Cleveland at the time of the South Improvement Company and who had sold out in the panic caused by it. It contained a review of the havoc which that scheme and the manipulation of the railroads by the Standard which followed it had caused in the refining trade in Pennsylvania, and it gave the affidavits of Mrs. B—— and of her secretary and others concerning the circumstances of her sale in 1878 (see Chapter VI). The affidavits filed by John D. Rockefeller, Oliver H. Payne and Henry M. Flagler in reply to the set presented by Sco-

* See Appendix, Number 43. Answer of William C. Scofield *et al.*

field, Shurmer and Teagle are curious reading. From the point of view of our present knowledge they deny a number of things now known to be true.*

It was not necessary, however, for the defendants to have presented their elaborate array of evidence to support the charge of intended monopoly. The character of the agreement itself was sufficient to prevent any judge from attempting to enforce it. The amazement was that the Standard Oil Company ever had the hardihood to ask for its enforcement. "That it should venture to ask the assistance of a court of equity to enforce a contract to limit the production and raise the price of an article of so universal use as kerosene oil," said the Chicago Tribune, "shows that the Standard Oil Company believed itself to have reached a height of power and wealth that made it safe to defy public opinion." This case is not the only one belonging to the period which goes to support the opinion of the Tribune.

Scofield, Shurmer and Teagle were now obliged to stand on their own feet. They could refine all the oil they wished, but they must make their own freight contracts, and they found rates when you worked with Mr. Rockefeller were vastly different from rates when you competed with him. The agent of the Lake Shore Railroad, by which most of their shipments went, told them frankly that they could not have the rates of the Standard unless they gave the same volume of business. The discrimination against them was serious. For instance, in 1880, when the Standard paid sixty-five cents a barrel from Cleveland to Chicago, Scofield, Shurmer and Teagle paid eighty. From April 1 to July 1, 1881, the Standard paid fifty-five cents and their rival eighty cents; from July 1 to November 1, 1881, the rates were thirty-five and seventy cents respectively, and so it went on for three years, when the firm, despairing of any change, took the case

* See Appendix, Number 44. Affidavit of John D. Rockefeller.

into court. This case, fought through all the courts of Ohio, and in 1886 taken to the Supreme Court of the United States, is one of the clearest and cleanest in existence for studying all the factors in the rebate problem—the argument and pressure by which the big shipper secures and keeps his advantage, the theory and defence of the railroad in granting the discrimination, the theory on which the suffering small shipper protests, and finally the law's point of view. The first trial of the case was in the Court of Common Pleas, and the refiners won. The railroad then appealed to the District Court (the present Circuit Court), where it was argued. So "important and difficult" did the judges of the District Court find the questions involved to be, that on the plea of the railroad they sent their findings of the facts in the case to the Supreme Court of the state for decision—a privilege they had under the law in force at that time.

These findings are elaborate, including some twenty-three propositions.* They have been confused by certain writers with the *opinion* on them given later by the Supreme Court; for instance, in an economic study recently published—"The Rise and Progress of the Standard Oil Company,"—the twelfth and thirteenth and part of the fourteenth proposition which the District Court sent up to the Supreme Court in its "findings of facts" are quoted separately, and the inference from the context is that the writer supposed he was citing part of the court's *opinion*. As the reader will see from what follows, the paragraphs in question are important, for, taken as quoted, they seem to show that the rebate the Standard received, and which Scofield, Shurmer and Teagle wanted, was on account of facilities it gave which the other refiners could not give:

"The court further find that prior to 1875 it was a question whether the Standard Oil Company would remain in Cleveland or remove its works to the oil-producing

* See Appendix, Number 45, Findings of Fact.

country, and such question depended mainly upon rates of transportation from Cleveland to market; that prior thereto said Standard Company did ship large quantities of its products by water to Chicago and other lake points, and from thence distributed the same by rail to inland markets; that it then represented to defendant the probability of such removal; that water transportation was very low during the season of navigation; that unless some arrangement was made for rates at which it could ship the year round as an inducement, it would ship by water and store for winter distribution; that it owned its tank-cars and had tank stations and switches, or would have, at Chicago, Toledo, Detroit and Grand Rapids, on and into which the cars and oil in bulk could be delivered and unloaded without expense and annoyance to defendant; that it had switches at Cleveland leading to its works at which to load cars, and would load and unload all cars; that the quantity of oil to be shipped by the company was very large, and amounted to ninety per cent. or more of all the oil manufactured or shipped from Cleveland, and that if satisfactory rates could be agreed upon it would ship over defendant's road all its oil products for territory and markets west and northwest of Cleveland, and agree that the quantity for each year should be equal to the amount shipped the preceding year; that upon the faith of these representations the defendant did enter into the contract and arrangement substantially as set forth in defendant's answer; that the rates were not fixed rates, but depended upon the general card tariff rates as charged from time to time, but substantially to be carried from time to time for about ten cents per barrel less than tariff rates, and, in consideration of such reduced rates as to bulk oil, the Standard Company agreed to furnish its own cars and tanks, load them on switches at distributing points, and unload them into distributing tanks, and was also to load and unload oil shipped in barrels, and without expense to defendant, and with, by reason thereof, less risk to defendant, which entered into the consideration, and was also to ship all its freight to points west and northwest of Cleveland, except small quantities to lake ports not reached by rail, and to so manage the shipments, as to cars and times, as would be most favourable to defendant; that defendant then agreed to said terms; that said agreement so made in 1875 has remained in force ever since.

"That, at a cost exceeding $100,000, said Standard Company had and constructed the terminal facilities promised and herein found; that, in fact, the risk of danger from fire to defendant, the expense of handling, in loading and unloading, and in the use of the Standard tank-cars is less (but how much the testimony does not show) than upon oil shipped without the use of such or similar terminal facilities; that said Standard Company commenced by shipping about 450,000 barrels a year over defendant's road, which increased from year to year until, in 1882, the year before filing the petition in this action, the quantity so shipped on defendant's road amounted to 742,000 barrels, equal to 2,000 barrels or one full train-load per day.

"That said arrangement was not exclusive, but was at all times open to others shipping a like quantity and furnishing like service and facilities; that it was not made or continued with any intention on the part of the defendant to injure the plaintiffs in any manner."

Now, as a matter of fact, other propositions in this same set from which the above are quoted, find that Scofield, Shurmer and Teagle offered the railroad exactly the same facilities as the Standard, a switch, loading racks, exemption from loss by fire or accident.* "The manner of making shipments for plaintiffs and for the Standard Oil Company was precisely the same, and the only thing to distinguish the business of the one from the other was the aggregate yearly amounts of freight shipped," said Judge Atherton, of the Supreme Court, who gave the decision on the findings of fact, and he held in common with his predecessors that a rebate on account of volume of business only was "a discrimination in favour of capital," and contrary to a sound public policy, violation of that equality of rights guaranteed to every citizen, and a wrong to the disfavoured person. "We hold, . . ." he said, "that a discrimination in the rate of freights resting extensively on such a basis ought not to be sustained. The principle is opposed to sound public policy. It would build up and foster monopolies, add largely to the accumulated power of capital and money, and drive out all enterprise not backed by overshadowing wealth. With the doctrine, as contended for by the defendants, recognised and enforced by the courts, what will prevent the great grain interest of the Northwest, or the coal and iron interests of Pennsylvania, or any of the great commercial interests of the country bound together by the power and influence of aggregated wealth and in league with the railroads of the land, driving to the wall all private enterprises struggling for existence, and with an iron hand thrusting back all but themselves?" Judge Atherton was scathing enough in his opinion

* See Appendix, Number 45.

of the contract between the Lake Shore and the Standard. Look at it, he said, and see just what is shown. In consideration of the company giving to the railroad its entire freight business in oil, they transport this freight about ten cents a barrel cheaper than for any other customer. "The understanding was to keep the price *down* for the favoured customer, but *up* for all others, and the inevitable tendency and effect of this contract was to enable the Standard Oil Company to establish and maintain an overshadowing monopoly, to ruin all other operators and drive them out of business in all the region supplied by the defendant's road, its branches and connecting lines."

Judge Atherton was particularly hard on the portion of the contract * which pledged the Standard to give the Lake Shore *all* its freight in return for the rebates, and for this reason: In 1883 a new road Westward was opened from Cleveland, the New York, Cincinnati and St. Louis. It might become an active competitor in transporting petroleum for customers other than the Standard Oil Company. It might establish such a tariff of rates that other operators in oil might successfully compete with the Standard Oil Company. To prevent this, the Lake Shore road, on the completion of the new road, entered into a tariff arrangement giving to it a portion of the Westward shipments of the Standard Oil Company, on condition of its uniting in carrying out the understanding in regard to rebates to the Standard Oil Company. "How peculiar!" exclaimed Judge Atherton. "The defendant, by a contract made in 1875, was entitled to all the freights of the Standard Oil Company, and yet, say the District Court, 'for the purpose of securing the *greater part* of said trade,' they entered into a contract to divide with the new railroad, if the latter would only help to keep the rates *down* for the Standard and *up* for everybody else."

* Number 20, Findings of Facts. See Appendix, Number 45.

Such a contract so carried out was, in the opinion of the court, "not only contrary to a sound public policy, but to the lax demands of the commercial honesty and ordinary methods of business."

Another fact found by the District Court incensed Judge Atherton. This was that the contract "was not made or continued with any intention on the part of the defendant to injure the plaintiffs in any manner." It does not "make any difference in the case," he declared. "The plaintiffs were not doing business in 1875, when the contract was entered into, and, of course, it was not made to injure them in particular. If a man rides a dangerous horse into a crowd of people, or discharges loaded firearms among them, he might, with the same propriety, select the man he injures and say he had no intention of wounding him. And yet the law holds him to have intended the probable consequences of his unlawful act as fully as if purposely directed against the innocent victim, and punishes him accordingly. And this contract, made to build up a monopoly for the Standard Oil Company and to drive its competitors from the field, is just as unlawful as if its provisions had been aimed directly against the interests of the plaintiffs." *

Having lost their case in the Supreme Court of the state, the Lake Shore now appealed to the Supreme Court of the United States, and the record was filed in November, 1886. It was never heard; the railroad evidently concluded it was useless, and finally withdrew its petition, thereby accepting the decision of the Supreme Court of Ohio restraining it from further discrimination against Scofield, Shurmer and Teagle.

This case, which was before the public constantly during the six or seven years following the breaking up of the Producers' Union, in which the Oil Regions presented no united

* Ohio State Reports, 43, pages 571–623.

BURST IN A PIPE LINE

front to Mr. Rockefeller, served to keep public attention on the ruinous effect of the rebate and to strengthen the feeling that drastic legislation must be taken if Mr. Rockefeller's exploit was to be prevented in other industries.

One other case came out in this war of individuals on the rebate system which heightened the popular indignation against the Standard. It was a case showing that the Standard Oil Company had not yet abandoned that unique feature of its railroad contracts by which a portion of the money which other people paid for their freight was handed over to them! This peculiar development of the rebate system seems to have belonged exclusively to Mr. Rockefeller. Indeed, a careful search of all the tremendous mass of materials which the various investigations of railroads produced shows no other case—so far as the writer knows—of this practice. It was the clause of the South Improvement contracts which provoked the greatest outcry. It was the feature of Mr. Cassatt's revelations in 1877 which dumfounded the public and which no one would believe until they saw the actual agreements Mr. Cassatt presented. The Oil Regions as a whole did not hesitate to say that they believed this practice was still in operation, but, naturally, proof was most difficult to secure. The demonstration came in 1885, through one of the most aggressive and violent independents which the war in oil has produced, George Rice, of Marietta, Ohio. Mr. Rice, an oil producer, had built a refinery at Marietta in 1873. He sold his oil in the state, the West, and South. Six years later his business was practically stopped by a sudden raise in rates on the Ohio roads—an advance of fully 100 per cent. being made on freights from Marietta, where there were several independent refineries, although no similar advance was made from Wheeling and Cleveland, where the Standard refineries were located. These discriminations were fully shown in an investigation by the Ohio State Legislature in 1879. From that time on Mr. Rice was in constant difficulty about rates. He seems to have taken rebates when he could get them, but he could never get anything like what his big competitors got.

In 1883 Mr. Rice began to draw the crude supply for his refinery from his own production in the Macksburg field of Southeastern Ohio, not far from Marietta. The Standard had not at that time taken its pipe-lines into the Macksburg field; the oil was gathered by a line owned by A. J. Brundred, and carried to the Cincinnati and Marietta Railroad. Now, Mr. Brundred had made a contract with this railroad by which his oil was to be carried for fifteen cents a barrel, and all other shippers were to pay thirty cents. Rice, who conveyed his oil to the railroad by his own pipe-line, got a rate of twenty-five cents by using his own tank-car. Later he succeeded in getting a rate of 17½ cents a barrel. Thus the rebate system was established on this road from the opening of the Macksburg field. In 1883 the Standard Oil Company took their line into the field, and soon after Brundred retired from the pipe-line business there. When he went out he tried to sell the Standard people his contract with the railroad, but they refused it. They describe this contract as the worst they ever saw, but they seem to have gone Mr. Brundred one better, for they immediately contracted with the road for a rate of ten cents on their own oil, instead of the fifteen cents he was getting, and a rate of thirty-five on independent oil. And in addition they asked that the extra twenty-five cents the independents paid *be turned over to them!* If this was not done the Standard would be under the painful necessity of taking away its shipments and building pipe-lines to Marietta. The Cincinnati and Marietta Railroad at that time was in the hands of a receiver, one Phineas Pease—described as a "fussy old gentleman, proud of his position and fond of riding up and down the road in his private car." It is probably a good description. Certainly it is evident from what follows that the receiver was much "fussed up" ethically. Anxious to keep up the income of his road, Mr. Pease finally consented to the arrangement the Standard demanded. But he was worried lest his immoral arrangement be dragged into court, and wrote to his counsel, Edward S. Rapallo, of New York City, asking if there was any way of evading conviction in case of discovery.

"Upon my taking possession of this road," the receiver wrote, "the question came up as to whether I would agree to carry the Standard Company's oil to Marietta for ten cents per barrel, in lieu of their laying a pipe-line and piping their oil. I, of course, assented to this, as the matter had been fully talked over with the Western and Lake Erie Railroad Company before my taking possession of the road, and I wanted all the revenue that could be had in this trade.

"Mr. O'Day, manager of the Standard Oil Company, met the general freight agent of the Western and Lake Erie Railroad and our Mr. Terry, at Toledo, about February 12, and made an agreement (verbal) to carry their oil at ten cents per barrel. But Mr. O'Day compelled Mr. Terry to make a thirty-five cent rate on all other oil going to Marietta, and that we should make the rebate of twenty-five cents per barrel on all oil shipped by other parties, and that the rebate should be paid over to them (the Standard Oil Company), thus giving us ten cents per barrel for all oil shipped to Marietta, and the rebate of twenty-five cents per barrel going to the Standard Oil Company, making that company say twenty-five dollars per day clear money on George Rice's oil alone.

"In order to save the oil trade along our line, and especially to save the Standard Oil trade, which would amount to seven times as much as Mr. Rice's, Mr. Terry verbally agreed to the arrangement, which, upon his report to me, I reluctantly acquiesced in, feeling that I could not afford to lose the shipment of 700 barrels of oil per day from the Standard Oil Company. But when Mr. Terry issued instructions that on and after February 23 the rate of oil would be thirty-five cents per barrel to Marietta, George Rice, who has a refinery in Marietta, very naturally called on me yesterday and notified me that he would not submit to the advance, because the business would not justify it, and that the move was made by the Standard Oil Company to crush him out. (Too true.) Mr. Rice said: 'I am willing to continue the 17½ cent rate which I have been paying from December to this date.'

"Now, the question naturally presents itself to my mind, if George Rice should see fit to prosecute the case on the ground of unjust discrimination, would the receiver be held, as the manager of this property, for violation of the law? While I am determined to use all honourable means to secure traffic for the company, I am not willing to do an illegal act (if this can be called illegal), and lay this company liable for damages. Mr. Terry is able to explain all minor questions relative to this matter." *

Mr. Rapallo, after consulting his partner and "representative bondholders," "fixed it" for the receiver in the following amazing decision:

"You may, with propriety, allow the Standard Oil Company to charge twenty-five cents per barrel for all oil transported through their pipes to your road; and I understand from Mr. Terry that it is practicable to so arrange the details that the company can, in effect, collect this direct without its passing through your hands. You may agree to carry all such oil of the Standard Oil Company, or of others, delivered to your road through their pipes, at ten cents per barrel. You may also charge all other shippers thirty-five cents per barrel freight, *even though they deliver oil to your road through their own pipes;* and this, I gather from your letter and from Mr. Terry, would include Mr. Rice." †

Now, how was this to be done "with propriety"? Simply enough. The Standard Oil Company was to be charged ten cents per barrel, less an amount equivalent to twenty-five cents per barrel upon all oil shipped by Rice. "Provided your accounts, bills, vouchers, etc., are consistent with the real arrangement actually made, you will incur no personal responsibility by carrying out such an arrangement as I suggest." Even in case the receiver was discovered nothing would happen to *him,* so decided the counsel. "It is possible that, by a proper application to the court, some person may prevent you, in future, from permitting any discrimination. Even if Mr. Rice should compel you, subsequently, to refund to him the excess charge over the Standard Oil Company, the result would not be a loss to your road, taking into consideration the receipts from the Standard Oil Company."

* Proceedings in Relation to Trusts, House of Representatives, 1888. Report Number 3,112, pages 575–576.

† See Appendix, Number 46. Letter of Edward S. Rapallo to General Phineas Pease, receiver Cleveland and Marietta Railroad Company.

Fortified by his counsel, Receiver Pease put the arrangement into force, and beginning with March 20, 1885, a joint agent of the Standard pipe-line and of the Cincinnati and Marietta road collected thirty-five cents per barrel on the oil of all independent shippers from Macksburg to Marietta. Ten cents of this sum he turned over to the receiver and twenty-five cents to the pipe-line. When Mr. Rice found that the rate was certainly to be enforced he began to build a pipe of his own to the Muskingum River, whence he was to ship by barge to Marietta. By April 26 he was able to discontinue his shipments over the Cincinnati and Marietta road. This was not done until a rebate of twenty-five cents a barrel had been paid to the Standard Oil Company on 1,360 barrels of his oil—$340 in all.

Mr. Rice, outraged as he was by the discrimination, was looking for evidence to bring suit against the receiver, but it was not until October that he was ready to take the matter into court. On the 13th of that month he applied to Judge Baxter of the United States Circuit Court for an order that Phineas Pease, receiver of the Cleveland and Marietta Railroad, report to the court touching his freight rates and other matters complained of in the application. The order was granted on the same day the application was made. It was specific. Mr. Pease was to report his rates, drawbacks, methods of accounting for discrimination, terms of contracts, and all other details connected with his shipment of oil. No sooner was this order of the court to Receiver Pease known than the general freight agent, Mr. Terry, hurried to Cleveland, Ohio, to meet Mr. O'Day of the Standard Oil Company, with whom he had made the contract. The upshot of that interview was that on October 29, twelve days *after* the judge had ordered the contracts produced, a check for $340, signed by J. R. Campbell, Treasurer (a Standard pipe-line official), was received from Oil City, headquarters of the

Standard pipe-line, by the agent who had been collecting and dividing the freight money. This check for $340 was the amount the pipe-line had received on Mr. Rice's shipments between March 20 and April 25. The agent was instructed to send the money to the receiver, and later, by order of the court, the money was refunded to Mr. Rice. But the Standard was not out of the scrape so easily.

Receiver Pease filed his report on November 2, but the judge found it "evasive and unsatisfactory," and further information was asked for. Finally the judge succeeded in securing the correspondence between Mr. Pease and Mr. Rapallo, quoted above, and enough other facts to show the nature of the discrimination. He lost no time in pronouncing a judgment, and he did not mince his words in doing it:

"But why should Rice be required to pay 250 per cent. more for the carriage of his oil than was exacted from his competitor? The answer is that thereby the receiver could increase his earnings. This pretence is not true; but suppose it was, would that fact justify, or even mitigate, the injustice done to Rice? May a receiver of a court, in the management of a railroad, thus discriminate between parties having equal claim upon him, because thereby he can accumulate money for the litigants? It has been repeatedly adjudged that he cannot legally do so. Railroads are constructed for the common and equal benefit of all persons wishing to avail themselves of the facilities which they afford. While the legal title thereof is in the corporation of individuals owning them, and to that extent private property, they are by the law and consent of the owners dedicated to the public use. By its charter and the general contemporaneous laws of the state which constitute the contract between the public and the railroad company—the state, in consideration of the undertaking of the corporators to build, equip, keep in repair and operate said road for the public accommodation, authorised it to demand reasonable compensation from everyone availing himself of its facilities, for the service rendered. But this franchise carried with it other and correlative obligations.

"Among these is the obligation to carry for every person offering business under like circumstances, at the same rate. All unjust discriminations are in violation of the sound public policy, and are forbidden by law. We have had frequent occasions to enunciate and enforce this doctrine in the past few years. If it were not so, the managers of railways in collusion with others in command of large capital could control the business of the country, at least to the extent that the business was dependent on railroad transportation for its success, and make and unmake the fortunes of men at will.

"The idea is justly abhorrent to all fair minds. No such dangerous power can be tolerated. Except in the modes of using them, every citizen has the same right to demand the service of railroads on equal terms that they have to the use of a public highway or the government mails. And hence when, in the vicissitudes of business, a railroad corporation becomes insolvent and is seized by the court and placed in the hands of a receiver to be by him operated pending the litigation, and until the rights of the litigants can be judicially ascertained and declared, the court is as much bound to protect the public interests therein as it is to protect and enforce the rights of the mortgagers and mortgagees. But after the receiver has performed all obligations due the public and every member of it—that is to say, after carrying passengers and freight offered, for a reasonable compensation not exceeding the maximum authorised by law, if such maximum rates shall have been prescribed, upon equal terms to all, he may make for the litigants as much money as the road thus managed is capable of earning.

"But all attempts to accumulate money for the benefit of corporators or their creditors, by making one shipper pay tribute to his rival in business at the rate of twenty-five dollars per day, or any greater or less sum, thereby enriching one and impoverishing another, is a gross, illegal, inexcusable abuse of a public trust that calls for the severest reprehension. The discrimination complained of in this case is so wanton and oppressive it could hardly have been accepted by an honest man having due regard for the rights of others, or conceded by a just and competent receiver who comprehended the nature and responsibility of his office; and a judge who would tolerate such a wrong or retain a receiver capable of perpetrating it ought to be impeached and degraded from his position.

"A good deal more might be said in condemnation of the unparalleled wrong complained of, but we forbear. The receiver will be removed. The matter will be referred to a master to ascertain and report the amount that has been as aforesaid unlawfully exacted by the receiver from Rice, which sum, when ascertained, will be repaid to him. The master will also inquire and report whether any part of the money collected by the receiver from Rice has been paid to the Standard Oil Company, and if so how much, to the end that, if any such payments have been made, suit may be instituted for its recovery." *

On December 18 George K. Nash, a former governor of Ohio, was appointed master commissioner to take testimony

* Proceedings in Relation to Trusts, House of Representatives, 1880. Report Number 3,112, pages 577–578.

and clear up the point doubtful in the judge's mind—to whom had the extra money paid by Rice been paid; the receiver declared that he never paid the Standard Oil Company any part of Rice's money. Mr. Nash summoned a large number of witnesses and gradually untangled the story told above. Mr. Pease spoke truly, he had never paid the Standard Oil Company any part of Mr. Rice's money. A joint agent of the railroad and the pipe-line had been appointed, at a salary of eighty-five dollars a month, sixty dollars paid by Pease and twenty-five dollars by the Standard, who collected the freight on independent shipments and divided the money between the two parties. It was from this agent that it was learned that, twelve days *after* Judge Baxter ordered Receiver Pease to bring his contracts into court, the money paid on Mr. Rice's oil had been returned by the Standard Oil Company.* While the investigation in regard to Mr. Rice's oil was going on, complaints came to Commissioner Nash from two other oil works at Marietta that they had been suffering a like discrimination for a much longer time. The commissioner investigated the cases and found the complaints justified. The Standard Oil Company had received $649.15 out of the money paid by one concern to the railroad for carrying its oil, and $639.75 out of the sum paid by another concern! Both of these sums were returned by the Standard.†

Of course the case aroused violent comment. In 1888 it came before the Congressional Committee which was investigating trusts, and an effort was made to explain the twenty-five cents extra as a charge of the pipe-line for carrying oil to the railway. Now, the practice in vogue in the Oil Regions

* See Appendix, Number 47. Testimony of F. G. Carrel, freight agent of the Cleveland and Marietta Railroad Company.

† See Appendix, Number 48. Report of the Special Master Commissioner George K. Nash to the Circuit Court.

then and now is that the *purchaser of the oil pays the pipe-line charge.* The railroad has nothing to do with it. Even if the Standard Oil Company puts a tax on railroads for allowing them to take oil carried by its pipe-lines—thus collecting double pay—the tax would not apply in Mr. Rice's case, for the oil came to the Cincinnati and Marietta road not through Standard pipes but through Mr. Rice's own pipes. This much Mr. O'Day was obliged to admit in 1888:

Q. But did that other oil which was in competition with you pass through your pipe?

A. No, sir.

Q. Did not they, therefore, on that oil which only passed over their railroad and not through your pipe-line, pay to you the same allowance or rebate that they did on your oil which did pass?

A. They did, but we returned it through the advice of our counsel, Mr. Dodd.

Q. Now, out of that sum how much did you get from the railroad out of what they had received from Mr. Rice?

A. We did not get any; that is, we did not retain any. The railroad company agreed to account to us for the oil that went over its lines, and they did make an accounting, to my recollection, of about $200, or something like that, on oil other than that which passed through the lines. Our counsel, Mr. Dodd, advised me that we could not do that business, and we refunded the money.

Soon after the report of the Congressional Committee was published John D. Rockefeller himself explained the case in an interview published in the New York World for March 29, 1890: "When the arrangement was reported to the officers of the company at New York," Mr. Rockefeller told the interviewer, "it was not agreed to because our counsel pronounced it illegal in so far as it embraced oil carried by the pipe-line. Some $250 had been paid to the pipe-line under this contract on oil which the line had not transported. This was refunded. We repudiated the contract before it was passed upon by the courts and made full recompense. In a business as large as ours, conducted by so many agents, some things are likely to be done which we cannot approve. We correct them as soon as they come to our knowledge. The public hears of the wrong—it never hears of the correction." In the Digest of Evidence made by the Industrial Commission in its report published in 1900 (page 158), it is stated that the money collected was refunded *before* suit was brought. The facts show that the statement in the report of the Industrial Commission that the money was refunded *before* suit was brought is wrong, and that, while Mr. Rockefeller is technically correct in stating that the Standard repudiated the contract before it was passed on by the courts, he should have added they did not repudiate the contract until *eight months after* it was made, and did not refund the money until *twelve days after* it became certain that the contract would be produced in court. He also does not explain why the Standard Oil Company did not return the money unjustly paid to them on the shipments of the other independent oil concerns of Marietta until exposure by Commissioner Nash's investigation made it inevitable.*

But it was not only manipulation of the railroads by the Standard Oil Company of which the public was complaining at this time. The policy of making it impossible for even small independent concerns to do business was attracting more and more attention. Indeed, there was going on in Buffalo, New York, simultaneously with these two cases, a most sensational trial, growing out of an indictment for the crime of conspiracy, by the Grand Jury of Erie County, New York, of three prominent members of the Standard Oil Company—H. H. Rogers, John D. Archbold and Ambrose McGregor—with two refiners with whom they were associated—H. B. Everest and C. M. Everest. The case is reported

* The documents from which the statements are drawn are all on file in the office of the Clerk of the United States Circuit Court for the Southern District of Ohio, Eastern Division.

in the next chapter at some length, because of the importance it has assumed in the popular controversy which has been going on for the last twenty years over "Standard methods," it being the case on which is based the often-repeated charge that Mr. Rockefeller, to win his point, has been known to burn refineries.

CHAPTER TWELVE

THE BUFFALO CASE

THE STANDARD BUYS THREE-FOURTHS OF THE VACUUM OIL WORKS OF ROCHESTER—TWO VACUUM EMPLOYEES ESTABLISH BUFFALO LUBRICATING OIL COMPANY AND TAKE WITH THEM AN EXPERIENCED STILLMAN FROM THE VACUUM—THE BUFFALO LUBRICATING OIL COMPANY HAS AN EXPLOSION AND THE STILLMAN SUDDENLY LEAVES—THE BUFFALO LUBRICATING OIL COMPANY IS SUED BY VACUUM FOR INFRINGEMENT OF PATENTS—MATTHEWS SUES THE EVERESTS OF THE VACUUM FOR DELIBERATELY TRYING TO RUIN HIS BUSINESS—MATTHEWS WINS HIS FIRST CIVIL SUIT—HE FILES A SECOND SUIT FOR DAMAGES, AND SECURES THE INDICTMENT OF SEVERAL STANDARD OFFICIALS FOR CRIMINAL CONSPIRACY—ROGERS, ARCHBOLD AND McGREGOR ACQUITTED—THE EVERESTS FINED.

VERY soon after Mr. Rockefeller began to "acquire" independent refineries, whose owners were loath to sell or go out of business, unpleasant stories began to be circulated in the oil world of the methods used in getting the offending plants out of the way. When freight discriminations, cutting off of crude supply, and price wars in the market failed, other means were tried, and these means included sometimes, it was whispered, the actual destruction of the plants. The only case in which this charge was made which ever came to trial was that of the Buffalo Lubricating Oil Company, Limited. For sake of clearness, a narrative of the case has been drawn from the testimony offered, no statements being admitted which were not brought out in the trials.

It seems that some time in 1879 the owners of the Vacuum

Oil Works, of Rochester, New York—H. B. and C. M. Everest, father and son—sold to H. H. Rogers, J. D. Archbold and Ambrose McGregor of the Standard Oil Company, for $200,000, a three-fourths interest in that concern. The purchase was not made for the gentlemen in whose names it appeared, but for the Standard. Thus, when on the witness-stand J. D. Archbold was questioned as to the real ownership of the stock which had been bought in his name, the examiner wanted to know whether the purchasers represented themselves or somebody else.

"Mr. Archbold," he asked, "you made the contract, did you not, with reference to the transfer of the seventy-five shares of the Vacuum Oil Company's stock by the Messrs. Everest?"

A. I bought the seventy-five shares, yes, sir.

. . . .

Q. Whom did you represent in that transaction?

A. I represented the shareholders of the Standard Oil Company.

Q. After this purchase was made did you continue to represent the purchasers in the management of the affairs of the Vacuum Oil Company?

A. I did.

Q. By virtue of power delegated to you, or by virtue of being a member of the board of directors or trustees of the Vacuum?

A. By the virtue of power delegated to me.

Q. By the purchasers?

A. By the purchasers.

The Vacuum manufactured principally lubricating oils used on harness and car wheels. It controlled several valuable patents and had been doing a prosperous business for a number of years. By the terms of the sale in 1879 the Everests remained as managers of the refinery, on a salary of $10,000 a year. They also contracted to enter into no outside oil business for ten years. The business policy of the Vacuum, including the fixing of salaries, was dictated by a board of directors made up of Messrs. Rogers, Archbold, McGregor and the two Everests. The meetings of this board were held at the office of the

Standard Oil Company, in New York or in Rochester, as convenient.

So far as can be inferred from the testimony, the works were well managed, the dividends large, and the employees well treated. In 1880 the salesman of the concern, J. Scott Wilson, decided to leave the Vacuum and go into business for himself. The decision seems natural, for until 1878 Mr. Wilson had carried on an independent oil business of one kind or another. He had been a partner in a refinery and understood making oils. He had been a jobber on his own account before going with the Everests, and as such had had a considerable clientele. Wilson told one of his fellow employees, Charles B. Matthews, of his decision, and asked him to go with him. Matthews had been with the Everests about the same length of time as Wilson—some two years. Previous to this engagement he had been a farmer, and his acquaintance with the Vacuum people had come about by his drilling on his farm for oil. Matthews was worth some $20,000, but he had had no experience in oil refining, for his duties at the Vacuum had been mainly looking after outside business—for instance, he had several times gone to New York to consult J. D. Archbold and H. H. Rogers concerning business matters, and particularly concerning patents owned by the Vacuum, of whose validity there was some doubt. For some time Matthews had been dissatisfied with his salary—he had asked for a raise, but had not got it—a fact which probably made him more favourable to Wilson's suggestion.

The two men decided finally to form a company and to build an oil refinery at Buffalo. Wilson said on the witness-stand that he did not want to handle the Vacuum processes in the new works, but to make only the oils with which he was familiar. Matthews, however, had convinced himself that the patents which covered certain of the Vacuum processes and apparatus were invalid, and insisted that they build at least one Vacuum still. The question of what steps the Vacuum might take to stop them was discussed, and according to Wilson's testimony Matthews remarked that he expected they would pay $100,000 or $150,000 to prevent their going into business. Matthews's remark was natural enough, considering the conditions under which outside refiners were forced to do business. It is probable that no man undertook any kind of independent oil business at that time, particularly oil refining, without considering the possibility of being driven to sell.

The new firm needed an experienced stillman accustomed to the Vacuum processes, and early in 1881 they asked one Albert Miller, a stillman in the Vacuum works, to join them. "If we have Miller," they told each other, "we can go to the customers of the Vacuum Oil Company and say to them: 'We have the same process and the same apparatus and the same oils as the Vacuum Oil Company, and we have their former superintendent, Mr. Miller, to manufacture the oils.'" Miller had been with the Everests for several years, having worked his way up from a labourer at two dollars a day to a position where, as stillman, he was paid by the hour, and earned from $1,200 to $1,400 a year. He and his wife had been thrifty, and had several thousand dollars in property. Miller thought there was money in the new venture, and consented to join Wilson and Matthews. The three set about carrying out their plans before they notified their employers of their intention to leave —Miller going so far as to order certain iron castings needed in the construction of their works, made after patterns owned by the Everests. He had these made at the foundry patronised by the Everests. He paid for them himself, and carried them away, presumably giving the impression that they were for his employers.

Early in March Matthews and Miller notified C. M. Everest, who was in charge, his father being in California, that they were going to leave and establish at Buffalo an independent oil refinery. Mr. Everest, surprised out of discretion by the news, told them plainly that although he had nothing against them personally, he should do all in his power to injure the proposed concern. He asked them where they expected to get oil, and they replied that they would get it from the Atlas Refining Company, an independent concern in Buffalo, which had its own pipe-line. "You will wake up some morning and find it is in the Standard," replied Mr. Everest. Apparently Mr. Everest's threat had little influence on the men, for they pushed the building of the works in Buffalo as rapidly as possible. On March 15 they signed an agreement to carry on the proposed business for five years, each man to put in $2,000. A month later the three men, with two relatives of Matthews, organised a stock company—the Buffalo Lubricating Oil Company, Limited—with a capital of $40,000.

Although Miller had gone to Buffalo the first of March with Matthews and Wilson, he returned frequently to Rochester to see his family. On several of these visits he saw C. M. Everest, who never failed to ask about the progress of the new concern, and to warn him that the Vacuum Company would never allow it to do business. "Don't you think, Miller," Everest said to him once, "that it would be better for you to leave those men and have $20,000 deposited to your wife's credit than to go to these parties?" Miller affirms that he answered that he had gone with the new firm in good faith, and thought he ought not to leave them.

About two months after the new firm began building, the elder Everest, who had been in California, returned to Rochester, and soon after had several interviews with Miller. He impressed on the man, as his son had done, that the Buffalo Lubricating Works would never succeed. He told him that the Vacuum meant to bring suit against them for infringing their patents, and would get an injunction and stop the works;

BLEACHING TANK

CONSTRUCTING AN IRON TANK FOR STORING OIL

OIL AGITATORS

FIVE-BARREL STILL
USED IN THE FIFTIES IN DISTILLING
CRUDE OIL AS A LUMINANT

that Miller would lose all the money he had put in. To save himself, Everest advised Miller to come back to the Vacuum. "But that would leave them in a pretty bad fix," Miller said. "That is exactly what I want to do," replied Everest. The fear that the new concern might be ruined through the hostility of the Vacuum, and he lose his savings, seems to have preyed on Miller's mind. He took his wife into his confidence, and she, too, became alarmed. He began to neglect his work in Buffalo. He was often away at nights. Matthews began to be worried by Miller's neglect and absence, and to watch the stations to find, if possible, where he went. Miller's question now became, how could he get away from the Buffalo firm? He had signed for the company a note for $5,000. He was under contract for a term of years. He discussed the question with the Everests, and they advised him to see his lawyer. On the seventh of June, according to H. B. Everest,* who went with him to help present the case, Miller did consult George Truesdale, a lawyer of Rochester, who had always handled his business. Mr. Truesdale afterwards told in court what occurred:

"Mr. Everest stated that Miller had left his employ, and got engaged with another oil concern in the City of Buffalo; that he desired to get back again; he wanted him to come back; and he said he supposed Miller had explained to me his situation, and the obligations he was under to the Buffalo company. I told him that he had made some statements to me about his contract with the parties in Buffalo; that he had spoken about being an endorser or party to the note made by, I think he said, Matthews and Wilson and himself, and I think another party—four or five of them had made, endorsed a note to raise money, done to start the Buffalo business, and that he had a contract or an arrangement with them to go into a company at Buffalo to manufacture oil, and that he wanted to know how he could get out of that arrangement. I stated what I had said to Miller, that he would, of course, be liable on the note, if he was *charged* properly when it became due, and that if he wanted to get out of that arrangement my advice to him had been to see if he couldn't get released; if they wouldn't release him

* Proceedings in Relation to Trusts, House of Representatives, 1888. Report Number 3,112, page 864.

or buy out his interest; then, if he couldn't do that, the only other way I saw was for him to leave them and take the consequences. I told him that I did not know the exact terms of his contract, but, if he had entered into a contract and violated it, I presumed there would be a liability for damages, as well as a liability for the debts of the Buffalo party. Mr. Miller and Everest both talked on the subject, and Mr. Everest says, 'I think there is other ways for Miller to get out of it.' I told him I saw no way except either to back out or to sell out; no other honourable way. Mr. Everest says, substantially, I think, in these words: 'Suppose he should arrange the machinery so it would bust up, or smash up, what would the consequences be?'—something to that effect. 'Well,' I says, 'in my opinion, if it is negligently, carelessly done, not purposely done, he would be only civilly liable for damages caused by his negligence; but if it was wilfully done, there would be a further criminal liability for malicious injury to the property of the parties, the company.' Mr. Everest said he thought there wouldn't be anything only civil liability, and said that would—he referred to the fact that I had been police justice, had some experience in criminal law—and he said that he would like to have me look up the law carefully on that point, and that they would see me again."

Miller's version of this interview is similar:

"I think Mr. Truesdale or myself, I am not positive which, asked the question what means I could take to get out of the company. H. B. says, 'There is a good many ways he could get out.' Either Mr. Truesdale or myself asked him how. 'Well,' he says, 'he can cut up something or do something to injure them; something of that kind, to get out'; H. B. said this. Mr. Truesdale spoke up and said, 'You must be very careful what you do or you will lay yourself criminally liable.' Mr. Everest says to me, 'There is ways that you can get out.' I says to him, 'You wouldn't want me to do anything, would you, to lay myself liable?' I think Mr. Truesdale spoke up and says, 'You must be very careful or you will end in state's prison,'—that is, I. There was considerable conversation I cannot just exactly remember; I have told all I recollect at present. Mr. Truesdale asked me if I had a contract with the Buffalo parties; I told him I had; 'Well,' he says, 'the best thing you can do is to stay there, then,' or something of that kind. I cannot say those were his exact words. H. B. Everest says, 'If he comes back with us, why, we will look after him.' I think Mr. Truesdale said that these men would be after me for leaving them. I think I told him the terms of the contract. . . . Mr. Everest says, 'They will have to catch Miller before they can do anything to him; we will take care of him.'" *

* Proceedings in Relation to Trusts, House of Representatives, 1888. Report Number 3,112, page 864.

In a talk with Miller a little while after this, C. M. Everest said to him: "You go back to Buffalo and construct the pipes so that they cannot make a good oil, and then, I think, if you would give them a little scare. You might scare them a little, they not knowing anything about the business, and you know how to do it." On account of Miller's neglect, the first still in the new refinery was not ready to be fired until June 15—it was an ordinary still, as was the second one built—the third only was built for the Vacuum process. As soon as the still was ready it was filled with some 175 barrels of crude oil and a very hot fire—"inordinary hot" was the droll description of the fireman—built under it. Miller, who superintended the operations, swore at the fireman once or twice because the fire was not hot enough, and then disappeared. While he was gone the brickwork around the still began to crack. The safety valve finally blew off, and a yellow gas or vapour escaped in such quantities that the superintendent of a neighbouring refinery came out and warned the fireman that he was endangering property. Miller was hunted up. He had the safety valve readjusted—it was thought by certain witnesses that he had it too heavily weighted—and ordered the fires to be rebuilt, hot as before. He again disappeared. In his absence the safety valve again blew off. The run of oil was found to be a failure. It was not a pleasant augury, but oil refiners are more or less hardened to explosions and no one seems to have thought much of the accident. Nobody was injured; nothing was burned, nothing but 175 barrels of oil spoiled; that, in an oil refinery, is getting off easy.

On the 23d of June Miller made the transfer of property advised by the Everests, talked over things with Truesdale, and a week later left the Buffalo Works suddenly on receipt of a telegram, and joined H. B. Everest at the Union Square Hotel in New York. Here Everest advised him to telegraph his wife to move at once to Rochester lest Matthews attach their household goods, and then proposed the two go to Boston. The only event of interest at the Union Square Hotel was an entirely casual meeting with H. H. Rogers, one of the directors of the Vacuum Oil Company. Mr. Rogers seems to have had no conversation with Miller other than to remark, in leaving, that he would see him the next day if he did not go to Boston. The men did, however, go to Boston, where they registered as "H. B. Everest and friend," and where several times, at least, Everest introduced Miller under an assumed name. They junketed about for some days on what Everest tried, with indifferent success, to persuade Miller was a pleasure excursion! While they were amusing themselves, Everest hired Miller at $1,500 a year to "do any fair job we put him at, either at Rochester or some other place." The job turned out to be a rambling one—a few weeks of semi-idleness in Boston—then nothing until September, when he undertook to supervise the drilling of a salt well in Leroy, New York. This lasted until February, 1882; then nothing until May, when, on the advice of H. B. Everest, who had returned to California, Miller went there: "Pack up, sell your property there and come on. Come right to my house and I will help you to get a place and show you how to raise fruit and be an independent man." Miller went, the Vacuum Oil Company paying his expenses. On his arrival he was put to work in a cannery. The Everests explained that they made this arrangement because they thought it would put Miller where he could not be brought back to trouble them any more.

In the meantime things were going badly with the Buffalo Lubricating Works. Miller's loss was a severe one. The men were all novices in making oil, save Wilson, and he was on the road, and they seem to have been unable to find a competent manager. The Everests soon succeeded, too, in getting Wilson out of the new firm by bringing a suit against him for damaging its business by unlawfully leaving it. The suit was withdrawn and the costs paid, when Wilson consented, in December, 1881, to leave the Buffalo Works. Wilson's loss was particularly serious, as he was a salesman of experience.

The suits for infringing the Vacuum patents and processes, which Everest at the start had warned Matthews would be brought, were begun in September, 1881—four separate suits within a year. Matthews, as has been said, had convinced himself that the patents were not valid, and some time in the spring of 1882 he saw H. H. Rogers in New York concerning the suits. "I told him I had come in to talk with him about the patent litigation, or suits that were begun by the Vacuum Oil Company against my company," Matthews said in his testimony. " 'Well,' he said, 'well, what about it?'—something like that. I told him that the product patent, that I well knew, was without merit, and that he knew it was without merit, and I could not see what object or good they could get out of it by bringing suit on that patent. And also the steam patent I considered was without value, and that he knew it was without value. He said that if one court did not sustain the patents they would carry along up until we got enough of it—that was the substance of that talk."

Matthews was evidently discouraged by the result of his talk with Mr. Rogers, for, meeting Benjamin Brewster, of the Standard Oil Company, he offered to sell the Buffalo Lubricating Works for $100,000. The offer was refused, and the suits against which Mr. Matthews protested were pushed. On the 21st of February, 1882, the Vacuum Oil Company filed a complaint in the United States Circuit Court of the Northern District of New York, asking that the Buffalo company be prevented from manufacturing lubricating oils, on the ground that the Vacuum Oil Company had a patent covering the process of manufacturing lubricating oils. The action was regarded as unfounded by the court, and was dismissed on July 16, 1884, "the ground being that the letters sued on in this cause are void." April 25, 1882, another action was commenced by the Vacuum Oil Company against the Buffalo company to obtain an injunction and an accounting for damages upon the ground that the Buffalo company was using an apparatus covered by a patent belonging to the Vacuum Oil Company, but this action also was dismissed March 17, 1885, upon the ground that the letters patent sued upon were "null and void." On February 23, 1883, the Vacuum Oil Company commenced still another action against the Buffalo company asking for an injunction to prevent the Buffalo company from using a label advertising "The Acme Harness Oil made by the Vacuum Process," because the Vacuum Company had long used a somewhat similar label advertising "The Vacuum Harness Oil manufactured by Vacuum Oil Company," but the judge in the case decided that the Vacuum Company had no more right to use labels than the Buffalo company. This decision has since been affirmed by the General Term of the Supreme Court. Still another action was brought against the Buffalo company April 25, 1882, for infringing a patent on a steam process, also a patent upon a fire test. This action resulted in a decree sustaining the fire-test patent, but declaring the steam patent void. The case was then referred to James Breck Perkins, of the Rochester bar, to decide the amount which the Buffalo company had infringed on this patent. Mr. Perkins on a number of different occasions took a large amount of proof there in behalf of the Vacuum Company upon which its counsel claimed that it was entitled to $12,000 damages upon the accounting. The Buffalo company submitted no proof in contradiction, but insisted that the whole proof showed nothing more than a purely technical infringement of the patent, and this view was sustained by Mr. Perkins in his report which awarded six cents damages against the Buffalo company.

The disappearance of Miller, the man on whom the firm

had depended for superintending building and refining, the withdrawal of Wilson, with whom the enterprise had originated and on which it had staked its hopes of finding a ready market, and the series of suits for infringement of patents, suits which cost Matthews thousands of dollars as well as much embarrassment and delay, were troubles brought on him, so he believed, as the result of a deliberate attempt on the part of the Vacuum Oil Company to make good C. M. Everest's threat to do all in his power to ruin the Buffalo Lubricating Works, and, in the spring of 1883, he brought a civil suit against the Everests for $100,000. While Matthews was working up his case he learned that Miller had returned from California, that he had left the Everests because he claimed they had "not treated him right," and that he was idle in Rochester. Miller seems to have left California chiefly because he had gotten it into his head that the information he had about the measures the Vacuum had taken to prevent the Buffalo Works carrying on their business was valuable. H. B. Everest testified that Miller once said to him after he was settled in California: "Mr. Everest, you have always been kind to me, and I shall do nothing to injure you, but I am going to bust the Standard." I said: "Al, how will you go to work to do that?" "More ways than one," he said; "they can't afford to let me loose," he said. "Sha'n't be bought off, either, unless I get something for it. It will cost them more than twenty-five or fifty thousand dollars before they get through with me." I said: "Al, I think you can make more money raising fruit in California than you can fighting the Standard." This conversation was held immediately after the Vacuum had paid Miller $1,000, in addition to the salary of $1,500 they gave him, and for no apparent purpose except to keep him quiet.

When Matthews learned of Miller's return he asked him to come to Buffalo, and evidently got from him then, for the first time, the story of the pressure the Everests had brought to bear on him to leave the Buffalo Lubricating Works, the "fixing" of the still at their advice so that something would "smash," the transfer of his property, his two years of semi-idleness on $1,500 a year and a bonus of $1,000, paid for a reason which can only be surmised, and his final breaking in California, because, as he claimed, he saw no settled employment in view and no prospect of the Everests doing more for him than they were, and, as they claimed, because he believed he could get a big sum from the Standard to keep silent. To all of this Miller made deposition in July, 1884.

The first civil suit was brought to trial early in March, 1885, and it resulted in the jury giving a verdict of $20,000 to Matthews for damages. The court set the sum aside, claiming that they had proved only $4,000 in damages and that he would not sustain an award of punitive damages. Matthews's counsel now obtained a stay of proceedings and finally a new trial. Now about this time Matthews secured evidence which emboldened him to give his suit a much wider range than he had at first intended. This was the testimony of the lawyer Truesdale, quoted above, that in his office Everest had suggested that Miller "arrange the machinery so that it would bust up or smash up." The explosion of June 15 was immediately construed as the result of this counsel. On the strength of this evidence Matthews instituted a second civil suit for damages of $250,000 caused by conspiracy to blow up the works of the Buffalo company, to entice away its employees, to bring unfounded suits against it, and to slander the company's product, and he added to the original defendants the three other directors of the Vacuum Works — H. H. Rogers, J. D. Archbold and Ambrose McGregor — and the Standard Oil Company of New York, the Acme Oil Company of New York and the Vacuum Oil Company. Matthews seems to have argued that, as Rogers, Archbold and McGregor were directors with the Everests in the Vacuum Oil Company, they had probably been consulted by the Everests concerning Miller, and could be included in the conspiracy, and, as the Vacuum, Standard Oil Company and Acme Oil Company were all concerns in the Standard Oil Trust, they, too, could be included. He also went before the Grand Jury of Erie County in opposition to the advice of his counsel and secured there an indictment of H. H. Rogers, J. D. Archbold, Ambrose McGregor and the two Everests for criminal conspiracy. The defendants succeeded in getting the indictment set aside the first time, but Matthews re-presented the case, and a second indictment was found of the same persons. It should be noted that Mr. McGregor was indicted only because he was a director of the Vacuum Works, his name not being mentioned in the evidence presented to the Grand Jury.

An indictment for conspiracy of three men of such prominence as Mr. Rogers, Mr. Archbold and Mr. McGregor riveted the attention of the whole country on the coming trial. It was apparent from the first that the Standard meant to put up a big fight to have the indictment quashed. They had, indeed, set a strong machinery at work immediately to get evidence on which to bring a counter charge of conspiracy; that is, that Matthews's intention in starting the Buffalo Lubricating Works was never to do business, but to force the Standard to buy him out at a big price. They at once set a detective to work on the case, one item of his instructions reading: "We have reason to believe that the suit is brought for the purpose of forcing the Standard to purchase the works of the Buffalo Lubricating Company, and Matthews has made certain statements to that effect; would like reports of any statements or admissions by him in relation to his objects in these suits." Under the direction of this detective, a man employed in Matthews's works for some months made daily reports of what he saw and heard there, copies of which were forwarded to the Standard office in New York. A detective was also put on Miller's track. Miller was now employed in a refinery in Corry, Pennsylvania, and here he was for a long time under espionage. The chief expression obtained from him was by luring him into a saloon one Sunday afternoon and getting him half drunk. While in this condition, the saloon-keeper testified, he said the Buffalo suit was a —— humbug, but there was money in it and that they (he and the persons who were drinking with him) might as well make it as anybody.

It was on May 2, 1886, that the trial began. The array of wealth and legal learning in the Buffalo court-room during the fourteen days' case set not only the town, but the country agape. There were not only the Standard men indicted for conspiracy—H. H. Rogers, J. D. Archbold, Ambrose McGregor—but Mr. Rockefeller himself was there, quiet, steady, watchful. The hostile said the accused and their counsel were disdainful of the proceedings—nobody charged Mr. Rockefeller with disdain. With him were other strong men of the concern, William Rockefeller, Daniel O'Day, J. P. Dudley. There was a great array of legal learning—five eminent lawyers—Wilson S. Bissell, a former law partner of ex-President Cleveland; W. F. Cogswell, of Rochester, counted then one of the ablest lawyers of the state; Theodore Bacon and F. G. Outerbridge, both of Rochester; Daniel Lockwood, famous in politics as well as law; and, of course, S. C. T. Dodd. This for the accused. For the people was the district-attorney of Erie County, George T. Quinby, with one assistant. For fourteen days witnesses were examined, and the above story was dragged from them by dint of questioning and cross-questioning. On May 10 the testimony for the prosecution ended, and the "people rested." The Standard lawyers immediately applied for the acquittal of Mr. Rogers, Mr. Archbold and Mr. McGregor, on the ground that no fact or circumstance had been proved that connected them in the slightest degree with the charge of conspiracy to lure Miller away or to destroy the Buffalo Works. The district-attorney combated the proposition vigorously. These gentlemen, he contended, owned three-fourths of the Vacuum Works; they were always present at directors' meetings; it was a fair presumption that they knew what was done to persuade Miller to leave the Buffalo Works; they must have known the moneys paid him while he was doing little work. Mr. Rogers had certainly threatened Matthews that he would carry up the patent suits until the Buffalo Works got enough of it. Judge Haight, however, advised the jury to acquit Mr. Rogers, Mr. Archbold and Mr. McGregor. "The indictment charges a conspiracy," the judge said. "It also charges certain overt acts. One of the acts charged in the indictment is the enticing away from the Buffalo company of a servant. Another of the acts alleged is an attempt to blow up or destroy the Buffalo Works, and another act that of bringing false suits against the corporation. So far as the agreement or combination to entice away a servant from the Buffalo company is concerned, I have not been able to recall any evidence which shows that either of these three defendants ever knew of it, ever heard of it, or ever took any part in it at all. So far as the charge of an attempt to blow up the Buffalo Works is concerned, I have been unable to recall any evidence that has been given in which either of these three defendants ever knew of it, ever heard of it, ever advised it, or ever took any part in it whatever. The only thing about which I have had any doubt was in reference to the maintaining of actions which have been brought upon patent rights which were formerly owned by the Everests, and by the Everests transferred to the Vacuum Oil Company, and it appears that two suits were brought upon patents, and that there was another suit, a third one, in reference to a trade-mark. It appears from the evidence that upon one occasion Mr. Matthews went to New York and had a talk with Mr. Rogers, and that his conversation has already been discussed and related in your hearing. The query in my mind was as to whether or not the inference could not be drawn, from this conversation, that Rogers did know of the bringing of these actions, acquiesced in their being brought, and in that way became a party to them; but, even conceding that the actions were brought with his knowledge and consent, I am inclined still to think that the evidence is hardly sufficient to warrant his conviction, for the reason that it does not appear that the actions were brought without probable cause; in other words, the bringing of an action and being defeated in the action is not of itself sufficient to authorise a jury to say that it was a false action. That standing alone is not sufficient to authorise a jury to say that it is a false action, but there must be shown in addition to that that there was a want of probable cause; in other words, that the party bringing the action knew and understood beforehand that he had no good cause of action. . . . I am inclined to the opinion that the evidence would not warrant his conviction upon that ground."

The acquittal of the three Standard gentlemen was followed by an application for the acquittal of the Everests, but the case with them was different. It had been proved conclusively that they threatened at the start to ruin the new concern, and that they had counselled Miller "to arrange the machinery so it would bust up or smash up"; there was a strong presumption that Miller, acting on this advice, had arranged for the explosion of June 15, though, as he claimed, he meant only to "give them a scare." The judge denied the application in their case, therefore, and the trial went on. The whole force of the defence was now thrown to proving that Matthews had gone into the Buffalo Lubricating Company merely to sell out. His offer to Mr. Brewster in 1882, his talk of making the Standard settle, were rehearsed. Two witnesses were produced also who told of seeking Matthews in 1885, after the criminal suit was brought, and of offering, on the ground that they knew the Standard defendants, to attempt to settle the affair. Matthews had told these men that if the Standard would give him $250,000 for his refinery, he would withdraw the civil suit, but that he could not touch the criminal suit, as it was in the hands of the district-attorney. The jury was not greatly influenced by the evidence produced to show that Matthews was a blackmailer. Evidently they concluded that, granting that the Everests had cause of complaint against the men for using their processes—they certainly had no just cause in the fact of the three men setting up in business for themselves—granting that the enterprise was started for blackmailing purposes—and there was no proof offered that it was—the Everests should have taken their case into the courts—not plotted the destruction of the refinery by any such underhand methods as they employed. Whatever the jury's process of reasoning, however, it is certain that on May 16 they brought in a verdict of "guilty as charged by the indictment."

The most strenuous efforts were made to set the verdict aside. The judge granted a stay, and an attempt to get a new trial was made, but unsuccessfully. The sentence was stayed until May, 1888. The statute provided a penalty of one year's imprisonment or $250 fine, or both. Efforts were at once made to soften the sentence. A petition signed by over forty "leading citizens" of Rochester, New York, the home of the Everests, was sent to Judge Haight, praying him, on account of the "untarnished fidelity and integrity" of the convicted men, to make the penalty as light as the court was authorised by law to fix. Six of the jurors were induced by Standard agents to sign a paper claiming that in their belief the jury in rendering its verdict of guilty did not mean to pronounce the Everests guilty of an attempt to blow up or burn the works of the Buffalo company, but guilty only of enticing Miller away, and they recommended that the sentence, therefore, be a fine and not imprisonment. District-Attorney Quinby offered to prove on a hearing for a new trial that the Standard's representatives used money in getting these affidavits. The result was that the two Everests were each fined $250. This sentence was made light, the judge explained, because of the civil suits brought to recover damages for the very same acts—a person could not be punished twice for the same offence.

The first civil suit referred to above resulted in an award by the jury of $20,000 to Matthews. The second civil suit was for $250,000, but before it was tried Matthews's business had become so involved by all this trouble that in January, 1888, it was put into the hands of a receiver. The defendants finally offered to settle the civil suits for $85,000. The judge ordered the receiver to acept the offer, on the ground that the Everests had already been declared guilty of criminal conspiracy and had been fined, and that a person could not be punished twice for the same offence!

It was not until June, 1889, that the receiver filed his account of the settlement of the affairs of the Buffalo Works. Of the $85,000 paid by the Standard, Matthews seems not to have gotten a cent. The entire sum went to settle the debts of the concern and pay the lawyers. The leading claimants among the lawyers were Thomas Corlett, Edward W. Hatch and Adelbert Moot, all of Buffalo. Their claims aggregated nearly $35,000. The receiver thought these fees exorbitant, and a referee was appointed by the court to take the testimony of the claimant as to their services. The testimony was voluminous, and the upshot was that the referee cut these claims to about $22,000. The final account filed by the receiver shows that the three gentlemen finally were paid about $15,000.

The large claims made by the lawyers and certain circumstances of the settlement have led the Standard, in later years, to advance a counter charge of conspiracy of much more serious nature than that which they depended on in the trial. This new charge makes Matthews's counsel his fellow conspirators, and alleges that at least two of them used important official positions to influence the verdict. In the present year (1904) the Standard's official organ, the Oil City Derrick, published a supplement containing the evidence on which this counter charge is based, and editorially accused the writer of bias in not using this material in the story of the Buffalo case which was published practically as it stands here in McClure's Magazine for March, 1904. It is true, as the Derrick claims, that through the courtesy of the Standard Oil Company this material was placed in the writer's hands before the article was published. It was not used because it was not thought it established the charge.

The points brought out in the evidence published by the Derrick which are held by the Standard to establish the charge of a conspiracy between Matthews and his counsel are the following: In the first place, they declare it a conspiracy because Corlett, who was called to the bench in January, 1884, and Hatch, who was called to the bench in January, 1886, were both in consultation with their successors after they became judges. That this is true there is no doubt whatever. Mr. Moot in his full statement of his services made to the referee refers again and again to consultations with Corlett and Hatch after they had given up the case. Hatch speaks freely in his statement to the referee of counselling with Quinby and Moot.* If there was an impropriety in what he did, he certainly made no effort to conceal it, nor did the referee, the court, or the receiver, to whom this statement was submitted, raise any question of impropriety. The counsel which both Judge Corlett and Judge Hatch gave Quinby and Moot they

* The Derrick published in a four-page supplement to the issue of April 23, 1904, the full text of both statements under the title "More of Tarbell's Tergiversations."

owed Matthews. They had been his counsel for years. They were obliged to give up his cases because of their election to the bench. They were debarred by their relation to the case, of course, from hearing it, but there was no reason why their knowledge and experience should not be drawn upon to a reasonable degree by the new attorneys. Certainly this is a universal practice in law courts. It is difficult to see how it could be otherwise. If either judge had used his position to influence his fellow judge who heard the case there would be a just criticism, but no such intimation has ever been made, to the writer's knowledge.

The second proof of conspiracy drawn from this testimony to the referee is the statements of both Hatch and Moot that they had no contracts for compensation and that they knew they would receive nothing if they lost. For instance, when Moot was examined by the referee he was asked:

Q. Did you have any contract or agreement as to how you should be compensated?

A. Not the slightest. I never had such a contract in my life, except that I should be liberally paid if I succeeded. If I did not succeed, the party being poor, my work would be without compensation. . . .

Q. Did you ever have any conversation with Matthews or with any officer of the company with reference to that?

A. No, sir. I feel very clear that I never had a conversation with a single member of this company about what we should receive for our services, except to this extent: Mr. Matthews once said, in referring to or commenting on these litigations, that they were like any other independent company, as I very well knew; that if the lawyers could not keep them alive with litigation, the Standard would beat them—we would not get anything.

Judge Hatch in his statement said: "Matthews and I or any one for his company never had any talk with respect to compensation for services at the time of their commencement or during their rendition. I knew, however, that the payment for services was largely contingent upon the success of the litigation, and the company was not able to pay much more than the actual expenses in the event they failed to succeed, and that we would get a very meagre compensation unless we succeeded in the actions. I think no conversation was ever had except Mr. Matthews stating that if we should succeed we should be well paid. I think he mentioned that once or twice."

It is not an unusual thing for lawyers to take cases they believe just, knowing that their compensation depends on their winning. Many clients with just cases would be deprived of counsel if they had to insure a fixed compensation, for not infrequently all that a client has is involved in a suit. The practice is so common among reputable lawyers that it certainly cannot be regarded as a proof of a conspiracy, unless there is a reason to suppose that they have taken a case of whose merits they themselves are suspicious. There is absolutely no evidence that Matthews's counsel were not convinced from the first that they had a strong case. Quinby, the district-attorney who tried the criminal case, certainly conducted it with a fire and a logic which nothing but conviction could have inspired. Moreover, it must be remembered that these attorneys never failed to convince the juries before whom they appeared of the merits of their case. Four juries, two grand juries and two petit juries gave unanimous verdicts of conspiracy against the defendants in the course of the litigation. A case backed by evidence which would convince such diversified bodies of men could hardly be called a speculation. Their claims were large, but lawyers are not proverbial for the modesty of their charges, and in the cases of Hatch and Moot, the two making the largest claims, the labour had been very great and had extended over long periods, as one can see who will examine the testimony published by the Derrick; and besides, exorbitant charges can hardly be construed as a proof of conspiracy.

This, then, in outline, is the history of the case on which are based all charges, so far as the writer knows, that the

Standard Oil Company has deliberately destroyed property to get rid of rivals. The case is of importance not only as showing to what abuses the Standard policy of making it hard for a rival to do business will lead men like the Everests, but it shows to what lengths a hostile public will go in interpreting the acts of men whom it has come to believe are lawless and relentless in pursuing their own ends. The public, particularly the oil public, has always been willing to believe the worst of the Standard Oil Company. It read into the Buffalo case deliberate arson, and charged not only the Everests, but the three co-directors, with the overt acts. They refused to recognise that no evidence of the connection of Mr. Rogers, Mr. Archbold and Mr. McGregor with the overt acts was offered, but demanded that they be convicted on presumption, and when the judge refused to do this they cursed him as a traitor. To-day, in spite of the full airing this case has had in the courts and investigations, Judge Haight is still accused of selling himself to a corporation, and Mr. Rogers is accused daily in Montana of having burned a refinery in Buffalo. As a matter of fact, no refinery was burned in Buffalo, nor was it ever proved that Mr. Rogers knew anything of the attempts the Everests made to destroy Matthews's business.

CHAPTER THIRTEEN

THE STANDARD OIL COMPANY AND POLITICS

OIL MEN CHARGE STANDARD WITH INTRENCHING ITSELF IN STATE AND NATIONAL POLITICS—ELECTION OF PAYNE TO SENATE IN OHIO IN 1884 CLAIMED TO ESTABLISH CHARGE OF BRIBERY—FULL INVESTIGATION OF PAYNE'S ELECTION DENIED BY UNITED STATES SENATE COMMITTEE ON ELECTIONS—PAYNE HIMSELF DOES NOT DEMAND INVESTIGATION—POPULAR FEELING AGAINST STANDARD IS AGGRAVATED—THE BILLINGSLEY BILL IN THE PENNSYLVANIA LEGISLATURE—A FORCE BILL DIRECTED AGAINST THE STANDARD—OIL MEN FIGHT HARD FOR IT—THE BILL IS DEFEATED—STANDARD CHARGED WITH USING MONEY AGAINST IT—A GROWING DEMAND FOR FULL KNOWLEDGE OF THE STANDARD A RESULT OF THESE SPECIFIC CASES.

THE cases described in the last two chapters naturally aroused intense interest in the Oil Regions. The two in Ohio demonstrated afresh the chief grievances which the oil men had against the Standard Oil Company since 1872—that they were securing rebates on their own shipments and drawbacks on those of their competitors. The Buffalo case demonstrated that when their ordinary advantages failed to get a rival out of the way they winked at methods which a jury called criminal. It was fresh proof of what the oil men had always claimed, that the Standard Oil Company was a conspiracy! At the same time that these cases were arousing their indignation anew there occurred in Ohio an affair which gave them new evidence of their old charge that the Standard was steadily intrenching itself in state and national politics in order to direct the course of legislation to suit itself. There had been many evidences of this, satisfactory enough to the initiated. There was no doubt that the investigation of 1876 and the first bill to regulate interstate commerce introduced at that time had been squelched largely through the efforts of two members of Congress, one of them directly and the other indirectly interested in the Standard—these were J. N. Camden of West Virginia, head of the Camden Consolidated Oil Company, now one of the constituent companies of the Standard Oil Trust, and H. B. Payne of Ohio, the father of the treasurer of the Standard, Oliver H. Payne. It had certainly used its influence to oppose the free pipe-line bill which the independent oil men had been fighting for since the early days of the industry. In 1878 and 1879, during the prosecution of the suits against the railroads and the Standard by the Petroleum Producers' Union, there had been incessant charge of the use of political influence to secure delay. It was a matter of constant comment in Ohio, New York and Pennsylvania that the Standard was active in all elections, and that it "stood in" with every ambitious young politician, that rarely did an able young lawyer get into office who was not retained by the Standard. The company seems to have taken a hand in politics even before the days of the South Improvement Company, for Mr. Payne once said in the United States Senate that when he was a candidate for the House of Representatives in 1871, "no association, no combination" in his district did more to bring about his defeat or spent so much money to accomplish it as the Standard Oil Company! *

But all of the examples they quoted were more or less poor in evidence. Of no one of them perhaps could they have produced satisfactory proof. Now, however, simultaneously with the three cases outlined in the last two chapters there

* Congressional Globe, September 12, 1888, pages 8520–8604.

came a case of bribery in an election which they held established their charge. The case was the familiar one of the election of H. B. Payne of Ohio to the United States Senate in January, 1884. Mr. Payne was at the time of his election the aristocrat *par excellence* of Cleveland, Ohio. He had birth and education, distinction of manner and mind. His fine old mansion still remains one of the most distinguished houses in a city of beautiful homes. He had been active in Democratic politics for many years—a member of the state Senate and a member of Congress, and he had been mentioned as the Democratic candidate for the presidency in 1880, receiving eighty-one votes on the first ballot. At the time of his election to the Senate he was a man seventy-four years old. Now Mr. Payne's son, Oliver H. Payne, was one of the thirteen original members of the South Improvement Company, and one of the rare Cleveland refiners who had a strong enough stomach to go into the Standard Oil Company when it swept up the oil trade of Cleveland in 1872, and he had gathered in his share of the spoils of that raid. Oliver Payne was proud of his father, and it was well known that he wanted to see him in the Senate of the United States, but there had been no movement to nominate him, and in 1883 he seems to have made up his mind to see what he could do.

A United States Senator was to be elected in Ohio in November. In October a new State Legislature was chosen, and the Democratic members were instructed for one of two candidates for the Senate, George H. Pendleton or General Durbin Ward, both men of prominence and long service in the public life of the state. Mr. Payne's name was not mentioned in the canvass. Nevertheless, hardly had the Legislature convened when there sprang up at the Neil House in Columbus an extraordinary Payne boom. Its backers were Senator Payne's own son, Oliver H. Payne, at that time treasurer of the Standard Oil Company, and Colonel Thompson, a prominent personage in the same concern. Their lieutenants were also members of the company in one capacity or another. Large sums of money were alleged to have been circulated. There was a rumour that Oliver Payne said the election cost him $100,000. It was claimed that it could be proved that a check for $65,000 had been cashed in Cleveland by one of the men most prominent in the Payne boom, and that the whole sum had been spent in Columbus.

A perfect uproar of indignation followed the announcement of Mr. Payne's choice. All over the state the Standard Oil Company was charged with the election. The Democratic press was particularly bitter:

Said the Butler County Democrat: "It was simply a question whether Pendleton, Ward, Thurman, Converse, Follett, Geddes, or any other capable and honest Democrat, should receive the compliments of a seat in the Senate, or that the Standard Oil Company should buy the place for Henry B. Payne. It was an honest and divided Democracy against a hydra-headed dictatorship of rich men on whose banner was inscribed 'Money Talks.'"

The Carroll County Chronicle in commenting on the election said: "It is a great mistake to suppose Standard Oil has captured the Democratic party of Ohio. It may have captured a score or two of men elected to the Legislature, but they are not the Democracy of Ohio by a long shot. When the British got General Benedict Arnold they imagined they had captured the United States army, but it was a mistake."

"The monopoly of the Standard Oil Company must be destroyed," declared the Columbus Times. "Its intrusion into political circles must be prevented. There must be no later acceptance of this outrage. Political purity and perpetuity permit no complacency. These pernicious foreign elements must be eradicated, and until they are no Democrat will enter the capitol of Ohio or of the nation. The rottenness that uncovered itself last night has not its confines in Ohio."

The comments were not confined to papers of the state. The New York Sun, under the head "Was Payne's Election Bought?" said:

"The subjoined communication from a source which we always respect is worthy of more attention than is usually bestowed upon the animated expressions of those whose preferences have not been realised:

"'It is now believed, and I believe, that the Standard Oil Company recently bought with money Ohio's seat in the Senate of the United States for Mr. Payne. Now, can the social respectability of a man make such a crime respectable? Or is there to be one standard of political morality for Republicans and another for Democrats? Or are Democrats expected to condemn corruption only when practised by Republicans, and to condone, defend, and cover it up when practised by Democrats, or when it is found only in the Democratic party? In my opinion there is no danger so threatening to free institutions as the sale and purchase of political power, and nothing more to be condemned.'"

Although these charges were kept up for two years neither the Standard Oil Company, Mr. Payne, nor the Legislature which had elected him noticed them. The scandal became one of the issues of the next campaign and was instrumental in making the next Legislature of Ohio Republican. As soon as the new Legislature convened at the opening of 1886 an investigation of the Payne case was ordered. Some fifty-five witnesses were examined, and the resulting testimony turned over to the Senate of the United States for its examination. The testimony did not prove the charge of bribery, the Ohio Legislature said, but it was of such a nature as to require the Senate's attention. The matter went to the Senate Committee on Elections, and in July, 1886, a majority reported against the further investigation asked by the state of Ohio.* Against this decision two members of the committee, Senators Hoar and Frye, protested:

"Is the Senate to deny to the people of a great state, speaking through their Legislature and their representative citizens, the only opportunity for a hearing of this momentous case which can exist under the constitution? We have not prejudged the case, nor do we mean to prejudge it. We sincerely trust that the investigation, which is as much demanded for the honour of the sitting members as for that of the Senate or the state of Ohio, may result in vindicating his title to his seat and the good name of the Legislature that elected him.

. . . .

* Report Number 1490, United States Senate, Forty-ninth Congress. This report, and Miscellaneous Documents Number 106, United States Senate, Forty-ninth Congress, 1886, contain the evidence of bribery collected by the Ohio Legislature and the majority and minority reports of the committee.

"How can a question of bribery ever be raised or ever be investigated if the arguments against this investigation prevail? You do not suppose that the men who bribe or the men who are bribed will volunteer to furnish evidence against themselves? You do not expect that impartial and unimpeachable witnesses will be present at the transaction? Ordinarily, of course, if a claim like this be brought to the attention of the Senate from a respectable quarter that a title to a seat here was obtained by corrupt means, the Senator concerned will hasten to demand an investigation. But that is wholly within his own discretion and does not affect the due mode of procedure by the Senate. From the nature of the case, the process of the Senate must compel the persons who conducted the canvass and the persons who made the election to appear and disclose what they know; and until that process issue, you must act upon such information only as is enough to cause inquiry in the ordinary affairs of life.

"The question now is not whether the case is proved; it is only whether it shall be inquired into. That has never yet been done. It cannot be done until the Senate issues its process. No unwilling witness has ever yet been compelled to testify; no process has gone out which could cross state lines. The Senate is now to determine, as the law of the present case and as the precedent for all future cases, as to the great crime of bribery—a crime which poisons the waters of republican liberty in the fountain—that the circumstances which here appear are not enough to demand its attention."

For three oppressive July days the Senate gave almost all of its time to a bitter debate on the report. The name of the Standard was freely used. "The Senate of the United States," said Senator Frye, "when the question comes before it as this has been presented, whether or not the great Standard Oil Company, the greatest monopoly to-day in the United States of America, a power which makes itself felt in every inch of territory in this whole republic, a power which controls business, railroads, men and things, shall also control here; whether that great body has put its hands upon a legislative body and undertaken to control, has controlled, and has elected a member of the United States Senate, that Senate, I say, cannot afford to sit silent and let not its voice be heard in an inquiry as to the truth of the allegation." The majority report was adopted, however, by a vote of forty-four to seventeen. "The most unfortunate fact in the history of the Senate," said Senator Hoar.*

For the time the matter rested, but only for the time. The failure to investigate rather intensified the convictions that Payne's seat was bought by the Standard Oil Company. In 1887 Mr. Payne voted against the Interstate Commerce Bill. "That is why he was put in the Senate," people said bitterly. The feeling became still more intense in 1888. The question of trusts was before Congress. The Republicans had come out with an anti-trust plank in their platform; the Democrats, in response to Mr. Cleveland's message, were declaring the tariff the greatest trust-builder in existence, and calling on their opponents for reform there if they were sincere in their anti-trust attitude. In this agitation the Standard Oil Company undoubtedly exerted its influence against all trust investigation and legislation. The charge became general that they were helping the Democrats. This is why they wanted a Democratic Senate. In September, 1888, when a phase of the question was before the Senate, Mr. Hoar, with his genius for asking far-reaching questions, said one day: "Is there a Standard Oil Trust in this country or not? . . . If there be such a trust, is it represented in the Cabinet at this moment? Is it represented in the Senate? Is it represented in the councils of any important political party in the country?"

It was the first time that Mr. Payne had been sufficiently aroused to reply. "There is nothing whatever to sustain the insinuation which the honourable Senator conveys. I make the declaration now for the first time, and it will be the last time I shall ever take notice of it. The Standard Oil Company is a very remarkable and wonderful institution. It has accomplished within the last twenty years of commercial enterprise what no other company or association of modern times

* Congressional Globe, July, 1886.

has accomplished, but, Mr. President, I never had a dollar's interest in that company. I never owned a dollar of its stock; I never rendered it any service, and that company never rendered me any service. On the contrary, when a candidate for the other House in 1871, no institution, no association, no combination in my district did more to bring about my defeat and went to so large an expense in money to accomplish it as the Standard Oil Company. . . .

"As a matter of fact, nine-tenths of the stockholders of the Standard Oil Company are now and always have been Republicans. Within my knowledge there are but two Democrats who have ever been stockholders in that company." Farther on Mr. Payne interpolated this irrelevant remark: "Not only are the majority Republicans, but they are very liberal in their philanthropic contributions to charities and benevolent works, and I venture the assertion that two gentlemen in that company have donated more money for philanthropic and for benevolent purposes than all the Republican members of the Senate put together."

Mr. Payne's denial was not sufficient to silence Senator Hoar. He returned to the attack. It was a "general public belief," he declared, that the Standard Oil Company was represented in the Cabinet and Senate. He called attention to the newspapers' charge to that effect, and declared that he had received many personal letters charging that the Standard was helping the Democrats. He asked for information when he asked his question; he made no charges. Mr. Whitney was the member of Mr. Cleveland's Cabinet to whom Senator Hoar referred, and he promptly, in a public letter, disclaimed all connection with the Standard Oil Company. Mr. Hoar said he "cheerfully accepted" the denial. As for Mr. Payne, he was not satisfied, and when Mr. Payne in heat replied to him, Senator Hoar closed his lips forever in a burst of biting sarcasm:

"A Senator who, when the Governor of his state, when both branches of the Legislature of his state complained to us that a seat in the United States Senate had been bought, when the other Senator from the state rose and told us that that was the belief of a very large majority of the people of Ohio without distinction of party, failed to rise in his place and ask for the investigation which would have put an end to those charges if they had been unfounded, sheltering himself behind the technicalities which were found by some gentlemen on both sides of this chamber, that the investigation ought not to be made, but who could have had it by the slightest request on his own part and then remained dumb, I think should forever after hold his peace. . . . I think few men ever sat in the Senate who would refrain from demanding an investigation under such circumstances, even if it were not required by the Senate itself. . . . There were Senators who thought that the admission of that Senator, the continuance of that Senator in his seat without investigation, indicated the low-water mark of the Senate of the United States itself." *

And there the Payne case rested. It was never *proved* that the Standard Oil Company had contributed a cent to his election. It was never *proved* that his seat was bought, but the fact that, in the face of such serious charges, rehearsed constantly for four years, neither Mr. Payne nor the Standard Oil Company had done aught but keep quiet, convinced a large part of the country that the suspicion under which they rested was less damaging than the truth would be. In the minds of great numbers this silence was a confession of guilt. The Payne case certainly aggravated greatly the popular feeling that the Standard Oil Company was using the legislative bodies of the country in its own interest.

This feeling was intensified in 1887 by a terrific battle between the oil producers and Standard forces in the Legislature of the state of Pennsylvania. Since the compromise of 1880 the body of the oil producers had been taking no concerted action against the Standard. But their inaction was not due to reconciliation to Standard domination. As a matter of fact they were almost as bitter in 1886 as they had been in 1878, when they formed the Union which for two years

* Congressional Globe, September, 1886, pages 8520–8604.

fought so good a fight. The specific complaint of the oil producers at this time was that they were being "robbed" by the National Transit Company—the big Standard pipe-line consolidation, which had secured by the series of manœuvres already outlined the monopoly of handling and transporting crude oil. If the oil producers had been making money at this time it is quite possible that they would have paid little attention to the profits of the National Transit Company. The service they got was about as perfect as any human machine could render, and they would probably have recognised this and been willing to pay high if they too had been prosperous. But the condition of the oil producer in these days was in glaring contrast to that of Mr. Rockefeller. They had piled up oil until there were in 1886 over 33,000,000 barrels on hand. Naturally this had driven prices down. The average price for the last years had been under a dollar a barrel. In 1886 it fell down to 71⅜, and everyone said it must go lower. Embittered and discouraged, the producers fell to comparing what they were getting out of the business with what Mr. Rockefeller was getting. It was not a consoling showing. The Standard Oil Trust had from its organisation in 1882 paid dividends on its $70,000,000 capital. In spite of the extraordinary outlay for tank-building and seaboard pipe-lines made from 1881 to 1884—$30,000,000 it is computed to have been—the trust paid 10½ per cent. in 1885, ten per cent. in 1886, and Standard Oil stock stood near 200! In contrast, the oil producer, in 1886, is estimated to have lost about six per cent. on his expenditures, and oil property depreciated one-third in value.*

Something was wrong. They could not charge the Standard with the price of oil. As long as over 33,000,000 barrels in stock lay on the market it could not rise. But they could

* See Appendix, Number 49. A statement from an oil-producer's stand-point for 1886.

JOHN D. ROCKEFELLER
By Eastman Johnson

and did complain of what it cost them to handle this oil, of storage and carrying charges, of the deductions for shrinkage and for loss by fire. If the Standard had not forced out every competing line, there would have been sufficient competition to have lowered these items—which at the present prices soon ate up the value of oil. And they fell to rehearsing the raids by which the various transporting companies which had fought themselves into independent positions had been forced into combination, their chief grievances being naturally the affair of the Tidewater. In this state of mind, and incited by the Buffalo, the Payne, and the Rice cases, it was natural enough that when suddenly, at the opening of 1887, a bill evidently intended to strike a blow at the Standard was introduced into the Legislature of Pennsylvania, the oil producers rushed pell-mell to support it. The opening sentence was enough for them. It was "An act to *punish* corporations." * This was what they had always sought, some way to *punish* Mr. Rockefeller for what they believed to be a conspiracy against their interests. The way in which the Billingsley Bill, as it was called from the name of its father, proposed to punish the Standard was to make it a criminal offence to charge in excess of certain rates it fixed—ten cents a barrel for gathering and delivering oil to storing points (the current rate was twenty cents); one-sixtieth of one per cent. per barrel a day for storage, with no storage charge for the first thirty days (one-half of one per cent. was the current rate); one-half of one per cent. shrinkage, instead of three per cent. Besides, the bill required the Standard to go to any well on application of the owner, it made the company liable for damage, and it required it to deliver oil of like kind and quality as that received.

The enthusiasm with which the bill was greeted was cooled a little by the announcement that as it stood it was

* See Appendix, Number 50. The Billingsley Bill.

unconstitutional—acts to punish being forbidden by the constitution of the state—as well as by an immediate realisation that the prices fixed for services were in nearly every case less than cost. The bill was immediately amended. When it came back it was at once apparent that, in spite of this preliminary hitch, a tremendous fight to carry it was being organised by the oil men. Then determination to push it grew in proportion to the Standard opposition. The Standard, indeed, realised immediately that unless a hard fight was made the bill would go through by popular clamour, and they turned their big lawyer, Mr. Dodd, against it, set their newspapers—the Oil City Derrick, Titusville Herald and Bradford Era, all of them by this time subsidised organs—to argue against it, and sent Mr. Scheide, one of the ablest of their pipe-line managers, to present their side at Harrisburg. They also secured the services of a well-known young Republican member of the Legislature, Wallace Delemater, of Crawford County, one of the counties in the Oil Regions, to organise an opposition to the bill in the Legislature.

In February a hearing was given the bill, Mr. Dodd presenting the Standard side. It is rare that so able a lawyer has to fight so weak a measure, and Mr. Dodd riddled it easily. As a matter of fact the Billingsley Bill was as bad as it could be. It was characterised by all sorts of constitutional, legal and practical difficulties. The pipe-line business was an interstate business, and this bill attempted to regulate it—which evidently it could not do. It could, of course, regulate Pennsylvania oil, but, by so doing, it created two classes of oil in the lines, a situation which would have been confusing and undesirable. It was evidently intended that the prices it fixed should apply to the 30,000,000 barrels of stocks on hand, but these were held under contract, and could not be touched. There were many other objections to the bill. Even Judge Heydrick, the able lawyer whom the oil men had engaged to defend it, was obliged to apologise for it at every point, and its most valiant supporter, Senator Lewis Emery, Jr., said frankly that the framer of the bill knew too little of the oil men's needs to be able to make a bill, and that this would have to be thoroughly revised.

In spite of all the reasonable, indeed overwhelming, objections to the Billingsley Bill, the oil men clung to it. Mass-meetings were held nightly from one end of the region to the other, petitions flooded the Legislature, a big delegation was kept constantly in Harrisburg lobbying for it. The support was intemperate, bitter, unreasonable. In March it was intensified by the knowledge that a self-constituted committee of leading oil men were in New York treating with the Standard in regard to certain of the abuses the bill aimed to cure. These men felt that the Standard was unjust in its dealings with the oil men, excessive in its charges, and arbitrary in its service, but they felt that the confusion the Billingsley Bill would bring into the business more than offset the grievances it righted, and they had gone to Mr. Rockefeller to see if matters could not be compromised. Now nothing could have more effectually added to the warlike spirit abroad in the Oil Regions at that moment than the suggestion of a compromise. Their cause was being "sold." It was "compounding with felony," and when, after a three days' sitting in New York, the committee came home with an agreement from the National Transit Company, making certain concessions—as two per cent. instead of three for shrinkage, twenty-five cents a day per 1,000 barrels, instead of forty, for storage, and with a promise that certain other points should be settled by joint committees—two of the leading members were hung in effigy in Titusville!

In April the final vote on the Billingsley Bill came. Harrisburg was alive with oil men determined that the bill should go through. The Standard was present, and if it had less of a *claque,* it had more of the "sinews of war." Indeed, it was charged later by Senator Lewis Emery that the leader of the Standard forces in the Senate received $65,000 for his services—a charge which, so far as the writer knows, has never been either proved or disproved. The bill came to a vote after a passionate wrangle. It was defeated eighteen to twenty-five. A storm of violent protest from the oil men's representatives followed the defeat, and the lobbies, the hotels, and even the streets of Harrisburg were scenes in the next hours of bitter quarrels and excited gatherings. When finally the oil men withdrew from the town it was with the understanding that they were to meet two weeks later in Oil City to organise a new protective association. The protests and resolutions passed at their final gatherings foreshadowed no intention of reviving the Billingsley Bill. Indeed, the bill itself had received scant attention from them in the violent campaign over its passage which they had carried on for three months. All their passion had been expended on the Standard. This was a question of whether the Standard Oil Company ruled the Legislature of Pennsylvania or whether the people ruled it—so declared the oil men; and when their bill was defeated they charged it was by bribery, and henceforth quoted the defeat of the Billingsley Bill along with the Payne case as proof of the corrupt power of the Standard Oil Company in politics. Their outbreak, for it was nothing else, was the culmination of their indignation and resentment at fifteen years of unfair play on the part of the Standard Oil Company, of resentment at the South Improvement Company, at forced combination of refineries and pipe-lines, at railroad rebates and drawbacks, at the immediate shipment outrages, at the Tidewater defeat. It was revolt against the incessant pressure of Mr. Rockefeller's pitiless steel grip. It was bitterness at the idea that it was he who was reaping all the profit of a business in which they were taking the chief risks, and if things went on as they were that it was he who always would. Out of their burst of passion was to grow a solid determined effort, but for the moment they were defeated, and the defeat, which really was merited, was another added to their series of just and unjust complaints against Mr. Rockefeller.

All of these bitter and spectacular struggles aroused intense public interest. The debate on the Interstate Commerce Bill was contemporaneous with them—the bill was passed in 1887, and had its effect. The feeling grew all over the country that whatever the merits of these specific cases, there was danger in the mysterious organisation by which such immense fortunes and such excessive power could be built up on one side of an industry, while another side steadily lost money and power. A new trial was coming to Mr. Rockefeller, one much more serious than any trial for overt acts, for the very nature of his great creation was to be in question. It was a hard trial, for all John D. Rockefeller asked of the world by the year 1887 was to be let alone. He had completed one of the most perfect business organisations the world has ever seen, an organisation which handled practically all of a great natural product. His factories were the most perfect and were managed with the strictest economy. He owned outright the pipe-lines which transported the crude oil. His knowledge of the consuming power of the world was accurate, and he kept his output strictly within its limit. At the same time the great marketing machinery he had put in operation carried on an aggressive campaign for new markets. In China, Africa, South America, as well as in remote parts of Europe and the United States, Standard agents carried refined oil. The Standard Oil Company had been organised to do business, and if ever a company did business it was this one. From Mr. Rockefeller himself, sitting all day in his den, hidden from everybody but the remarkable body of directors and heads of departments which he had "acquired" as he wiped up one refinery and one pipe-line after another, to the humblest clerk in the office of the most remote marketing agency, everybody worked. There was not a lazy bone in the organisation, nor an incompetent hand, nor a stupid head. It was a machine where everybody was kept on his mettle by an extraordinary system of competition, where success met immediate recognition, where opportunity was wide as the world's craving for a good light to cheer its hours of darkness. The machine was pervaded and stimulated by the consciousness of its own power and prosperity. It was a great thing to belong to an organisation which always got what it wanted, and which was making money as no business in the country had ever made it.

What more, indeed, could Mr. Rockefeller ask than to be let alone? And why not let him alone? He had the ability to keep together the wide-spread interests he had acquired—not only to keep them together, but to unify and develop them; why not let him alone? Many people even in the Oil Regions were inclined to do so, some because they feared him—rumour said Mr. Rockefeller was vindictive and never forgot opposition; others because they were canny and foresaw that they might want his help one day; still others because criticism of success is an ungracious business and arouses a suspicion that the critic may be envious or bitter. But there were a few people, as there always are, whom no cowardice, no self-interest, no fear of public opinion could keep quiet, and these people insistently urged that the Standard Oil Company was a menace to the commerce of the country. We have been and are being wronged, they repeated. We have a right to do an independent business. Interference to drive us out is conspiracy. Let Mr. Rockefeller succeed in the oil business and he will attack other industries; he will have imitators. In fifty years a handful of men will own the country.

Mr. Rockefeller handled his critics with a skill bordering on genius. He ignored them. To see them, to answer them, called attention to them. He was too busy to answer them. "We do not talk much—we saw wood." This attitude of serene indifference is supremely wise. It belittles the critic and it gives the outsider who watches the game a feeling that a serenity so high must come from an impregnable position. There is no question but many a mouth opened to testify against the Standard Oil Company has been closed by Mr. Rockefeller's policy of silence. Only the few irreconcilables withstood his sphinx-like attitude, and yearly, from the compromising of 1880, these warnings and accusations were louder and more fierce. Probably the greatest trial Mr. Rockefeller has ever had has come from the persistency with which the few malcontents kept him before the public. They interfered with two of his great principles—"hide the profits" and "say nothing." It was they who had ruined the South Improvement Company; it was they who had indicted him for conspiracy and compelled him to compromise in 1880. It was they who now, after the splendid pipe-line organisation was completed and his market machinery was in order, kept up their agitation and their cursing. Their work began to tell. The feeling grew that the Standard Oil Company, or Trust, as it was by this time generally called, must be looked into. Even those who, dazzled by Mr. Rockefeller's achievement, were inclined to overlook its ethical side and to refuse to consider to what aggregation of power and abuse it might lead, began to feel that it would be quite as well to have the matter thrashed out, to have it settled once for all, whether the thing had been so bad in its making and was so dangerous in its tendencies as the "oil-shriekers" pretended. In the House of Representatives, when the question of ordering an investigation of trusts by the Committee on Manufactures was up in 1887, the liveliest concern was shown as to whether the Standard Oil Company, "the most important case" of all, would escape. More than one member asked to be assured before consenting to the investigation that the Standard would be put on the rack. The same interest was shown in the Senate of New York State, where an investigation was ordered for February, 1888. It was certain indeed now that Mr. Rockefeller would not be allowed much longer to work in the dark. He was to be dragged into the open, much as he might deplore it, to explain what his trust really was, to prove to a suspicious and hostile public that he had a right to exist.

CHAPTER FOURTEEN

THE BREAKING UP OF THE TRUST

EPIDEMIC OF TRUST INVESTIGATION IN 1888—STANDARD INVESTIGATED BY NEW YORK STATE SENATE—ROCKEFELLER'S REMARKABLE TESTIMONY—INQUIRY INTO THE NATURE OF THE MYSTERIOUS STANDARD OIL TRUST—ORIGINAL STANDARD OIL TRUST AGREEMENT REVEALED—INVESTIGATION OF THE STANDARD BY CONGRESS IN 1888—AS A RESULT OF THE UNCOVERING OF THE STANDARD OIL TRUST AGREEMENT ATTORNEY-GENERAL WATSON OF OHIO BEGINS AN ACTION IN QUO WARRANTO AGAINST THE TRUST—MARCUS A. HANNA AND OTHERS TRY TO PERSUADE WATSON NOT TO PRESS THE SUIT—WATSON PERSISTS—COURT FINALLY DECIDES AGAINST STANDARD AND TRUST IS FORCED TO MAKE AN APPARENT DISSOLUTION.

THERE was no characteristic of Mr. Rockefeller and his great corporation which from the beginning had been more exasperating to the oil world than the secrecy with which operations were conducted. The plan of the South Improvement Company had only been revealed to those who signed an agreement to keep secret all transactions they might have with it. The purchase in 1874 and 1875 by the Standard Oil Company of Lockhart, Frew and Company of Pittsburg, of Warden, Frew and Company of Philadelphia, and of Charles Pratt and Company of New York was so thoroughly concealed that Mr. Rockefeller, five years after it occurred, dared make an affidavit that it had never occurred! * Men who entered into running arrangements with Mr. Rockefeller were cautioned "not to tell their wives," and correspondence between them and the Standard Oil Company

* See Appendix, Number 44.

was carried on under assumed names! Whenever the subject of the relations between the various companies came up in a lawsuit or an investigation, a candid and straightforward answer was always avoided by both Mr. Rockefeller and the men known to be associated with him in some way. For instance, in 1879, when H. H. Rogers was before the Hepburn Committee, an effort was made to find out what relation the firm of Charles Pratt and Company, of which he was a member, sustained to the Standard Oil Company. Mr. Rogers's testimony was a masterpiece of good-natured evasion,* and all that the examiners could get, though they returned again and again to the inquiry, was that Charles Pratt and Company worked "in harmony" with the Standard Oil Company.

When ex-Governor Nash of Ohio was investigating the relations of the Cleveland and Marietta Railroad and the National Transit Company, try his best he could not find out anything definite. In his report Mr. Nash said: "I have purposely referred to the parties who entered into this arrangement with Receiver Pease and his freight agent, J. E. Terry, as the parties represented by O'Day and Scheide, for the reason that I have not been able to ascertain who or what the parties are." That they were officers of the National Transit Company he had evidence, but what relation had the National Transit Company to the Standard Oil Company? Was it a part of it? Mr. Nash was unable to find from Mr. O'Day, closely as he might question him.†

In the Buffalo case, when John D. Rockefeller was on the stand, he was put through a questioning in regard to the relations of the persons concerned in the suit to the Standard Oil Trust, whose existence he admitted. Mr. Rockefeller answered all the questions his lawyers would allow, but at the end the

* See Appendix, Number 51. Extracts from testimony of H. H. Rogers.

† See Appendix, Number 48.

plaintiffs had gained little or nothing, and there was a strong impression, from the attitude of his lawyers rather than from that of Mr. Rockefeller, that an effort was making to conceal the nature of the agreement or charter or whatever it was under which the companies involved were working. Naturally enough this attitude inspired resentment and aggravated the feeling that this secrecy meant evil-doing. When the epidemic of trust investigation broke out in 1888, and the Standard Oil Trust was brought up for examination, there was a general public demand to have the matter cleared up. The first investigation of importance took place in February, 1888, in New York City, and by the direction of the Senate of New York State. A list of more than a score of trusts was in the hands of the committee, and, with the limited time at their disposal, it was certain that they could not look into more than half a dozen. There seems to have been no hesitation about including the Standard Oil Trust. "This is the original trust," wrote the committee. "Its success has been the incentive to the formation of all other trusts or combinations. It is the type of a system which has spread like a disease through the commercial system of this country."

There were several things the committee wanted to know about the Standard Oil Trust, and its president was summoned for examination. (1) What was it? Was it an organisation recognised by any law of the land? Long ago men had decided that partnerships, corporations, companies, in which men united to do business, must be regulated by law and subjected to a certain amount of publicity, if the public good was to be protected. Was the Standard Oil Trust within or without the law? (2) By the testimony of its own members, in other years the Standard Combination controlled from eighty to ninety per cent. of the oil business of the country. Was this supremacy due in any measure to special privileges, such as discrimination in railroad rates? (3) Was its power used to manipulate production and prices, and to prevent men outside entering the oil business?

It was to learn these things that the commission summoned Mr. Rockefeller. Flanked by Joseph H. Choate, present Ambassador to the Court of King Edward and the most eminent lawyer of the day, and S. C. T. Dodd, a no less able if a less well-known lawyer, Mr. Rockefeller submitted himself to his questioners. In no case where he has appeared on the stand can his skill as a witness be studied to better advantage. With a wealth of polite phrases—"You are very good," "I beg with all respect"—Mr. Rockefeller bowed himself to the will of the committee. With an air of eager frankness he told them nothing he did not wish them to know. The committee had a desire to begin at the beginning. It evidently had heard that a short-lived organisation, called the South Improvement Company, had given Mr. Rockefeller his whip-hand in the oil business as far back as 1872, enabling him in three months' time to raise his daily capacity as a refiner from 1,500 to 10,000 barrels, and so they asked Mr. Rockefeller:

Q. There was such a company?
A. I have heard of such a company.
Q. Were you not in it?
A. I was not.*

It is a perfectly well-known fact that Mr. Rockefeller owned 180 shares in the South Improvement Company, of which he was a director; that, when a public uprising caused the destruction of the company, he was one of the two men who tried to save it; also that the Standard Oil Company of Ohio was the only concern which profited by the short-lived conspiracy.

Another staggering bit of testimony concerned railroad rates. Asked if there had been any arrangements by which

* Report on Investigation Relative to Trusts, New York Senate, 1888 pages 419-420.

the trust or the companies controlled by it got transportation at any cheaper rates than was allowed to the general public, Mr. Rockefeller answered: "No, sir." As a matter of fact, the three great oil-carrying systems of the country—the Central, Erie and Pennsylvania—had all of them, for much of the period between 1872 and 1888, granted to Mr. Rockefeller rebates calculated to keep freight rates down for the Standard Oil Company and up for its competitors. Contracts and agreements to this effect are easily accessible to any one caring to investigate the quality of Mr. Rockefeller's "no." "No," said Mr. Rockefeller, "we have had no better rates than our neighbours," and then, with that lack of the sense of humour which, ethical qualities aside, is his chief limitation, he hastened to add: "But, if I may be allowed, we have found repeated instances where other parties had secured lower rates than we had."

Later in the day the committee, which seems to have known something of Mr. Rockefeller's former contracts with the railroads, returned to the subject, and the following colloquy, worthy of the study of all witnesses interested in how not to tell what you know, took place:

Q. Has not some company or companies embraced within this trust enjoyed from railroads more favourable freight rates than those rates accorded to refineries not in the trust?

A. I do not recall anything of that kind.

Q. You have heard of such things?

A. I have heard much in the papers about it.

Q. Was there not such an allegation as that in the litigation or controversy recently disposed of by the Interstate Commerce Commission, Mr. Rice's suit; was not there a charge in Mr. Rice's petition that companies embraced within your trust enjoyed from railroad companies more favourable freight rates?

A. I think Mr. Rice made such a claim; yes, sir.

Q. Did not the commission find that claim true?

A. I think the return of the commission is a matter of record; I could not give it.

Q. You don't know it; you haven't seen that they did so find?

A. It is a matter of record.

Q. Haven't you read that the Interstate Commerce Commission did find that charge to be true?

A. No, sir; I don't think I could say that. I read that they made a decision, but I am really unable to say what that decision was.

Q. You did not feel interested enough in the litigation to see what the decision was?

A. I felt an interest in the litigation; I don't mean to say that I did not feel an interest in it.

Q. Do you mean to say that you don't know what the decision was? that you did not read to see what the decision was?

A. I don't say that; I know that the Interstate Commerce Commission had made a decision; the decision is quite a comprehensive one, but it is questionable whether it could be said that that decision in all its features results as I understand you to claim.

Q. You don't so understand it? Will you say, as a matter of fact, that none of the companies embraced within this trust have enjoyed more favourable freight rates than the companies outside of your trust? Will you say, as a matter of fact, that it is not so?

A. I stated in my testimony this morning that I had known of instances where companies altogether outside of the trust had enjoyed more favourable freights than companies in this trust; and I am not able to state that there may not have been arrangements for freight on the part of companies within this trust as favourable as, or more favourable than, other freight arrangements; but, in reply to that, nothing peculiar in respect to the companies in this association; I suppose they make the best freight arrangements they can." *

The committee had a vague idea that refineries outside of the Standard Combination had had a hard time to live, and asked if the trust had sought in any way to make the operations of outsiders so unprofitable that they would either have to come in or go out of the business.

"They have not; no, sir, they have not," replied Mr. Rockefeller.

"And they have lived on good terms with their competitors?"

* Report on Investigation Relative to Trusts, New York Senate, 1888, pages 420–421.

"They have, and have to-day very pleasant relations with those gentlemen."

It would have been interesting to have heard the comments of a number of gentlemen trying to carry on an independent business in 1888 on that answer: of the refiners in Oil City and Titusville, at that time preparing to carry their troubles to the Interstate Commerce Commission; of George Rice and others at Marietta, Ohio; of H. H. Campbell, of the Bear Creek Refining Company at Pittsburg; of Scofield, Shurmer and Teagle at Cleveland.

If all of Mr. Rockefeller's testimony had been of the nature of the above, the investigation would have been worth little to the people who demanded it. But when it came to the questions which, after all, it was most essential to have answered at that moment, Mr. Rockefeller, after some skirmishing, gave the committee as frank testimony as is on record from him. The information wanted was in regard to the organisation of the Standard Oil Trust. As pointed out in a previous chapter, there had been some kind of an agreement adopted in 1882, binding together the varied interests which controlled the oil business. But what it was, where it was kept, by what authority it lived, nobody knew. For six years it had succeeded in hiding itself. What was the understanding which had made a trust of a company? The committee asked to know. Mr. Rockefeller and his counsel were the soul of amiability under the demand. They had only one request, and Mr. Choate made it persuasively:

"If the committee please," he said, "I do not arise to make an objection to a request of the committee; we think that it is very proper that the committee should be made acquainted with this document and everything pertaining to it in order to advise them as to the nature and operation of this trust; at the same time, there are private interests and controversies involved which might be seriously prejudiced by a public exposition of its details, and therefore, in producing it, we, without asking the committee to make any promise or to commit themselves at all, request that while they make whatever use of it they please, it shall not be in all its details made a matter of public record

or exhibition unless in their final judgment, after consideration of the matter, they shall consider it necessary. There are very important private interests involved that ought not, under the guise of a public investigation, to be interfered with."

The committee examined the document and concluded to include it in its report.* Like all great things, it was simplicity itself—an agreement which anybody could understand, by which some fifty persons holding controlling interests in corporations, joint stock associations, and partnerships of different states, placed all their stock in the hands of nine trustees, receiving in return trust certificates. These nine trustees themselves owned a majority of the stock and had complete control of all the property. Mr. Rockefeller, when questioned, stated that one of the trustees was a responsible officer in almost every refinery or organisation in the trust; that the trustees, as a body, knew by reports and correspondence, and by frequent consultation in New York with active promoters of each concern, just how the business was going on. "We all know how the business goes," said Mr. Rockefeller; "we get reports once in thirty days showing what it has cost for everything."

The trustees evidently ran the entire great combination under the agreement. But consider the anomaly of the situation. Thirty-nine corporations, each of them having a legal existence, obliged by the laws of the state creating it to limit its operations to certain lines and to make certain reports, had turned over their affairs to an organisation having no legal existence, independent of all authority, able to do anything it wanted anywhere; and to this point working in absolute darkness. Under their agreement, which was unrecognised by the state, a few men had united to do things which no incorporated company could do. It was a situation as puzzling as it was new. The committee in reporting on what it discovered did nothing to solve the puzzle. It simply sounded a warning:

* See Appendix, Number 52. The Trust Agreement of 1882.

"The actual value of property in the trust control at the present time is not less than one hundred and forty-eight millions of dollars, according to the testimony of the trust's president before your committee. This sum in the hands of nine men, energetic, intelligent, and aggressive—and the trustees themselves, as has been said, own a majority of the stock of the trust which absolutely controls the one hundred and forty-eight millions of dollars—is one of the most active and possibly the most formidable moneyed power on this continent. Its influence reaches into every state and is felt in remote villages, and the products of its refineries seek a market in almost every seaport on the globe. When it is remembered that all this vast wealth is the growth of about twenty years, that this property has more than doubled in value in six years, and that with this increase the trust has made aggregate dividends during that period of over fifty millions of dollars, the people may well look with apprehension at such rapid development and centralisation of wealth wholly independent of legal control, and anxiously seek out means to modify, if not to prevent, the natural consequence of the device producing it, a device of late invention, namely, the aggregation of great corporations into partnerships with unbounded resources and a field of operations quite as extended as its resources. So much for the nature of the Standard Oil Trust. The committee regret that they are not able to make a more complete and satisfactory report as to the method of its operations and its effect upon public interests.

"The brevity of the time within which the investigation was required to be made rendered it impossible for your committee to do more than examine the persons most prominent in the management of its affairs. Its cause was thus presented to the most favourable light possible, and it is only fair to conclude that nothing was left unsaid by them that could be said in its favour. No witness came forward to accuse it of the great offences commonly laid to its charge. No proofs were made of its rapacity or of the greed with which it lays hold of every competitive industry, except such as might be drawn from the fact that it is the almost sole occupant of the field of oil operations, from which it has driven nearly every competitor. No witness appeared to prove its power over railroad and transportation companies and to wring from already impoverished lines better terms than other shippers, except such as might be drawn from the admission of its officers, made with hesitation, that this wealth and the amount of its business enabled it to obtain better terms than its poorer competitors." *

The New York Senate made its investigation of trusts in February, 1888. In March the Committee on Manufactures of the House of Representatives began a similar inquiry. This committee, like the earlier one, made the Standard its princi-

* Report on Investigation Relative to Trusts, New York Senate, 1888, pages 9–10.

pal subject. Fully 1,000 pages of a report of 1,500 pages are devoted to Mr. Rockefeller's creation—five times the space given to the Sugar Trust, ten times that given to the Whiskey Trust. The testimony was wide in range. Indeed, from the volume alone, a pretty complete history of the Standard Oil Company up to 1888 could be written. Here are found the South Improvement Company charter and contracts in full. Here is Mr. Cassatt's testimony, taken in the case of the Commonwealth of Pennsylvania *vs.* the Pennsylvania Railroad, showing the character of the rebates the Standard Combination was able to secure from the railroads at that time. Here is a partial history of the growth of the Standard pipelines. Many personal histories of refiners driven out of business by the conditions brought about by railroad discriminations; full accounts of the war of the producing element on the Standard; all of the testimony in the Buffalo case, where two refiners were found guilty of conspiring to ruin an independent refining concern; the reports of the Interstate Commerce Commission in the cases of George Rice; and much interesting explanation of various matters by leading Standard Oil officials appear in the report.

Mr. Rockefeller was on the stand, and one item of his testimony affords a curious comparison. On the 28th of February, when before the New York Senate committee, Mr. Rockefeller was asked if he was not a member of the South Improvement Company.

"I was not," he replied.

On the 30th of the April following, when before the House Committee, the following colloquy took place:

Q. I want the names particularly of gentlemen who either now or in the past have been interested with you gentlemen who were in the South Improvement Company?

A. I think they were O. T. Waring, W. P. Logan, John Logan, W. G. Warden, O. H. Payne, H. M. Flagler, William Rockefeller, J A. Bostwick, and—*myself.*

It was in this investigation that Henry M. Flagler gave

explanations of various operations of the Standard, which have been quoted in the course of this narrative, notably explanations of the South Improvement Company, of the ten-cent rebate secured from all the railroads in 1875, of the purchase of the Empire Transportation Company, of the rebate on other people's shipments enjoyed in 1878 by the American Transfer Company. Some of Mr. Flagler's testimony in this investigation compares as curiously with affidavits of his made in 1880 as does that of his great chief. For instance, in 1880 Mr. Flagler swore that "the Standard Oil Company owns and operates its refineries at Cleveland, Ohio, and also a refinery at Bayonne in the state of New Jersey. That at no other place in the United States does the said Standard Oil Company *own,* operate, or control any refinery or refineries." * But in this investigation the following colloquy took place:

Q. When did the Standard Company of Ohio first enter into an alliance with other refineries?

A. If you mean (by) an alliance, Mr. Gowen, I should say never.

Q. I am only endeavouring to aid your friends in getting at what they want. Here, I notice, they propose to prove by you—I will give it in this way—that on account of the disastrous condition of the refining business, the Standard, on October 15, 1874, entered into an alliance with a number of Pittsburg refineries.

A. That is more correctly stated by saying that the Standard Oil Company *purchased* the refineries owned by the parties in Pittsburg.

Q. Who were they?

A. Lockhart, Frew and Company, I think, was the company. Wait a moment. It was the Standard Oil Company of Pittsburg, it being a corporation, and Warden, Frew and Company, of Philadelphia, and, I should say, Charles Pratt and Company, of New York.

Q. Any others?

A. That is all.

Q. All those gentlemen, Warden, Frew and Company, and the Standard Oil Company of Pittsburg, Charles Pratt and Company, of New York, are now associated with you as parties interested in the present Oil Trust?

* Affidavit of Henry M. Flagler in the case of the Standard Oil Company *vs.* William C. Scofield *et al.*, in the Court of Common Pleas, Cuyahoga County, Ohio, 1880.

A. They are stockholders. The property formerly owned by them was at that time purchased by the Standard Oil Company.

Q. When you speak of purchasing their interest, you do not exclude them from their interest? They united with you and remained as your associates in the business?

A. If it was not from the fact that ours was a corporation, we might call it a copartnership.

Q. They becoming interested in yours, and you in theirs?

A. Yes, sir.

Q. And you simply used your name to represent the joint ownership, as it was a corporation?

A. Yes, sir.*

Full as the testimony on the Standard Oil Trust gathered by the Federal committee of 1888 is, its report touched but one point, and that was its organisation. To the committee it seemed that the agreement under which the trust operated was such as to make it exempt from the anti-trust legislation which was then contemplated by Congress. The legislation proposed was directed against "combinations to fix the price or regulate the production of merchandise or commerce." Now a mass of testimony had been presented showing that, from the starting-point of the Standard's history with the South Improvement Company, its aim has been to regulate the output of refined oil so as to fix the price, but this testimony, the committee saw clearly enough, did not apply to the trust which it was investigating. For—so swore the trustees—they had nothing to do with the business operations of the separate concerns. They simply held the stock of the various corporations, exercised their right as stockholders, received and distributed the dividends. Each company did its own business in its own way. The trustees were not responsible for it. There was something humorous to those familiar with the oil world, in the idea of J. D. Rockefeller, William Rockefeller, J. D. Archbold, Henry H. Rogers, Charles Pratt, H. M. Flagler, Ben-

* Proceedings in Relation to Trusts, House of Representatives, 1888. Report Number 3,112, page 770.

jamin Brewster, W. H. Tilford and O. B. Jennings, having nothing to do, as trustees of the Standard Oil Trust, but to receive and divide dividends, engrossing and interesting a task as that undoubtedly was. But, as a matter of fact, nothing else could be settled on them by anything in the testimony. For instance, in 1887 there was an alliance formed between the Oil Producers' Protective Association and the Standard for limiting the production of crude oil (a movement of which we shall hear more later). This certainly was in restraint of trade. But, on examination, the committee found the contract had been signed by the Standard Oil Company of New York. The trustees had nothing to do with it! Taking up, point by point, the conditions of which the oil producers complained, not one of them could be fixed on the trust. It had made no agreements, signed no contracts, kept no books. It had no legal existence. It was a force powerful as gravitation and as intangible. You could argue its existence from its effects, but you could never prove it. You could no more grasp it than you could an eel. Certainly the Committee on Manufactures was justified in confining its report to pointing out the fact that the Standard Oil Trust agreement was a shrewd and slippery device for evading responsibility.

And there the investigations of 1888 ended. There had been much noise over them, and for what good? So asked the discontented oil public. It simply had secured the form of an agreement which could no more be touched by legislation than human greed. It was characteristic that the oil public, intent on immediate remedies, should be discouraged. If they had applied to their cause the same patience and foresight Mr. Rockefeller did to his, they would have realised that, as a matter of fact, a respectable first step had been taken toward their real goal, a goal which has not by any means been reached—that is, a legal form of organisation for corporations doing interstate business which would enable the public to know promptly if they were securing special privileges or were restricting trade. This first step was in securing the famous trust agreement. That was now in the hands of people given to thinking about things, and something came of it, even more quickly than the philosophical observer of public events might expect, and in this wise:

In 1887 there was elected to the attorney-generalship of Ohio a lawyer, something under forty years of age, named David K. Watson. Two years later Mr. Watson was a candidate for re-election. One day, while busy with his campaign, he came out of his office in the state-house on the public square in Columbus, and, crossing the street, stopped, as he often did, at a book-shop to look over new publications. He happened there on a small yellow leatherette volume entitled "Trusts." It was written by William W. Cook, of the New York bar, and cost fifty cents. Mr. Watson bought the book and spent the evening reading it. At the end he found the Standard Oil Trust agreement. It was the first time he had ever seen it. He read it carefully and saw at once that, if it was a bona fide agreement, the Standard Oil Company of Ohio was and had been for seven years violating the laws of the state of Ohio by taking the affairs of the company from the directors and placing them in the hands of trustees, nearly all of whom were non-residents of the state. M.r Watson knew on the instant that, if this were a bona fide agreement and he were re-elected attorney-general of Ohio, it would be his duty to bring an action against the Standard Oil Company of the state. He laid the little book away until he knew the result of the election.

A few weeks later Mr. Watson was re-elected attorney-general. He at once began a search into the authenticity of the documents in Mr. Cook's little volume. He sent for the reports of the investigations by the committees of the New York Senate and of Congress. He read the testimony word for word. But he still doubted the correctness of the document,

DAVID K. WATSON

Attorney-General of Ohio from 1887 to 1891. Mr. Watson brought suit against the Standard Oil Company in May, 1890, in the Supreme Court of Ohio.

FRANK S. MONNETT

Attorney-General of Ohio from 1895 to 1899. Mr. Monnett brought suit against the Standard Oil Company in 1897 in the Supreme Court of Ohio.

LEWIS EMERY, JR.

Independent oil operator and refiner. Leader in movement for free pipe-line bill and anti-discrimination laws. Founder of the United States Pipe Line.

GEORGE RICE

Plaintiff in numerous cases brought against the Standard Oil Company. Prominent independent witness in various State and congressional investigations.

fearing that, even if it were in the main correct, there might be some loophole by which the Standard Oil Company could escape. Now, in reading the report of the House investigations, Mr. Watson had been particularly impressed with the clearness and directness of the questions put by one of the members of the investigating committee, Mr. Buchanan, of New Jersey. He accordingly went to Washington, inquired from a friend if Mr. Buchanan could be relied upon, and, receiving the assurance of his high character, sought an interview with him. "Was the Standard trust agreement as published in the committee's report *bona fide?"* was the inquiry. "Yes," said Mr. Buchanan. "But why do you ask?" "Because if it is," replied Mr. Watson, "I believe the Standard Oil Company of Ohio has violated the laws of the state, and on my return to Columbus I shall file an action in *quo warranto* against it in the Supreme Court of the state."

"You would not *dare* do that, would you?" exclaimed Mr. Buchanan.

"I was young then," Mr. Watson told the writer in describing this interview, "and I supposed it was expected of a public officer to perform his duty. So I explained to Mr. Buchanan that there was a statute in Ohio which required an attorney-general to bring suit against any corporation which he had reason to believe was violating the laws of the state; that I had no personal feeling against the Standard Oil Company, but I meant to enforce the law against it as I would against any other company which I believed to be violating the law."

"I admire your courage," said Mr. Buchanan, "but I would not do it."

On May 8, 1890, Mr. Watson filed his petition in the Supreme Court of Ohio.* The petition averred that, in viola-

* The full style of the case was: The State of Ohio on the Relation of David K. Watson, Attorney-general, Plaintiff, against the Standard Oil Company, Defendant.

tion of the law of Ohio, the Standard Oil Company had entered into an agreement by which it had transferred 34,993 shares out of 35,000 to the trustees of the Standard Oil Trust, most of whom were non-residents of the state; that it was these trustees who chose the board of directors of the Standard Oil Company of Ohio, and directed its policy, and prayed that, on account of this violation of law, the company should be "adjudged to have forfeited and surrendered its corporate rights, privileges, powers and franchises, and that it be ousted and excluded therefrom, and that it be dissolved."

The petition came on the trust like a thunderbolt. There had been already more or less erratic and ill-advised anti-trust legislation in various states, but it had been framed in ignorance of the actual organisation of the trust, and carried out with a crude notion that the trust, in spite of the fact that it was already thoroughly intrenched in the business life of the country, could be destroyed by a hostile act of a Legislature. Mr. Watson's suit was something very different. It was an application of recognised laws to admitted facts. It brought the Standard Oil Company face to face with several legal propositions it did not like to meet. After a long delay an answer was filed by the Standard. To Mr. Watson's joy, the one thing he feared—the denial of the correctness of the agreement—made no part of this answer. It admitted the agreement, but it denied that the Standard Oil Company of Ohio was a party to it. The agreement was signed by the individual stockholders of the Standard Oil Company, not by the company in its corporate capacity. The Standard Oil Company of Ohio had nothing to do with the Standard Oil Trust. True, certain of its stockholders had turned over their stock to the nine trustees, but the company did its business as before, discharging all its duties as its charter required. This was the essential point of the defendant's answer. This, and the claim that if the court should hold that the action of the stockholders in becoming parties to the agreement in their individual capacity was a corporate act of the Standard Oil Company, even then the charter should not be forfeited, since the law barred an act committed more than five years before a petition was filed.

Anticipating that the trust would get together a strong array of counsel to defend its attacked member, Mr. Watson retained his personal and professional friend, John W. Warrington, an eminent lawyer of Cincinnati, to assist him. They were opposed by Joseph H. Choate, S. C. T. Dodd and Virgil P. Kline of Cleveland.

But, while the preparation for the argument of the case was going on, the courageous young attorney-general was beset on all sides for an explanation. *Why* had he brought the suit? What was the influence which had controlled him? Men in power took him aside to question him, incapable, evidently, of believing that an attorney-general could be produced in Ohio who would bring a suit solely because he believed it was his duty. Some suggested that some big interest, hostile to the Standard, was behind him; others said the suit was suggested by Senator Sherman, then interested in his anti-trust bill. Along with this speculation came the strong and subtle restraining pressure a great corporation is sure to exert when its ambitions are interfered with. From all sides came powerful persuasion that the suit be dropped. Mr. Watson has never made public the details of this influence in any documentary way, but the accounts he at the time gave different friends of it led to so much gossip in Ohio that in 1899 the attorney-general of the state, F. S. Monnett, made detailed charges of six deliberate attempts to bribe Mr. Watson to withdraw the suits.* But one bit of documentary proof of the efforts to reach the attorney-general ever reached the public—that

* See annual report of the attorney-general to the governor of the state of Ohio, 1899.

came out without his knowledge or consent, Mr. Watson claims, seven years after the suit was brought. It is interesting enough as evidence of the character of the pressure Mr. Rockefeller can set in motion when he will. Among Mr. Rockefeller's Ohio friends was the late Marcus A. Hanna, who was even then a strong factor in the Republican party of the state. A few months after the suit was brought he wrote Mr. Watson a letter of remonstrance. Many of Mr. Watson's friends saw this letter at the time and felt deep indignation over its contents. In 1897, when Mr. Hanna was a candidate for the United States Senate, an enterprising newspaper man of Ohio recalled that during 1890 it was common gossip in Ohio that Mr. Hanna had written the attorney-general a letter asking him to withdraw his suit against the Standard Oil Company. The correspondent sought Mr. Watson, who, so he avers, let him read the letter through, although he refused to allow him to copy it for publication. "No one could read it and ever forget it," said the correspondent; but to reinforce himself he sought persons who were associated with Mr. Watson at the time—yes, they remembered the letter perfectly. Certain of them said that they could never forget some of its expressions. Between them they pieced up the following portions of the letter which they declared correct and which the correspondent published in the New York World for August 11, 1897:

"I noticed some time ago that you had brought suit to take away the charter of the Standard Oil Company. I intended at the time to write you about it, but it slipped my memory. A few days ago while in New York I met a friend, John D. Rockefeller, and he called my attention to the fact that you had brought the suit, but did not ask me to influence you in any way."

. . . .

"I have always considered you in the line of political promotion," said Hanna, and then went on to intimate that unless the suit against the Standard was withdrawn, Watson would be the object of vengeance by the corporation and its friends forever

GROUP OF CLEVELAND CITIZENS

Who called on John D. Rockefeller at his residence, "Forest Hill," on July 25, 1896, to thank him for his gift of park lands to the city. Mr. Rockefeller is in the centre of the group, the late Senator Marcus A. Hanna in the right lower corner, and Governor Myron T. Herrick in the centre of the top row.

after. As if to clinch his threat and argument, Hanna wrote: "*You have been in politics long enough to know that no man in public office owes the public anything.*"

. . . .

The letter concluded with a reference to the present Secretary of State, John Sherman. Hanna wrote: "I understood that Senator Sherman inspired and instigated this suit. If this is so I will take occasion to talk to him sharply when I see him."

The letter was written on the typewriter and letter-heads of Hanna's business office in Cleveland.

Having secured this much, the correspondent, thinking it possible Mr. Watson might have answered Mr. Hanna's letter, undertook a bit of original investigation. He sought the files of the attorney-general's official correspondence for 1890, and the following is what he found. This letter certainly is evidence enough of the sort of letter Mr. Hanna had written even if the above restoration is not absolutely accurate:

HON. MARK HANNA, December 13, 1890.
Cleveland, Ohio.

My dear Sir:—Your communication of the 21st ult. came to hand. The delay in answering it has been caused largely by my being ill for several days. I did not intend that bringing the action to which you refer in your letter should be an attack on my part on "organised capital," for I am aware that great business transactions require the union and concentration of moneyed interests, and fully appreciate what has been done in that direction, yet I cannot but feel that I am justified in bringing the suit against the Standard Oil Company, and believe that there are many things relating to the case which, if you understood, would cause you to entertain different views concerning it and my relation to it. Let me impress one thing on you with special particularity, and you may depend absolutely on its truthfulness. Senator Sherman never suggested or encouraged this suit, either directly or indirectly. This must be understood in its broadest sense. The report probably arose from the fact that the action was brought shortly after the Senator made his great speech in support of his anti-trust bill. You will hardly receive my statement with favour, I fear, but I am alone responsible for the action. No one encouraged me to bring it or knew that it would be brought until I determined to do so, and it is unfair to other persons to charge them with suggesting it or encouraging it. With the highest appreciation of your personal friendship, I am, with great respect,

Truly yours,

DAVID K. WATSON.

The part which the terse phrase attributed to Mr. Hanna,

"NO MAN IN PUBLIC OFFICE OWES THE PUBLIC ANYTHING,"

played in the Senatorial campaign of 1897 is familiar to those who follow politics. It was kept standing for days in black-faced capitals at the head of the opposition newspapers in Ohio, and remained a potent weapon in the hands of Mr. Hanna's enemies to the time of his death.

Whatever the pressure Mr. Watson encountered, it had no effect on his purpose. He quietly went ahead, presented his brief, and, when the time came, he and Mr. Warrington argued the case. The following proposition from the brief presented by Mr. Watson and Mr. Warrington show tersely the line of their argument:

"Where the manifest object of an agreement is to unite corporations, partnerships and individuals into, or include them in a common enterprise, and control them through an agency unknown to the law of their creation, and all the officers, directors and stockholders of such corporations sign the agreement, and, in furtherance of its provisions, transfer their stock to such agency, permit the corporate executive agencies to make such transfers on the corporate books, submit without objection to the domination of the agency to which the stock is so transferred in the selection of directors and officers, and in the management of the corporate affairs and business suffer the corporate earnings to go to such agency and be placed and mingled with the earnings of the other parties in the combination so created, and, after deductions for uses of the combination, be divided as part of such common earnings among the persons interested, in such case the corporations become and are—or at least will be treated by the courts as—parties to such agreement and actors in its performance, although their corporate names are withheld therefrom. Such proceedings constitute actual corporate conduct, if not formal corporate action, on the part of each corporation.

"An agreement is in violation of law and void which in effect creates a partnership between corporations, or where its probable operation and effect—much more where its inevitable tendency—is to create a substantial monopoly, or is in restraint of trade or otherwise injurious to the public.

"Where a corporation, either directly or indirectly, submits to the domination of an agency unknown to the statute, or identifies itself with and unites in carrying out an agreement whose performance is injurious to the public, it thereby offends against the law of its creation and forfeits all rights to its franchises, and judgment of ouster should be entered against it.

"Even if the statute which prescribes a time within which an action against a corporation for forfeiture of its charter shall be commenced, be applicable to a case of this kind, yet, where the offences or acts committed or omitted by a corporation for which forfeiture of its charter is sought at the suit of the state, are concealed, or are of such character as to conceal themselves, such offences and acts as against the state are frauds, and such statute does not begin to run until the frauds are discovered."

Joseph H. Choate appeared for the defence. The most eminent lawyer in the country, his argument must have been anxiously awaited by Mr. Watson. Curiously enough, as it seems to the non-legal mind, Mr. Choate began his plea by a *prayer for mercy.* Whatever the sins of the Standard Oil Company of Ohio, pleaded Mr. Choate, do not take away its charter. Mr. Choate then proceeded with a strong argument in which he claimed "absolute innocence and absolute merit for everything we have done within the scope of the matters brought before the court by these pleadings."

The argument did not convince the court of the innocence of the Standard in the questions at issue. The court showed, out of the mouth of the trust agreement itself, that the Standard Oil Company of Ohio was "managed in the interest of the Standard Oil Trust—irrespective of what might be its duties to the people of the state from which it derives its corporate life." The court gave as its opinion that an act of a majority of the stockholders of a corporation affects the property of a company in the same way that a resolution by the board of directors affects it. "By this agreement," said the court, "indirectly, it is true, but none the less effectually, the defendant is controlled and managed by the Standard Oil Trust, an association with its principal place of business in New York City, and organised for a purpose contrary to the policy of our laws. Its object was to establish a virtual monopoly of the business of producing petroleum, and of manufacturing, refining and dealing in it and all its products, throughout the entire country, and by which it might not merely control the production, but the price, at its pleasure. All such associations are contrary to the policy of our state and void.

. . . .

"Much has been said in favour of the objects of the Standard Oil Trust and what it has accomplished. It may be true that it has improved the quality and cheapened the cost of petroleum and its products to the consumer. But such is not one of the usual or general results of a monopoly; and it is the policy of the law to regard, not what may, but what usually happens. Experience shows that it is not wise to trust human cupidity where it has the opportunity to aggrandise itself at the expense of others. The claim of having cheapened the price to the consumer is the usual pretext on which monopolies of this kind are defended." *

From all this the court decided the Standard Oil Company deserved punishment. The charter was not taken away—the statute of limitations being advanced as a reason for this leniency, although, as Mr. Watson and Mr. Warrington showed, the statute of limitations could hardly be pleaded in this case, when the state had been kept in ignorance by the concealment of the agreement. The company was allowed to live, but it was ousted from the privilege of entering into the trust agreement, from the power of recognising the transfer of the stock, and from the power of permitting the trustees to control its affairs. It was also ordered to pay the costs of the action.

The judgment of the court was not rendered until March 2, 1892, almost two years after the filing of the petition. As soon as it was received Virgil P. Kline, the chief counsel of the Standard Oil Company of Ohio, went to New York for consultation with the trustees. Five days later he wrote to

* History of Standard Oil Case in the Supreme Court of Ohio, 1897–1898. Part I, pages 27–28. Original opinion of the court.

Judge Spear, the chief justice of the Ohio Supreme Court, saying: "Decisive steps will be taken at once not only to release the Standard Oil Company from any relations to the trust, but to terminate the entire trust." But there were "practical difficulties" in the task. The company pleaded for a "temporary recognition," and he asked an interview where he could explain the situation. This was granted, and on the 16th of March Mr. Kline explained to the judges in chambers, to Mr. Watson, and to his successor in office, the situation of the company. The trustees had all but seven shares of its stock. Trust certificates had been issued for these ten years before. The Standard Oil Company did not know who held these certificates, and could only know through the trustees, therefore the trust certificates must be transferred back, the owners hunted up, and each one induced to make an exchange. A system must be devised for doing this. Anybody could see this would take time. The court was friendly in the matter, and Chief Justice Spear gave to Mr. Kline an informal note granting an extension. "The court is not disposed to change its order at this time," the chief justice wrote, "but, so long as those in control appear to be engaged, as now, in an honest effort to dissever the relations of the company with the trust, and liquidate and wind up the affairs of the trust, the court will not be disposed to interfere." Thus time was gained.

While Mr. Kline was securing time, the trustees were pushing a liquidation scheme. On March 11 the following notice was mailed to all holders of Standard Oil Trust certificates, and was published in a newspaper in each state where a Standard Oil Company had been organised:

NOTICE

A special meeting of the holders of Standard Oil Trust certificates will be held at the office of the trust, Number 26 Broadway, in the City of New York, on Monday, March 21, 1892, at eleven o'clock A.M., for the purpose of voting upon a resolution to terminate the trust agreement, in accordance with the terms of said agreement, and to take such further action as may be thereby rendered necessary.

H. M. FLAGLER, *Secretary*.

The meeting was held as called. Mr. Rockefeller was in the chair, and Mr. Dodd, who had drawn the trust agreement, now presented the resolution which was to dissolve it. The remarks with which Mr. Dodd introduced his resolution denied every point which the courts had charged against the combination:

"Something over ten years ago," said Mr. Dodd, "a few individuals owning stocks in a number of corporations engaged in transporting and refining oil, entered into an agreement by which their stocks were placed in the hands of trustees, and certificates were issued by said trustees showing the amount of each owner's equitable interest in the stocks so held in trust. This was not done in order to vest the voting power in the hands of a few persons, because the persons chosen as trustees then held, and always have held, the voting power by virtue of their absolute ownership of a majority of the stocks. It was not done to reduce competition, because the companies whose stocks were placed in trust were not competing companies, and could not be so long as their stocks were owned by these few persons. It was not done to limit production or to increase prices, but, on the contrary, was done to increase production, cheapen cost of manufacture, and to lower prices, and it has been successful in that object far beyond the anticipations of those who originated the plan. It was called a trust, because it was a trust in the sense in which the word was then understood. It vested a fiduciary obligation in a few for the benefit of many, and the trustees thus created have faithfully observed the trust confided in them.

"Other persons, however, found this trust plan a convenient one, and it is alleged that it has been adopted for and adapted to purposes quite different from those which actuated the framers of this trust. Whether these allegations be true or false, it is true that a trust is now defined to be a combination to suppress competition and to reduce production, and to increase prices. Public opinion has not unwisely been aroused against combinations for such purposes, and legislation of more or less severity, and rather more or less peculiarity, has been directed against them in seventeen or eighteen states of the Union. All such arrangements are now miscalled trusts, and all trusts are popularly supposed to partake of the same nature. For this reason, if for no other, it should be seriously considered whether this trust should not be terminated. So long as it exists, misconception of its purposes will exist.

"But another reason exists which seems to make it desirable to dissolve this trust.

Some two years ago a *quo warranto* issued in the name of the state of Ohio against the Standard Oil Company, a corporation of the state of Ohio, setting forth this trust agreement and alleging that that corporation, by becoming a party thereto, had done an act beyond its power, and thereby had forfeited its charter. The defendant corporation denied that it was a party to the agreement, and alleged that the agreement was on its face, and plainly, an agreement only between individuals, owners of corporate stocks, relating to their personal property, and was neither made by the corporation nor for the corporation. The court, however, held that the agreement was a corporate agreement, and decreed, among other things, that the corporation must cease to permit trustees to vote upon stocks held in trust.

"As this agreement was not entered into as a corporate agreement, and as this decision gives it an effect quite different from the intent of the parties who entered into it, it seems better to end it." *

It is probable that Mr. Dood had foreseen from the first just such an attack on his agreement as had come, for he had put into that instrument a paragraph providing for a dissolution, and it was in accordance with that article that the trust was now dissolved. The trustees were to continue to exist—under a new name: "Liquidating trustees." The property they had to take care of was vastly in excess of what it had been ten years before. Then the capital of the thirty-nine constituent companies was $70,000,000. These companies had been combined until they had been reduced to twenty, and their combined capital was now $102,233,700.† Property of about $20,000,000 in excess of the capital was held by the trustees. Mr. Dodd's resolution provided for the division of this property, and for the transfer of the trust certificates back to the corporations to which they belonged. The individual holders of the trust certificates were to get in exchange a proportionate share in each of the twenty companies. "A will not get stock in one corporation and B in another; each will get his due

* Proceedings of meeting dissolving trust. History of Standard Oil Case in the Supreme Court of Ohio, 1897–1898. Part I, pages 80–81.

† See Appendix, Number 53. List of constituent companies of the Standard Oil Trust, with assets and capitalisation in 1892.

proportion in the stocks of all," said Mr. Dodd. All of this change would make no difference with the management of affairs. Mr. Dodd assured the stockholders: "Your interests will be the same as now. The various corporations will continue to do the same business as heretofore, and your proportion of the earnings will not be changed."

The trustees went about liquidating at once, but it was not until the following November that the immense number of certificates held by them personally were exchanged. The process followed can be easily illustrated by Mr. Rockefeller's case. When the trust was ordered dissolved Mr. Rockefeller held 256,854 of the 972,500 shares of Standard Oil Trust which were out. He turned over to an attorney an assignment of this amount, with instructions to secure from each of twenty companies in the trust stock certificates for the portion belonging to him. The corporate stocks were turned over to Mr. Rockefeller, and the assignment of certificate, a properly framed and numbered document, was turned over to the liquidating trustees. This assignment of legal title, for all practical purposes, was the same thing as the trust certificate. It enabled the trustees to collect dividends from the various companies and pay them just as they had before. The documents showing the formal procedure in the case of Mr. Rockefeller's stocks are printed in the Appendix.*

At the end of the first year, after the dissolution of the trust, 477,881 shares were uncancelled. At the end of the second year it was the same; at the end of the third, 477,881 were still out. At the end of the fourth, 477,881. The dissolution of the trust seemed to have come to a stand-still. Mr. Dodd was right; things were going on as they did before; dividends were issued exactly as before. Nor was there any indication of an inten-

* See Appendix, Number 54. Forms of Mr. Rockefeller's certificate of holdings in the Standard Oil Trust, with assignment of legal title which took its place in 1892.

tion on the part of the liquidating trustees to change this state of things. If the monopolistic power of the Standard Oil Trust was to be broken, it was evidently not to be by any order of dissolution by the courts. Something more powerful than the courts was at work, however. The spirit of individualism was beginning to reassert itself in the oil industry—a new war for independence had been begun, was indeed well under way even before the state of Ohio made the dissolution of the trust necessary.

CHAPTER FIFTEEN

A MODERN WAR FOR INDEPENDENCE

PRODUCERS' PROTECTIVE ASSOCIATION FORMED—A SECRET INDEPENDENT ORGANIZATION INTENDED TO HANDLE ITS OWN OIL—AGREEMENT MADE WITH STANDARD TO CUT DOWN PRODUCTION—RESULTS OF AGREEMENT NOT AS BENEFICIAL TO PRODUCERS AS EXPECTED—PRODUCERS PROCEED TO ORGANISE PRODUCERS' OIL COMPANY, LIMITED—INDEPENDENT REFINERS AGREE TO SUPPORT MOVEMENT—PRODUCERS AND REFINERS' COMPANY FORMED—LEWIS EMERY, JR.'S, FIGHT FOR SEABOARD PIPE-LINE —THE UNITED STATES PIPE LINE—STANDARD'S DESPERATE OPPOSITION—INDEPENDENT REFINERS ALMOST WORN OUT—THEY ARE RELIEVED BY FORMATION OF PURE OIL COMPANY—PURE OIL COMPANY FINALLY BECOMES HEAD OF INDEPENDENT CONSOLIDATION—INDEPENDENCE POSSIBLE, BUT COMPETITION NOT RESTORED.

JOHN D. ROCKEFELLER'S one irreconcilable enemy in the oil business has always been the oil producer. There is no doubt that Mr. Rockefeller has sincerely deplored this. And well he might, for he learned in his first great raid on the industry in 1872 that the producers aroused and united made a powerful and dangerous foe.

No doubt, if it had been practical, Mr. Rockefeller would have begun at the start to take over oil production as he did oil refineries and pipe-lines, and thus would have gotten his enemy out of the way; but during the first fifteen years of his work it was not practical. The oil fields were too vast and undefined. It not being practical to own the oil fields, and yet essential that those who did own them, and of whose oil he aspired to be the only buyer, should be kept sufficiently satisfied not to interfere with his domination or to attempt to handle the oil for themselves, Mr. Rockefeller, whenever he had the chance, sought to persuade the producers to do what he would have done had he owned the oil fields—that was, to keep the supply of crude oil short.

"The dear people," he said once when asked by an investigating committee if his monopoly of oil refining and oil transportation had not prevented the producer from getting his full share of the profits—"the dear people," he said, "if they had produced less oil than they wanted, would have got their full price; no combination in the world could have prevented that, if they had produced less oil than the world required." *

It is quite possible that if Mr. Rockefeller had been able to convert the majority of the producing body to this theory, and the supply of crude oil producers would have forgotten consequently high, the oil producers would have forgotten their resentment at his early raids and would have relapsed into indifference toward his control. Material prosperity is usually benumbing in its effects. There always has been a factor in the great game playing in the Oil Regions, however, which not even Mr. Rockefeller could match. Nature has been in the oil game, and she has taken pains to prevent the only situation which would have enabled Mr. Rockefeller to reconcile the oil producers. Again and again when it seemed as if the limits of oil production were set, and when Mr. Rockefeller and his colleagues must have believed that they would soon have the industry sufficiently well in hand to pay the producers a satisfactory price for crude oil, their calculations have been upset by the discovery of a great deposit of oil which flooded the market and put down the prices. This happened so often between Mr. Rockefeller's first public appearance in the business and the time when he completed his control of transportation, refineries and markets,

* Report on Investigation Relative to Trusts, New York Senate, 1888, page 445.

that the yearly production of crude oil had risen from five and a half million barrels to thirty million barrels, and instead of a half million barrels above ground in stocks there were in 1883 over thirty-five million barrels, in 1884 nearly thirty-seven million, in 1885 thirty-three and a half million. The low price for crude which these vast stocks caused, the high charges for gathering, transporting and storing, all services out of which the Standard was making big profits, the fact that the profit on refined oil steadily increased in these years—the result of the overthrow of independent refiners and pipe-lines—while the profit on crude steadily diminished, were facts which the oil producers brooded over incessantly, and the more bitterly because they felt they could do nothing to help themselves. Every enterprise looking to relief which they had undertaken had, for one reason or another, failed. They had no faith that relief was possible. The Standard would never allow any outside interest to get a foothold. It was the bitterness which this conviction caused which was at the bottom of the outburst over the Billingsley Bill described in Chapter XIII. The Billingsley Bill was defeated, as it deserved to be, but the work done was by no means lost. For the first time since 1880 the Oil Regions were aroused to concerted action. The support of the Billingsley Bill had been a spontaneous movement, a passionate, unorganised revolt against the tyranny of the Standard, but it served to bring into action men who for six long years had been saying it was no use to resist, that Mr. Rockefeller's grip was too strong to be loosened. It revived their confidence in united action and steeled them to a determination to take hold of the industry and force into it again a fair competition in handling oil.

On the very night after the defeat of the bill (April 28, 1887) the oil men who had gathered in Harrisburg to support the measure, angry and sore as they were, arranged to call an early meeting in Oil City and organise. The meeting was held. It was large, and it was followed by others. In a very short time 2,000 oil men were enrolled in a Producers' Protective Association, and thirty-six local assemblies were holding regular meetings throughout the region. There were several important points about the new association, aside from the enthusiasm and determination which animated it:

(1) It was a secret order.

(2) Its membership was composed entirely of persons outside of and opposed to the Standard Oil Trust, one of its by-laws reading: "No person connected with the Standard Oil Company or any of its allies, as partners, stockholders, or employees, and friendly thereto, shall be elected to membership; and members becoming such shall be liable to expulsion."

(3) It proposed "to defend the industry against the aggregations of monopolistic transporters, refiners, buyers and sellers" by *handling its own oil.*

Hardly had the Producers' Protective Association been organised before Mr. Rockefeller had an opportunity to try his plan for conciliation. An independent movement had been started in the summer of 1887 by certain large producers in favour of a general "shut-down," its object, of course, being to decrease the oil stocks. The president of the Producers' Association, Thomas W. Phillips, who at that time was the largest individual producer in the oil country, his production averaging not less than 6,000 barrels a day, was called into consultation with the leaders of the "shut-down" movement. Mr. Phillips promptly told the gentlemen interested that he would not join in such an undertaking unless the Standard went into it. He pointed out that the Standard owned a large proportion of the 30,000,000 barrels of oil above ground. They had bought it at low prices. If the production was shut down prices would go up and the Standard would reap largely on the oil they owned. The producers would, as usual, be standing all the loss.

The upshot of the council was that the Producers' Protective Association took hold of the shut-down movement, its representative seeking an interview with the Standard officials as to their willingness to share in the cost of reducing the production. Here was a chance for Mr. Rockefeller to apply his theory of handling the oil producers—conciliate them when possible—encourage them in limiting their production. The oil men's representatives were met half-way, and an interesting and curious plan was worked out; the producers were to agree to limit their production by 17,500 barrels a day. They were to do this by shutting down their producing wells a part or all of the time and by doing no fresh drilling for a year. If they would do this the Standard agreed to sell the association 5,000,000 barrels of oil at sixty-two cents, and let them carry it at the usual rates as long as they wanted to. Whatever advance in price came from the shut-in movement the producers were to have on their oil, and it was to be shared by them according to the amount each shut in his production. Mr. Phillips, before agreeing to this arrangement, demanded that provision be made for the workingmen who would be thrown out of employment by the shut-down, and he proposed that the association set aside for their benefit 1,000,000 barrels of the oil bought from the Standard, and that the Standard set aside another million; all the profits above sixty-two cents and the carrying charges on the 2,000,000 barrels were to go to the workingmen. A memorandum covering the above points of the agreement was drawn up, and it was accepted by the two interests represented.*

Mr. Rockefeller's reason for signing the contract he gave

* See Appendix, Number 55. Agreement of 1887 between the Standard Oil Company and producers.

to the New York State Trust Investigating Committee four months later:

Q. . . . What was the inducement for the Standard Oil Trust to enter into such an agreement as that?

A. The inducement was for the purpose of accomplishing a harmonious feeling as between the interests of the Standard Oil Trust and the producers of petroleum; there was great distress throughout the oil-producing region; as an instance of that distress there was an outcry that our interest was getting a return, that theirs was not in the business, and we did not know, as a matter of fact, that the oil-producing interest was abnormally depressed, and we felt it to be to the interests of the American oil industry that a reasonable price should be had by the producer for the crude material, and we wanted to co-operate to that end.

Q. By advancing the price of the crude material you necessarily advance the price of the refined?

A. Yes, sir.*

The shut-down went into effect the first of November, 1887. The effect on stocks and the market was immediate—stocks fell off at the rate of a million barrels a month, and prices rose by January, 1888, some twenty cents. But at the end of the year, though oil was higher and stocks considerably less, the benefits of the shut-down had not been conspicuous enough to produce that "harmonious feeling" Mr. Rockefeller so much desired; not sufficient to distract the minds of the producers from the idea they had in forming their association, and that was a co-operative enterprise for taking care of their own oil. Throughout 1888 and 1889 two schemes, known as the Co-operative Oil Company, Limited, and the United Oil Company, Limited, were under consideration. By the end of the latter year it looked as if something could be done with the second, and it was turned over by the executive board of the association to a special committee, of which H. L. Taylor, of the Union Oil Company, one of the largest and oldest producing concerns of the Oil Regions, was chairman. How Mr. Taylor had succeeded in

* Report on Investigation Relative to Trusts, New York Senate, 1888, page 449.

getting into the Producers' Protective Association it is hard to say, for it was he and his partner, Mr. Satterfield, who in 1883 had tried to throw the Tidewater Pipe Line into the hands of the Standard Oil Company, and who, when that unworthy scheme failed, had sold their stock to the Standard, thus giving that company its first holdings in the Tidewater.* The independents had forgotten or overlooked this fact, for Taylor was a member of the Producers' Protective Association and prominent in its councils.

The special committee, of which Mr. Taylor was chairman, went actively to work. Lawyers were employed to consider the safest form of organisation for a company doing an interstate pipe-line business and carrying on refineries. Certain German capitalists, owners of tank-steamers and interested in foreign marketing agencies, were brought into the scheme. Things were going well, when suddenly the committee found the chairman cooling toward the enterprise. Then came the rumour that Mr. Taylor and his partners—Mr. Satterfield and J. L. and J. C. McKinney—had sold the Union Oil Company to the Standard. A meeting of the executive board was at once called, Messrs. Taylor and J. L. McKinney both being present. They acknowledged the truth of the report and were promptly informed their resignations would be accepted.

The rumour of the secret desertion of strong members of the Producers' Protective Association, while holding positions of trust, soon spread through the Oil Regions. It was a staggering blow. It took from them one of the largest single interests represented. It deprived them of men of ability on whom they had depended. It introduced a fear of treachery from others. It brought them face to face with a new and serious element in the oil problem—*the Standard as an oil producer*. Up to 1887, the year of the organisation of the

* See Chapter IX.

Producers' Protective Association, Mr. Rockefeller had not taken his great combination into oil production to any extent, and wisely enough from his point of view. It was a business in which there were great risks, and as long as he could control the output by being its only buyer, why should he take them? Now, however, the situation was changing. A number of sure fields had been developed—Bradford, Ohio, West Virginia. Their value was depressed by overproduction. Mr. Rockefeller had money to invest. The producers were threatening to disturb his control by a co-operative scheme. It was certain that he had not yet produced a "harmonious feeling." It was not sure he would. If he failed in that they might one day even shut off his supply of oil, as they had done in 1872, and Mr. Rockefeller, with great foresight, determined to become a producer. In 1887 he went into Ohio fields. Soon after he began quietly to buy into West Virginia. When he learned, in 1890, from Mr. Taylor and his partners, that a co-operative company of producers was on foot, he naturally enough concluded that the best way to dismember it was to buy out the largest interest in it. The Union Oil Company saw the advantage of being a member of the Standard Oil Trust, and sold. In this one year, 1890, over 40,000 shares of Standard Oil Trust certificates were issued to oil-producing companies,* as follows:

For stock of Union Oil Company	18,249	shares
" " " Forest Oil Company	17,378	"
" " " North Pennsylvania Oil Company	2,647	"
" " " Midland Oil Company	2,000	"
	40,274	"

There was general consternation in producing circles, and if there had not been a number of men in the organisation who realised that the life of the independent effort was at

* Plaintiff's Exhibit Number 52 in the case of James Corrigan *vs.* John D. Rockefeller in the Court of Common Pleas, Cuyahoga County, Ohio, 1897.

stake, and who turned all their strength to saving it, the association would undoubtedly have gone to pieces. Chief among these men were Lewis Emery, Jr., and C. P. Collins, of Bradford, Pennsylvania; J. W. Lee and David Kirk, of Pittsburg; A. D. Wood, of Warren; Michael Murphy, of Philadelphia; Rufus Scott, of Wellsville; J. B. Aiken, of Washington; R. J. Straight, of Bradford; Roger Sherman and M. W. Quick, of Titusville. They urged an immediate meeting of the General Assembly, at which a plan for co-operative action should be adopted and at once put into force.

On January 28, 1891, the General Assembly convened at Warren, Pennsylvania. The whole miserable story of the co-operative plan which the executive board had worked out, and its destruction by the desertion of the Union Oil Company, came out. It was at once evident that, instead of disheartening the Assembly, it was going to harden their determination and spur them to action; that they would not leave Warren until they had something to work on. The session lasted three days, and before finally adjourning it had adopted a drastic plan, framed by a committee of nine, of which Mr. Quick was chairman. This plan aimed, so the resolution adopted by the Assembly stated, *to cut off the supplies of the producers' oil from the Standard Trust!* This was to be accomplished by forming a limited partnership, whose subscribers should all be trusted members of the Producers' Protective Association (only persons having no affiliation with the Standard Oil Company were members of the Producers' Protective Association, it will be remembered), and which should aim to take care of the crude oil from the wells of the producers who went into the movement, furnish it local transportation, and find a market for it either by building independent refineries or by alliance with those already in existence.

From Warren the delegates went home to work for the

MICHAEL MURPHY

The present President of the Pure Oil Company.

DAVID KIRK

The first President of the Pure Oil Company.

JAMES W. LEE

The chief counsel of the Pure Oil Company. President of the company from 1897 to 1901.

THOMAS W. PHILLIPS

A leader in the independent movement, which resulted in the Pure Oil Company.

new scheme. J. W. Lee and J. R. Goldsborough, the secretary of the association, at once made a tour of the Oil Regions to explain the project and solicit subscriptions. The response was immediate. In a few weeks over 1,000 producers had subscribed to the new company, which was at once organised as the Producers' Oil Company, Limited, its capital being $600,000.

But it is one thing to organise a company, and another to do business. Where were they to begin? Where to set foot? The only thing of which they were sure was a supply of crude oil, and in order to take care of that they began operations by putting up four iron tanks at Coraopolis, Pennsylvania, near the rich McDonald oil field. But they must have a market for it, and their first effort was to ship it abroad. At Bayonne, New Jersey, on the border of the territory occupied by the Standard's great plant, stands an independent oil refinery, the Columbia Oil Company. The Columbia has "terminal privileges," that is, a place on the water-front from which it can ship oil—an almost impossible privilege to secure around New York harbour. The Producers' Oil Company now obtained from Hugh King, the president of the Columbia, the use of his terminal. They at once had fifty tank-cars built, and prepared to ship their crude oil, but the market was against them, stocks were increasing, prices dropping. The railroad charged a price so high for running their cars that there was no profit, and the fifty tank-cars were never used in that trade. A futile effort to use their crude oil as fuel in Pittsburg occupied their attention for a time, but it amounted to nothing. It was becoming clearer daily that they must refine their oil. The way opened to this toward the end of their first year.

In and around Oil City and Titusville there had grown up since 1881 a number of independent oil refineries. They had come into being as a direct result of the compromise made in 1880 between the producers and the Pennsylvania Railroad, a clause of which stipulated that thereafter railroad rates should be open and equal to all shippers. The Pennsylvania seems to have intended at first to live up to this agreement, and it encouraged refiners in both the Oil Regions and Philadelphia to establish works. At first things had gone very well. There were economies in refining near the point where the oil was produced, and so long as the young independents had a low rate to seaboard for their export oil they prospered. But in 1884 things began to change. In that year the Standard Pipe Line made a pooling arrangement with the Pennsylvania Railroad, by which rates from the Oil Regions were raised to fifty-two cents a barrel, an advance of seventeen cents a barrel over what they had been getting, and in return for this raise the Standard agreed to give the railroad twenty-six per cent. of all the oil shipped Eastward, or pay them for what they did not get. This advance put the independents at a great disadvantage. In September, 1888, another advance came. Rates on oil in barrels were raised to sixty-six cents, while rates on oil in tanks were not raised. The explanation was evident. The railroad owned no tank-cars, but rented them from the Standard Oil Company. It refused to furnish these tank-cars to the independents, but forced them to ship in barrels, and now advanced the price on oil in barrels. This second advance was more than the refiners could live under, and they combined and took their case to the Interstate Commerce Commission, a hearing being given them in Titusville in May, 1889. No decision had as yet been rendered, and they in the meantime were having a more and more trying struggle for life, and their exasperation against the Standard was increasing with each week. When, therefore, the representatives of the Producers' Oil Company proposed a league with the independent refiners they were cordially welcomed.

We have oil in tanks at Coraopolis, said the producers, plenty of it, but we have no market. If we build a pipe-line from our tanks to Oil City and Titusville and give you pipage at fifteen cents a barrel, five cents less than the Standard charges, will you enter into an agreement with us to take our oil for five years? The refiners saw at once the possible future in such an arrangement, and in a short time they had gone individually into a company to be called the Producers' and Refiners' Company, with a capital of $250,000, of which the Producers' Oil Company held $160,000, and whose object was the laying of a pipe-line from the fields in which the producers were interested to the refineries at Oil City and Titusville. The new plan was carried out with the greatest secrecy and promptness. Before the Standard men in the region realised what was going on, a right of way was secured and the pipe was going down. On January 8, 1893, the first oil was run. Here, then, was the first link in a practical co-operative enterprise—independent producers and refiners of oil joined by a pipe-line of which they were the owners.

While this enterprise was being carried out in Western Pennsylvania, in the northern part of the state a still more ambitious, independent project was under way, nothing less than a double pipe-line, one for refined and the other for crude oil, from the Oil Regions to the sea. This plan had originated with Lewis Emery, Jr., one of the most implacable and intelligent opponents Mr. Rockefeller's pretensions have ever met. Mr. Emery sympathised with the idea that there was no way for the producer to get his share of the profits in the oil business except by handling the product entirely himself. In his judgment a pipe-line to the seaboard was the first important link in such an attempt, and in 1891, on his own responsibility, he set out to see what hopes there were of securing a right of way. The Columbia Oil Company, through whom the Producers and Refiners were exporting, favoured such a scheme. It was certain many producers would go into it; but on all sides there was much scepticism about the Standard allowing a line to go through. Mr. Emery's first idea was a line from Bradford to Williamsport, on the Reading road. He consulted the railroad officials. They would be glad of the freight, they told him, and a preliminary contract was drawn up. The contract was never completed. Mr. Emery returned to find out why. "If we give you this contract," the Reading officials told Mr. Emery, "we shall disturb our relations with the Standard Oil Trust. We cannot do it."

Turning from the Reading, he projected a new route, a pipe-line from Bradford to the New York, Ontario and Western Railway near Hancock, New York, thence by rail to the Hudson River, and from there by water to New York harbour. The New York, Ontario and Western officials welcomed the proposal. It gave them a new and valuable freight. But the pipes must cross the Erie road near both its terminals. Mr. Emery saw the president of the road. "Yes," the president told him, "we are disposed to assist all progress. Go ahead." Thus encouraged, he sent his men into the field to get the right of way. They had made a good beginning before the project was known, but as soon as it was rumoured there appeared promptly on the route surveyed a number of men known to be Standard employees. They, too, wanted a right of way, the same as Mr. Emery wanted. They bought strips of land across his route, they bought up mortgages on farms where rights had already been acquired, and, mortgage in hand, compelled farmers to give them rights. It was an incessant harassing by men who never used the rights acquired—who did not want them save to hinder the independent project. This sort of hindrance by the Standard was certain, whatever route was taken, and Mr. Emery went ahead undismayed, and in September, 1892, organised his company—the United States Pipe Line Company—with a capital of $600,000. Among the incorporators were representatives of the independents' interests, both in New York and in the Oil Regions, and much of the stock was soon placed in the hands of the men who were interested in the independent concerns described above.

It looked very much as if the United States Pipe Line were to be laid. Now, the strength of the Standard Oil Trust had always been due to its control of transportation. An independent pipe-line, especially to the seaboard, was considered rightly as a much more serious menace to its power than an independent refinery. The United States Pipe Line could not be allowed, and prompt and drastic measures were taken to hinder its work. There is no space here for an account of the wearisome obstructive litigation which confronted the company, for the constant interference, even by force, which followed them for months. It culminated when an attempt was made to join the pipes laid to each side of the Erie tracks near Hancock, New York, the Eastern terminal of the pipe-line. Mr. Emery, relying on the promise of the Erie's president to allow a crossing, sent his men to the railway to connect the pipes. Hardly had they arrived before there descended on them a force of seventy-five railroad men armed for war. These men took possession of the territory at the end of the pipes and intrenched themselves for attack. The pipe-line men camped near by for three months, but they never attempted to join the pipes. Mr. Emery had concluded, on investigation, that the Erie officials, like the Reading, had found that it would be unwise to disturb their relations with the Standard, and while his men were keeping attention fixed on that point he was executing a flank movement, securing a right of way from a point seventy miles back to Wilkesbarre, on the Jersey Central. This new movement was executed with such celerity that by

June, 1893, the United States Pipe Line had a crude line 180 miles long connecting the Bradford oil fields with a friendly railway, and a refined line 250 miles long connecting the independent refiners of Oil City, Titusville, Warren and Bradford with the same railway.

With the completion of the refined line a question of vital importance was to be settled: Could refined oil be pumped that distance without deteriorating? The Standard had insisted loudly that it could not. When the day came to make the experiment an anxious set of men gathered at the Wilkesbarre terminal. They feared particularly that the oil would lose colour, but, to their amazement, not only was the colour kept, but it was found on experiment that the fire test was actually raised by the extra agitation the oil had undergone in the long churning through the pipes. A new advance had been made in the oil industry—the most substantial and revolutionary since the day the Tidewater demonstrated that crude oil could be pumped over the mountains. This new discovery, it is well to note, was not the work of the Standard Oil Trust, but it was accomplished in the face of their ridicule and opposition by men driven to find some way to escape from their hard dealings.

The success of the United States refined line aroused the greatest enthusiasm among the independent interests. It gave them access to the seaboard, and there was immediate talk of a closer union between them. Why should the Producers' and Refiners' Pipe Lines not be sold to the United States Line and completed to Bradford? By the spring of 1894 the project seemed certain of realisation.

The new movement was serious. Let this consolidation take place, and the producers had exactly what they had set out in 1887 to build up—a complete machine for handling the oil they produced. As the undertaking grew in solidity and completeness, the war upon it grew more systematic and determined. It took two main lines—discrediting the enterprise in the eyes of stockholders so that they would sell the stock to Standard buyers, the object being, of course, to get control of the companies; cutting the refined market until the refiners in the alliance should fail, or, becoming discouraged, sell. The work of discrediting the enterprise was turned over to the Standard organs in the Oil Regions, chief among which is the Oil City Derrick. Since 1885 the editor of this interesting sheet has been a picturesque Irishman, Patrick C. Boyle by name. Mr. Boyle's position as editor and proprietor of the Derrick is due to the generosity of the Standard Oil Trust, and he has discharged his allegiance to his benefactor with a zeal which, if it has not always contributed to the enlightenment of the Oil Regions, has, materially, to its gaiety. Mr. Boyle now turned all his extraordinary power of vituperation on three of the independents whose activity was particularly offensive to him—Mr. Emery, Mr. Wood and Mr. Lee—and he went so far that each of the three gentlemen finally sued him for libel. They all got judgments. In Mr. Emery's case, Mr. Boyle, after signing a bond of $5,000 to keep the peace—which bond he was obliged later to pay, with half as much more in costs—published the following retraction:

TO THE PUBLIC

For many years past there have appeared in the editorial and news columns of the Oil City Derrick various articles reflecting on the business, social and political character and integrity of Lewis Emery, Jr.

P. C. Boyle, the editor of the Derrick, was indicted and convicted for the publication of certain of such articles, and civil suit for damages was instituted by Mr. Emery against P. C. Boyle for damages for such publications.

The litigation has now been adjusted, and Mr. Boyle voluntarily retracts *in toto* all matters and things which he has said derogatory to the character, standing, or responsibility of Lewis Emery, Jr., published by him or under his direction in the past.

Mr. Boyle is fully satisfied that such articles have been published under a misapprehension of the facts, and is satisfied that Mr. Emery has been wronged, and should be vindicated, and this retraction is freely made as such.

Many of the articles have been republished in various papers in this country and Europe, and it is the desire of Mr. Boyle that this retraction shall be as freely and fully printed and published as were the original articles reflecting on Mr. Emery.

(Signed) P. C. BOYLE.

It is a satisfaction to the writer to be able to help gratify Mr. Boyle's laudable desire to have this document well circulated!

Although the greater part of the Oil Regions never took Mr. Boyle himself seriously, the conviction that his attacks were inspired, that this was the Standard's way of saying to the producers that their enterprise would not be allowed to live, gave a sinister look to what he said. More damaging still was the quiet confidence with which the solid men of the Standard smiled at the independent effort. What were their puny hundreds compared to the millions of the trust? What was a band of scattered "oil-shriekers" against the cold-blooded deliberation of Mr. Rockefeller's solid phalanx? The oil men were conscious enough of the inadequacy of their capital and their organisation, but they hung on, many of them because their blood was up, and they preferred spending their last cent to yielding; others on the principle which Mr. Phillips confesses held him, "that God sometimes chooses the weak things of the world to confound the mighty"; or that "one might chase a thousand, and two put ten thousand to flight."

The efforts which the Standard made to discredit the independent companies and their leaders were accompanied by a persistent, though quiet, attempt of Standard agents to buy in all the stock in the Producers' Oil Company and the United States Pipe Lines which timid, indifferent, or financially embarrassed stockholders could be induced to give up. The movement began to be rumoured and caused no little uneasiness in independent circles. How much would the Standard get? What would they do with it? They were soon to find out.

Before the use to be made of the stock developed, however, the Standard turned against the independents the most powerful and cruel weapon it wields—its control of the markets. The refiners were to be driven from the combination. The extent to which cutting was carried on for two years, beginning with the fall of 1893, is clear from a comparison of prices. In January of 1893 crude oil was selling at 53½ cents a barrel and refined oil for export at 5.33 cents a gallon. Throughout the year the price of crude advanced until in December it was 78⅜ cents. Refined, on the contrary, fell, and it was actually eighteen points lower in December than it had been twelve months before. Throughout 1894 the Standard kept refined oil down; the average price of the year was 5.19 cents a gallon, in face of the average crude market of 83¾ cents *—lower than in January, 1893, with crude at 53½ cents a barrel!

This much for the New York end of the export business. In Germany, where the export oil of the independents all went, it being handled there by one dealer, Herr Poth, whose depot was Mannheim, on the Rhine, prices were cut at every point which the independent oil reached. It was a matter of life and death to keep the foreign market they had devel-

* The following table shows the variation from 1890 to 1897 in price of crude oil per barrel of 42 gallons, and the price of refined oil per gallon in barrels in New York:

	1890		1891		1892		1893	
	Jan.	Dec.	Jan.	Dec.	Jan.	Dec.	Jan.	Dec.
Crude...	1.05⅛	67½	74⅛	59¼	62½	53¼	53½	78⅜
Refined..	7½	7¼	7.42	6.44	6.45	5.45	5.33	5.15

	1894		1895		1896		1897	
	Jan.	Dec.	Jan.	Dec.	Jan.	Dec.	Jan.	Dec.
Crude.....	80	91⅜	98⅝	1.43⅝	1.45¾	97⅞	88⅛	65
Refined....	5.15	5.61	5.87	7.77	7.85	6.35	6.13	5.40

oped, and for twenty months the independent refiners met the demand of their export agents and foreign dealers for lower prices with cut cargoes. For twenty months they lost money on every barrel they sold. Oil was sold by the Titusville refiners as low as 1.98 cents a gallon. The Lewis Emery works at Bradford sold one cargo at 1.07 cents net, and many at or below two cents. Had it not been for the union with pipe-lines such prices would have been impossible, but all through the struggle in the market the United States Pipe Line and the Producers' and Refiners' lines carried oil at cost or below. The pipe-lines were heavily in debt to the Reading Iron Works, but that company stood by them valiantly, extending their notes until the struggle was over and the pipe-lines able to meet them.

Such a situation could not go on forever, evidently. It had come apparently to be a question of how long the refiner had money to lose, and, as month after month the independents saw their bank accounts diminishing, and no relief in sight, the courage of a few began to ooze. Finally, late in 1894, a committee of the Western refiners, consisting of John Fertig of Titusville, H. P. Burwald of Titusville and S. W. Ramage of Oil City, went to New York to consult the Standard. Is there no hope of a better market? Is there any chance for us? None whatever, they were told, except to sell. We will buy the refineries and the stock of the independent concerns, but that is all we can do. The committee came home to report. The situation was hopeless, they said, and, as for them, they should sell. As they represented three of the largest concerns in the Union, and all carried stock in the allied enterprises, their withdrawal seemed at the moment a death-blow. It was a glum and beaten body of men which listened to the report, surrender written in every line of their faces.

Now Mr. Lee and Mr. Wood, two active men of the Producers Oil Company, had been invited to the meeting of the refiners. They realised fully that if the refiners pulled out of the Union now, the independent effort would in all probability go to pieces, and before a vote to sell could be taken Mr. Lee was on his feet. In an impassioned speech he pleaded for one more effort. He pointed out the fact that the abnormal condition of the oil market could not remain, that crude oil was steadily rising, and that no monopoly could permanently hold down a manufactured product in the face of the rising raw product. The Standard had done this for nearly two years—but it was contrary to the laws of nature that they do it for two years more. He told them that already conditions were better in Germany; that Mr. Emery had recently gone with Herr Poth, their foreign buyer, to several members of the German government, and presented to them the discrimination in prices of oil practised in the empire, oil from one and a half to three cents higher on the Elbe than on the Rhine, at points where freights were the same. He told the refiners of the interest that had been taken by the government in their case, and how they said, "Go home, gentlemen, and this shall stop," and that it had stopped. If criminal underselling can be checked in Germany, Mr. Lee argued, we can keep our market. He reminded the refiners that it was not merely a business they were establishing; it was a cause they were defending—the right of men to work in their own way without unlawful interference. The honour not only of themselves but of the Oil Regions was at stake. They were struggling for great principles. They were demonstrating that pluck, patience, and energy and brains can conquer any combination that ability and unscrupulousness can devise. "Do not give in," pleaded Mr. Lee. "Hold on, and we will go to the producers, lay your plight before them, and raise money to keep up the fight."

Aroused by his plea, all of the refiners, excepting Messrs.

Fertig, Burwald and Ramage, who had seen the Standard, decided to make another effort if the producers would help them out. In the next few days the leading men of the independent alliance worked with fury to call the Oil Regions into a mass-meeting. They travelled from assembly to assembly exhorting to action; they circulated dodgers announcing the gathering, and finally, in January, 1895, ran special trains to Butler, the rallying place. There was no lack of enthusiasm and blunt talk at the Butler mass-meeting. All the bitterness and determination of the region poured forth against the Standard, and when a resolution was offered by David Kirk, one of the most active and forceful of the independents, to raise money to form a new company, to be called the Pure Oil Company, its immediate object being to take care of the refiners in the tight place where they were, it went through with a whoop, and in a few moments $75,000 had been subscribed. A few days later this sum was raised to $200,000.

The objects of the company, as set forth in its prospectus issued at this time, were:

> To maintain and uphold the inherent right to do business, the right to transport and market the producer's own product, and his right to the just reward of his labour and capital invested.

Another clause of the prospectus is interesting:

> To prevent any interference of that monopoly which has obtained control of the oil business, the voting power of one-half of the stock of the Pure Oil Company is placed by the owners in the hands of five champions of this right of independence, who are bound by the terms of a permanent trust bond to vote only for such men and measures as shall forever make this company INDEPENDENT, so that no sales of interest will carry with them any power to jeopardise the policy or existence of the company, or the investments of its remaining members.

The Pure Oil Company had been organised none too soon. It was but a few months after it was well under way before a hurried meeting of the independents was called in New York. With scared faces the members learned that the German dealer, who for four years had been handling ninety per cent. of their export oil, had sold to the Standard marketing concern, the Deutsche-Amerikanische Company. Consternation was great. The independents had depended on the loyalty of Herr Poth as they did on that of each other. He had been enlisted in their cause by Mr. Emery, who, with the tragic earnestness which had characterised his entire struggle for independence, had asked him for an oath of loyalty, and, hand on his heart, Herr Poth had pledged his faith. In every respect he had served them loyally. His desertion was inexplicable and disheartening. Later they learned the truth, that Herr Poth had been informed, by what he supposed to be reliable authority, that the American independent interests had sold to the Standard. Believing that this would cut off his supply, he had turned over his concern to the Deutsche-Amerikanische. A few weeks later Herr Poth died suddenly. The story goes in independent circles that when he learned the truth he literally died of grief, believing he had perjured himself.

Herr Poth's sale left the independents in serious shape. They had cargoes of oil ready for Europe and no tankage in Europe to take it—nobody there to sell it. A meeting was at once called in Pittsburg to raise money, and in a few days Mr. Emery and Mr. Murphy went abroad, and, as quickly as such work could be done, they secured privileges in Hamburg and Rotterdam to erect tanks and establish marketing stations. The Pure Oil Company was in Europe. Once more the independents had been driven to depend on themselves, and once more they had proved sufficient to the emergency. But war was by no means over. With the establishment of the Pure Oil Company came the foreshadowing of a still closer union of the companies. At all hazards this was to be prevented. The Standard determined to play the stock of the Producers' Oil Company, Limited, and the United States Pipe Line, which it had been picking up quietly.

Already one attempt had been made to get into the former concern through one of the most conspicuous and successful producers of the oil country—Colonel John J. Carter, of Titusville, the president of the Carter Oil Company. Colonel Carter owned 300 shares of the stock of the Producers' Oil Company, Limited, and had been elected a member on it; according to the rules governing limited partnership in Pennsylvania, a stockholder must be elected to membership before he can vote his stock. In February, 1894, when a union of the pipe-lines had first been voted, he suddenly appeared in court and got an injunction against the sale. In the hearings on the injunction there came out a fact in regard to Colonel Carter which aroused a storm of wrath against him among the independents. The Standard Oil Company owned sixty per cent. of the Carter Oil Company! A harder fact was to be digested. On April 11, 1894, the company met in Warren, Pennsylvania. Colonel Carter was present and voted not only his 300 shares, but 13,013 more! Where had he got them? There was but one conclusion, and it proved to be true—the 13,013 belonged to the Standard Oil Company. They had been *loaned* to Mr. Carter; there was a form of transfer, but no sale, not even a price having been decided on—evidently in the hope that he, with a few other stockholders who were disaffected, would control the meeting and prevent the union of the pipe-lines. The attempt failed, for the Carter-Standard faction succeeded in getting together only 21,848 shares, while the independents held 30,560. The bitterness over this attack aroused terrible excitement. More than one member of the Warren meeting shouted "traitor" at Colonel Carter, and when the news of what happened reached the Producers' Protective Association there was a general demand that he be expelled from the Titusville assembly. It was done promptly, Mr. Carter not being given even a hearing.

The Standard took back its 13,013 shares and patiently went on picking up more. By January, 1896, they held 29,764 shares, enough, with Colonel Carter's 300, to give them a clean majority. Colonel Carter appeared at 26 Broadway at this opportune moment and offered to buy the stock at 100. Mr. Archbold and his colleagues thought it worth 150. (They are said to have paid as high as 220 for some of it.) Mr. Carter, in his frank colloquial testimony when on the witness-stand, described the conversation over the price:

> "Mr. Archbold says, 'I don't know, John, but what you are asking us to sell that stock too cheap. Don't you think it is worth more money?' I says, 'Not to me, it is not.' I says, 'I am willing to start in on this thing and put it on a paying basis and pay par for it.' 'Well,' he says, 'I guess that we will have to think that thing over,' and it dropped right there."

There were several interviews between Mr. Archbold, Mr. Rogers and Mr. Carter. They wanted to know how he proposed to run the Producers' Oil Company if he obtained a majority of the stock. "If I run that pipe-line," Mr. Carter reports himself as saying, "I am going to run it according to law and business principles. Any man that wants oil of me, and has the money to pay for it, shall have it."

"Will you let Mr. Emery have some oil if he wants it?" asked Mr. Rogers. "Yes, I will." "And all the outside refiners?" "Yes, I will. I shall make no discrimination against the outside refiner and in favour of the Standard Oil Company, or *vice versa*."

The Standard Oil seems to have been convinced that Colonel Carter was their friend—they probably never had any doubt of their ability to manage *him,* and it is evident from the Colonel's testimony that *he* never had any doubt about his own ability to manage both independents and

Standard—and the sale was made at 100, Colonel Carter giving his check for $297,640 on the Seaboard Bank.

Stock in hand, Colonel Carter went back to the Oil Regions to take possession. It was not so easy as he anticipated. The secretary refused to transfer the stock. He sought the president, Mr. Lee. What took place Colonel Carter himself told later on the witness-stand:

"Senator Lee and myself retired to my room in the hotel and we had quite a preliminary conversation on the situation and in regard to the Producers' Pipe Line. Then I stated to him my ownership of the majority of the stock of the Producers' Oil Company, Limited, and stated furthermore that I purchased it from the National Transit Company; that my desire was to stop all contention on the part of the producers and myself, to run the business on a business principle, so that the stock belonging to the various members and myself might pay something, instead of dragging its slow length along as it had been for the past six years. I told him, furthermore, that I was perfectly willing that he should elect what portion of the directors that his stock would warrant him, and I would elect those that I could. The Senator replied then: 'You propose to take charge of the association?' 'Yes,' I said; 'I did.' The Senator then stated emphatically that I could not do it; he would not permit it; if he had to spend the whole capital of the company he would resist it. . . . He gave me to understand emphatically that there was not anything except the management of the company by himself and his associates that would be tolerated, and I told him then I was sorry that I would have to go into court and determine my rights in court. That was about all, but it is only fair, furthermore, to say that at the time the Senator was rather warm, and I presume I was warm in the collar myself. I stated to him plainly that if there was any attempt to eject me from a legally constituted meeting in which I was there, I would resist it if I killed the man that attempted to put me out."

Mr. Carter's cool announcement that he meant to run the company "from a business stand-point, and not from the stand-point of a gadfly"—there seems to be a doubt about its being the producers who had played the part of the gadfly—exasperated the independents to the last degree, and in June, 1896, they met the colonel in court. His ownership of a majority of the company's stock was admitted, but it was urged by the independents that the Producers' Oil Company was a limited partnership, and that under the Pennsylvania law no one owning stock can become a member without being elected by a majority in number and value of the interests. Colonel Carter had been elected member on only 300 shares. Both the lower and supreme courts sustained the independents, and Colonel Carter found himself an owner of a majority of the concern's stock without the right of control. Under those circumstances neither he nor the Standard wanted the stock, and the company bought it below par.

The winning of the Carter case gave encouragement that a similar suit brought by the Standard pipe-lines against the United States Pipe Line might fail. As already noted, the Standard began to buy into that company as soon as it was under way, and by the summer of 1895 they had collected 2,613 shares. In August of that year the annual meeting of the company was held, and the agent of the Standard Oil Company who had been buying the stock, J. C. McDowell, presented himself prepared to vote. He was stopped at the door by Michael Murphy, the present president of the Pure Oil Company, and told emphatically that they considered that he was sent there by the Standard Oil Company to spy on their actions; that, legal or illegal, they would throw him out if he crossed the threshold. Mr. Murphy is well known to be a man of his word, and as he was backed by young and athletic independent stockholders, Mr. McDowell discreetly withdrew. Naturally a suit followed, but this time the independents lost. The United States Pipe Line, being a corporation, was obliged to recognise the Standard interest in the concern and eventually to allow them a director on its board.

The humiliation and disgust over this result shook the independents' interests to their foundation. There perhaps was never a period of more heart-breaking discouragement for many of the men than when they saw their dearest hopes frustrated, and a Standard representative in their councils. This defeat came, too, when they were smarting under a continued and intolerable interference by the Standard with the extension of their pipe-lines to the seaboard. That both the crude and refined lines should ultimately reach the sea had, of course, been the intention from the first. But it was not until 1895 that the company felt firm enough in its finances to push the extension. The route laid out was from Wilkes-barre to Bayonne, New Jersey, by way of Hampton Junction, on the Jersey Central Railroad. By this course two railroads were to be crossed, the Pennsylvania and the Delaware, Lackawanna and Western. Under both of them ran the pipe-lines of the Standard and the Tidewater, and the United States Pipe Line officials believed they had an equal right to go under, but they took it for granted they would be opposed, and prepared for it. Looking over the titles of the land along the Pennsylvania, Mr. Emery, the president of the company, who was personally directing the extension, found one for an acre; the owner did not know of his possession and was glad to sell it. This gave the United States people a crossing, but even then they were obliged to carry on a long litigation in the courts before they were free to use their right.

Coming to the Delaware, Lackawanna and Western, they decided to test their position by laying a pipe. It was promptly torn out. A farm over which the railroad passed was then purchased and preparations made to lay the pipe in a roadway under the tracks. As this road was some seventeen feet below the rails, any claim that there was possible danger from the oil seemed feeble. Knowing that the point was watched, Mr. Emery tried strategy. Taking fifty men with him he went in the night to the culvert under which he meant to cross, laid his pipes four feet under ground, fastened them down with heavy timbers, piled rocks on them, anchored them with chains, established a camp on each side of the track, and prepared for war. They soon had it. First,

LAYING A SIX-INCH PIPE LINE, CAIRO, WEST VIRGINIA

with a body of railroad men armed with picks and bars, who invaded the camp. "I told the boys," said Mr. Emery in describing the incident to the Industrial Commission in 1899, "to take the men by the shoulders and the seat of the pants, and take them out and lay them down carefully, which they did." The next day two wrecking-cars, with 250 men, came down the road and charged the camp, but again they were routed. The matter was taken by mutual agreement into court, and while Mr. Emery was before the justice of the peace, two locomotives were run down and the camp attacked with hot water and coals!

By this time the whole countryside was aroused. The unfairness of the thing was so patent that even the railroad employees engaged in it did not hestitate to say, in excuse of their employers, that it was the Standard Oil Company which was at the bottom of the opposition! As for the inhabitants, they offered any aid they could give. The local G. A. R. sent forty-eight muskets to the scene of war. Mr. Emery bought eighteen Springfield rifles, the camp was barricaded, and for seven months the pipes were guarded while the courts were deciding the legal title to the crossing.

This interim was employed by the pipe-line people in an attempt to get a free pipe-line bill through the New Jersey Legislature. If this could be done they could go under the Delaware, Lackawanna and Western without its consent. The bill was introduced in February, 1896, J. W. Lee, Hugh King and Lewis Emery, Jr., all appearing before the committee to argue for it. At first there seemed to be no opposition to it. Everybody agreed it was a just and proper measure. Then, suddenly, within a few days of the end of the session, a violent opposition sprang up. Trenton became alive with lobbyists—men well enough known to politicians. The newspapers came out boldly with the charge that the railroads and Standard were going to defeat the bill. Its friends

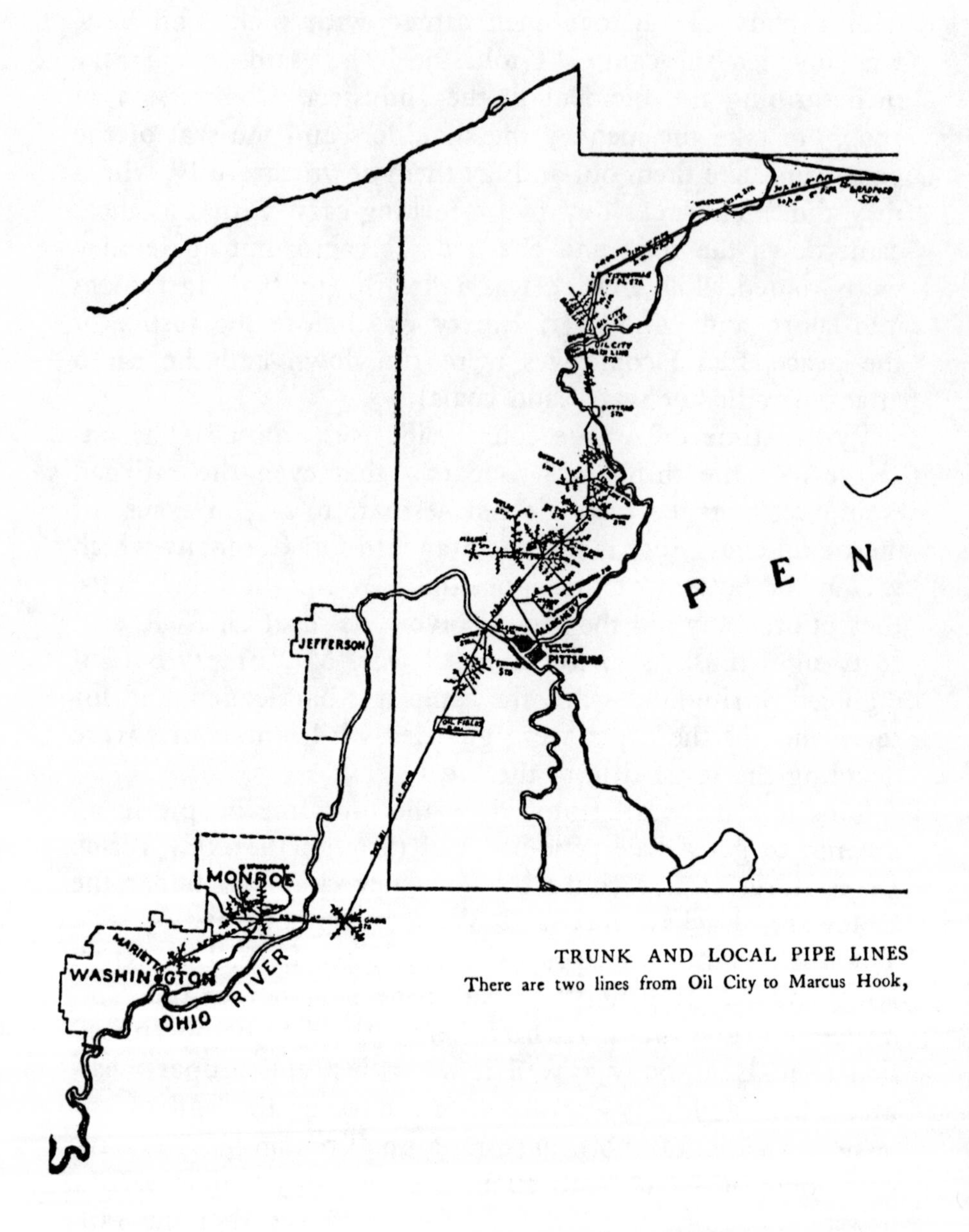

TRUNK AND LOCAL PIPE LINES
There are two lines from Oil City to Marcus Hook,

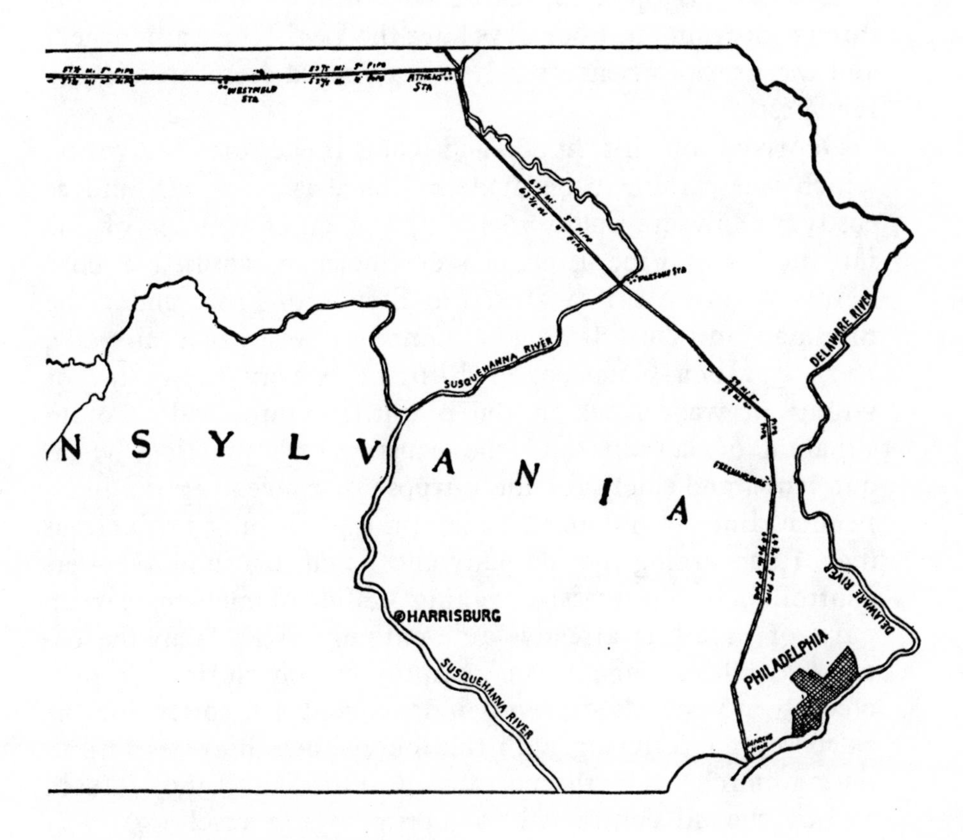

OF THE PURE OIL COMPANY.
near Philadeiphia, one for crude and one for refined oil.

could not believe it, nor did they until they found, the morning it was to be presented, that the Senator having it in charge had disappeared, taking with him the bill and everything concerning it. Four days later the Legislature adjourned, and the precious Senator, when next heard from, was in the far West!

Deprived of this hope, and condemned to a litigation which was certain to be made as long, as vexatious, and as costly as lawyers could make it, the chief counsel of the United States Pipe Line, Roger Sherman, advised a bold move—to bring suit against the Standard Trust under the Sherman anti-trust law. The summons was issued in July, 1897, by John Cunneen, of Buffalo. A very pretty list of wrongs it was of which the plaintiff complained: the instigation of lawsuits and the causing of injunctions without cause, and solely for the purpose of preventing the independent line from doing business; the publishing of libellous matter concerning the company and its officers in newspapers controlled by the trust; engaging bodies of men to tear up parts of pipe-line already laid; enticing away from the enterprise officers, agents and employees; chartering or purchasing any vessels carrying independent oil, solely for the purpose of interfering with the independent market; intimidating merchants by threats of underselling until they refused to buy the oil contracted for; criminal underselling solely for destroying the plaintiff's markets.

It was a serious case Mr. Sherman made out, and the evidence he collected was elaborate and detailed. But, for a sad reason, it was never to come to trial. Less than two months after the summons was issued Mr. Sherman died suddenly in New York City. The shock of his death was such that the independent companies had no heart for the suit, but allowed it to lapse.

There was nothing now but the slow course of Jersey justice for the United States Pipe Line, and for four long years it dragged itself through the courts. Twice it won, but at last, in 1899, the decisions of the lower courts were reversed and the pipe-line had to come up. Ordered out of New Jersey, the independents had to turn back to Pennsylvania. In that state there is a free pipe-line bill. Philadelphia is a shipping point. Luckily for the company, Mr. Murphy had, some time before this, and in anticipation of a defeat in New Jersey, bought on his own responsibility the land for a terminal at Marcus Hook, on the Delaware. This terminal he now sold to the company at the nominal price he had paid for it, and the United States Pipe Line was started again from Wilkesbarre to the sea. Finally, on May 2, 1901, after nine years of struggle in the face of an interference intolerable and unjust, after a quarter of a million dollars spent in litigation, in useless surveys, in laying and pulling up pipes, in loss of business, the first refined oil ever piped from the Oil Regions to the seaboard reached Philadelphia.

Mr. Emery, in telling his story of the difficulties of the United Pipe Line to the Industrial Commission in 1899, did not hesitate to attribute them to the Standard Oil Trust. John D. Archbold made a "general denial": "We have not at any time had any different relations with reference to any obstruction or effort at obstruction of their line *than would attach to any competitor in a line of business engaging against another.*" * "We asked our friends on the railroad and in the New Jersey Legislature to look after our interests, of course," a Standard official told the writer in discussing this case. "That was our right." Mr. Boyle, the editor of the Derrick, took the stand before the Industrial Commission that the Standard Oil Trust's opposition to the United States

* See Appendix, Number 56. John D. Archbold's statement to the Industrial Commission concerning the Standard's opposition to the building of the United States Pipe Line.

Pipe Line was merely fair competition, as justifiable as offering a higher price for land which your competitor is after.

From the Standard point of view it is evident that all this is legitimate business. They do not wish the United States Pipe Line to reach New York. They say to their friends of the Delaware, Lackawanna and Western, and in the Legislature of New Jersey: "These people are our competitors." Apparently neither the Delaware, Lackawanna and Western nor the New Jersey Legislature can afford to forget who are the competitors of the Standard Oil Trust. When the case becomes public and clamour is raised against such methods, the Standard disclaims all responsibility. It was the railroad who fought the pipe-line!

It was not only from without that trouble came upon these men. There were the inevitable internal struggles. They saw their stockholders diminish from discontent and timidity. One of their staunchest members withdrew because of his disbelief in the wisdom of a majority action, and twice they were robbed by death of their most valued members. In December, 1895, A. D. Wood, of Warren, died. Mr. Wood had been one of the most inspiring members in the independent work, and there was nobody left who could do what he had been doing there. In 1897 the chief counsel, Roger Sherman, died. He had conducted the enormous and vexatious litigation of the various concerns with consummate skill, and there was nobody to take his place. Mr. Emery, overwhelmed by the death of Roger Sherman and worn out by his six years of work and worry over the United States Pipe Line, fell ill and was obliged to resign. On every side it was fight and loss and despair, and yet these men hardened under it. Not only hardened, they expanded. Ten years after the unorganised uprising which brought them together in 1887 and forced from them the resolution to take care of their own product, what had they? A company of nearly

600 individual oil producers organised on a business basis, and connected by pipe-lines with some dozen individual oil refineries. For transporting this oil they had pipe-lines carrying both crude and refined from the Oil Regions to within fifty miles of the sea, and for markets they had those they had themselves worked up in the United States and Europe. They had something more. In spite of the continued hostility of the Standard they had the conviction that there was a future for their venture; but they saw clearly that to realise it they must get themselves into still more compact form—that their holdings must be put into the hands of trustees in a single company if they were to be free from the danger of the eventual dominance of the Standard. Now, in November, 1895, as we have seen, the independents had incorporated in New Jersey a marketing concern called the Pure Oil Company. After months of discussion it was decided to enlarge the capital of this company to $10,000,000, $2,000,000 in preferred and $8,000,000 in common stock, and put into this concern all their interests. There was opposition to the consolidation from some of the strongest interests concerned, but finally the idea prevailed, and in 1900 a majority of the stock of the Producers' Oil Company, the Producers' and Refiners' Company, and the United States Pipe Line was turned over to the Pure Oil Company.

The purpose of the combination was frankly stated to be the maintenance of the independence of the company. This was to be effected in the following way: the holders of 16,000 shares of stock—more than a majority—vested the voting power of these shares in fifteen persons for twenty years, and it was agreed that one-half of all shares thereafter subscribed should be transferred to those same trustees. Shares can be sold and transferred, but this transfer does not give the purchaser any right other than provided in the trust agreement. Any trustee may be summarily removed by three-fifths of the trustees, together with three-fifths of the shareholders in trust. It certainly looks as if the Pure Oil Company has devised an organisation which will effectually preserve its independence so long as its shareholders desire that independence. Mr. Archbold, in describing this voting trust of the Pure Oil Company to the Industrial Commission, called it "iniquitous." It is difficult to understand just how it is iniquitous, unless it is because of its success so far in keeping the Standard out of its councils. It is not a secret arrangement. It aims at no monopoly, at no restraint of trade. It claims only to be a device for protecting its obvious right to handle its own product. Of course, if we admit that the oil business belongs to the Standard, as Mr. Rockefeller claims, then the Pure Oil Company is certainly in the wrong!

As it stands to-day, the independents have a good showing for their fight. They have fully 900 stockholders, most of them producers. They handle a daily production of 8,000 barrels of crude oil; operate 1,500 miles of crude pipe-line and 400 miles of refined; are allied with some fourteen refineries, in some of which all the by-products of oil, as well as naphtha and illuminating oils, are produced; own one tank-steamer, the Pennoil, with a capacity of 42,000 fifty-gallon barrels, and charter several others; own oil barges on the Rhine, the Elbe and the Baltic; have fully equipped stations in Europe at Hamburg, Mannheim, Riesa, Stettin and Dusseldorf, in Germany; Rotterdam and Amsterdam, Holland; London and Manchester, England; and, in the United States, New York and Philadelphia. With conservative and loyal management, there seems to be no reason that the Pure Oil Company should not become a permanent independent factor in the oil business. Such a thing is worth the best efforts of the men who have made it. Their courageous and persistent struggle no doubt seems to most of them as of purely personal and local meaning. All they asked was to get a fair share of the profits in their business. They knew they did not get it, and they believed it was because there was not fair play on the part of the railroads and the Standard Oil Company. Aroused, they each fought for the particular thing which would give them relief. They only combined because driven to. They have become a strong organisation almost solely because of the persistent opposition of the Standard Oil Trust. The Standard's efforts to break up the Producers' Protective Association by buying out the biggest producers precipitated a co-operative company for handling oil. Its efforts to drive out the independent refineries by the manipulation of the railroads drove the producers and refiners to combine. The heavy charges for handling oil by the Standard pipe-line and by the railways drove these independents to build a seaboard pipe-line for both refined and crude, and to demonstrate that refined as well as crude could be pumped to the sea in pipes. The buying out of their foreign agents forced them to develop their own market in Europe. The secret buying in of their stock, and the combined effort to force the Standard directors on them, compelled them into their present close trust organisation. It looks very much as if in trying to make way with several small scattered bodies Mr. Rockefeller had made one strong, united one.

But while the experience of the Pure Oil Company demonstrates that it is possible to-day to build up an independent oil business if men have the requisite patience and fighting quality, it by no means follows that the success of the Pure Oil Company has restored competition in the oil business or that by its success the public is getting any marked reduction in the price of oil. That the control of that price—within limits—is now and has been almost constantly since 1876 in the hands of the Standard Oil Company is demonstrated, the writer believes, by the figures and diagrams of the next chapter.

CHAPTER SIXTEEN

THE PRICE OF OIL

EARLIEST DESIGNS FOR CONSOLIDATION INCLUDE PLANS TO HOLD UP THE PRICE OF OIL—SOUTH IMPROVEMENT COMPANY SO INTENDS—COMBINATION OF 1872–1873 MAKES OIL DEAR—SCHEME FAILS AND PRICES DROP—THE STANDARD'S GREAT PROFITS IN 1876–1877 THROUGH ITS SECOND SUCCESSFUL CONSOLIDATION—RETURN OF COMPETITION AND LOWER PRICES—STANDARD'S FUTILE ATTEMPT IN 1880 TO REPEAT RAID OF 1876–1877—STANDARD IS CONVINCED THAT MAKING OIL TOO DEAR WEAKENS MARKETS AND STIMULATES COMPETITION—GREAT PROFITS OF 1879–1889—LOWERING OF THE MARGIN ON EXPORT SINCE 1889 BY REASON OF COMPETITION—MANIPULATION OF DOMESTIC PRICES EVEN MORE MARKED—HOME CONSUMERS PAY COST OF STANDARD'S FIGHTS IN FOREIGN LANDS—STANDARD'S VARIOUS PRICES FOR THE SAME GOODS AT HOME—HIGH PRICES WHERE THERE IS NO COMPETITION AND LOW PRICES WHERE THERE IS COMPETITION.

IT is quite possible that in keeping the attention fixed so long on Mr. Rockefeller's oil campaign the reader has forgotten the reason why it was undertaken. The reason was made clear enough at the start by Mr. Rockefeller himself. He and his colleagues went into their first venture, the South Improvement Company, not simply because it was a quick and effective way of putting everybody but themselves out of the refining business, but because, everybody but themselves being put out, they could control the output of oil and put up its price. "There is no man in this country who would not quietly and calmly say that we ought to have a better price for these goods," the secretary of the South Improvement Company told the Congressional Committee which examined him when it objected to a combination for raising prices.

Four years after the failure of the first great scheme, a similar one went into effect. What was its object? J. J. Vandergrift, one of the directors of the Standard Oil Company at that time, questioned once under oath as to what they meant to do, said: "Simply to hold up the price of oil —to get all we can for it." Nobody pretended anything else at the time. "The refiners and shippers who are in the association intend there shall be no competition." "It is a struggle for a margin." "The scope of the association is an attempt to control the refining of oil, with the ultimate purpose of advancing its price and reaping a rich harvest in profits." These are some of the comments of the contemporary press. The published interviews with the leaders confirm these opinions. Mr. Rockefeller, always discreet in his remarks, denied that the scheme was to make a "corner" in oil; it was "to protect the oil capital against speculation and to regulate prices." H. H. Rogers was more explicit: "The price of oil to-day is fifteen cents per gallon" (March, 1875). "The proposed allotment of business would probably advance the price to twenty cents. . . . Oil to yield a fair profit should be sold for twenty-five cents per gallon."

What was the exact status of this refining business out of which it was necessary to make more in the year 1871, when the first scheme to control it was hatched? The simplest and safest way to study this question is by means of the chart of prices on pages 194 and 195.* On this chart the line A shows the variation in the average monthly price, per gallon, of export oil in barrels in New York from 1866 to June 1, 1904. The line B shows the average monthly price, per gallon, of crude oil in bulk at the wells. A glance at the chart

* Adapted from chart printed in Volume I of Report of Industrial Commission, and brought up to date.

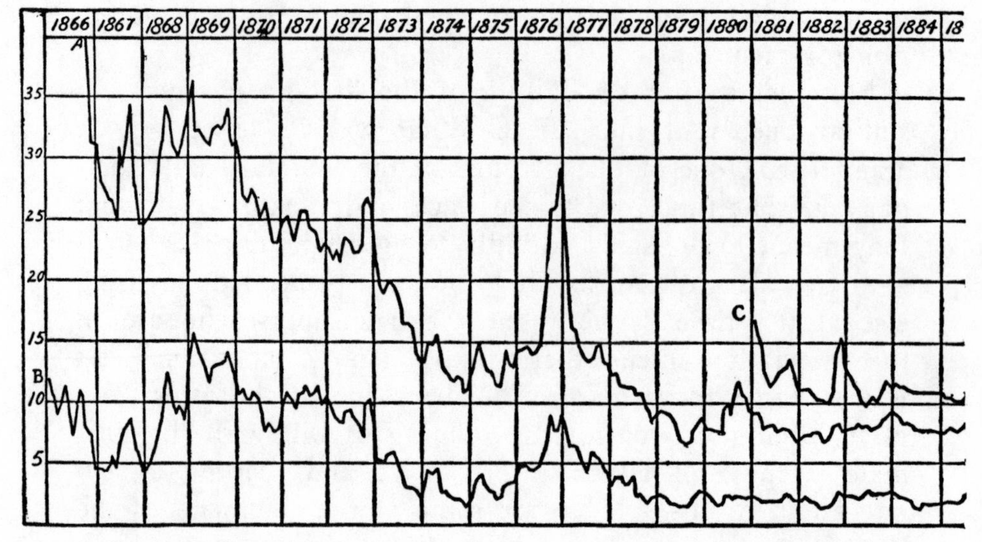

CHART SHOWING PRICE OF

The above chart is adapted from one published in the Report of the Industrial Commission, Volume cents. The dates are placed at the top. The figures on which the export and crude lines are based are are from the Oil, Paint and Drug Reporter.

A shows the variations in the price per gallon of refined oil for export in barrels in New York. The B shows the variations in the price per gallon of crude oil in bulk at the wells.

C shows the variations in the price per gallon of water-white oil (150° test) in barrels in New York.

The margin or difference between the price of crude and refined is easily calculated. Thus at the end the margin was therefore twenty cents.

will show the difference or margin between the two prices. It is out of this difference that the refiner must pay the cost of transporting, manufacturing, barrelling and marketing his product, and get his profits. Now in 1866, the year after Mr. Rockefeller first went into business, he had, as this chart shows, an average annual difference of 35 cents a gallon between what he paid for his oil and what he sold it for. In 1867 he had from 26½ to 20 cents; in 1868, from 20 to 22½; in 1869, from 21 to 18; in 1870, from 20 to 15.*

There were many reasons why this margin fell so enormously in these years. All of the refiners' expenses had rapidly decreased. In 1866 but two railroads came into the oil coun-

* See Appendix, Number 57. Tables of yearly average prices of crude and refined.

THE PRICE OF OIL

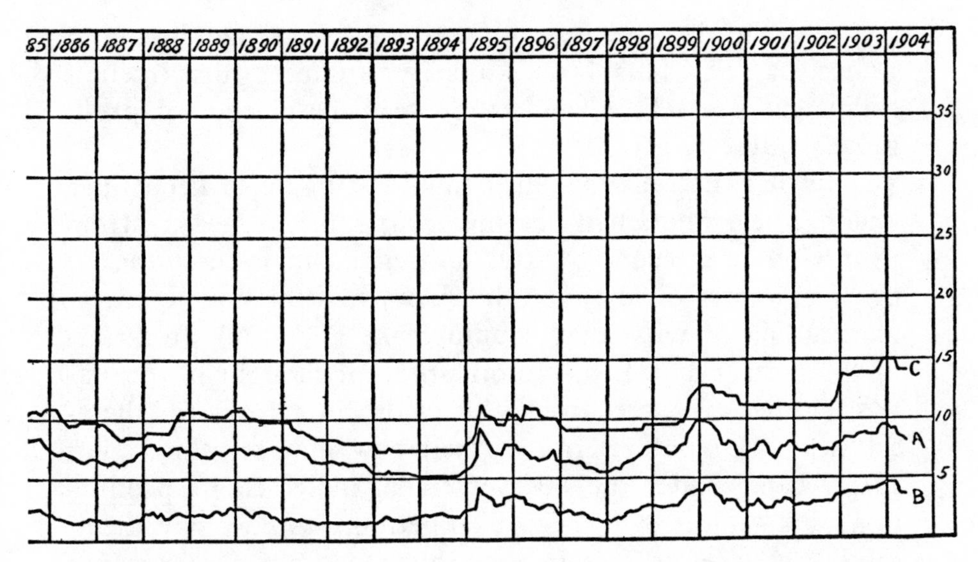

OIL FROM 1866 TO 1904.

1, 1900, and is brought up to date. The figures at the right and left stand for the price per gallon in those taken from the "Oil City Derrick Hand-Book." Those on which the water-white line is based

price of barrels varies slightly, but is usually estimated at 2½ cents per gallon.

This is the usual domestic oil.
of 1876 the crude line shows the price of crude to be about nine cents—the price of refined about twenty-nine;

try; by 1872 there were four connections, and freights fell in consequence. In 1866 carrying oil from the wells by pipe-lines was first practised with success, by 1872 all oil was gathered by pipes, thus saving the tedious and expensive operations of teaming. Tank-cars for carrying crude oil in bulk had replaced barrels and rack-cars. The iron tank, holding 20,000 barrels, was used instead of the wooden tank holding 1,000 barrels. On every side there had been economies, and because of them the margin had fallen. But not only were the expenses coming down; so were the profits. The money which had been made in refining oil had led to a rapid multiplication of refineries at all the centres. In 1872 there was a daily refining capacity of about 46,000 barrels in the country, and the daily consumption of that year had been but 15,000 barrels. This large capacity produced the liveliest competition in selling, and every year the margin of profit grew smaller.

Now it is natural that men should struggle to keep up a profit. The refiners had become accustomed to making from twenty-five per cent. to fifty per cent., and even more, on every gallon of oil they put out. They had the same extravagant notion of what they should make as the oil producers of those early days had. No oil producer thought in the sixties that he was succeeding if his wells did not pay for themselves in six months! And as their new industry slowly but surely came under the laws of trade, increased its production, was subjected to severe competition, as they saw themselves, in order to sustain their business, forced to practise economies and to accept smaller profits, they loudly complained. There was never a set of men who found it harder to accept the limitations of economic laws than the oil producers of Pennsylvania. The oil refiners showed the same dislike of the harness, and in 1871, as we have seen, Mr. Rockefeller and a few of his friends combined to throw it off. What they proposed to do was simply to get all the refineries of the country under their control, and thereafter make only so much oil as they could sell at their own interpretation of a paying price.

There was not enough profit in the margin of 1871. Now what was the profit? According to the best figures accessible of the cost of oil refining at that day, the man who sold a gallon of oil at 24¼ cents (the average official price for that year) made a profit of not less than 1¼ cents—52½ cents a barrel.* Josiah Lombard, a large independent refiner of New York City, when questioned by the Congressional

* Figures used in computing this profit are from the Oil City Derrick of the period, and from practical oil refiners of that day.

Committee which, in 1872, looked into Mr. Rockefeller's scheme for making oil dearer, said that his concern was making money on this margin. "We could ship oil and do very well." A. H. Tack told the Congressional Committee of 1888, which was trying to find out why he had been obliged to go out of the refining business in 1873, that he could have made twelve per cent. on his capital with a profit of ten cents a barrel. Scofield, Shurmer and Teagle, of Cleveland, made a profit of thirty-four cents a barrel in 1875, and cleared $40,000 on an investment of $65,000. Fifty-two cents a barrel profit then was certainly not to be despised. The South Improvement Company gentlemen were not modest in the matter of profits, however, and they launched the scheme whose basic principles have figured so largely in the development of the Standard Oil Trust.

The success which Mr. Rockefeller had in getting the refiners of the country under his control, and the methods he took to do it, we have traced. It will be remembered that for a brief period in 1872 and 1873 he held together an association pledged to curtail the output of oil, but that in July, 1873, it went to pieces.* It will be recalled that three years after, in 1875, he put a second association into operation, which in a year claimed a control of ninety per cent. of the refining power of the country, and in less than four years controlled ninety-five per cent.† This large percentage Mr. Rockefeller has not been able to keep, but from 1879 to the present day there has not been a time when he has not controlled over eighty per cent. of the oil manufacturing of the country. To-day he controls about eighty-three per cent.

Now it is generally conceded that the man or men who control over seventy per cent. of a commodity control its price—within limits, very strict limits, too, such is the force of economic laws. In the case of the Standard Oil Company

* See Chapter IV. † See Chapter V.

the control is so complete that the price of oil, both crude and refined, is actually issued from its headquarters.

Now, with the help of the chart, let us see what Mr. Rockefeller and his colleagues have been able to do from 1872 to 1904 with their power over the price of oil. The first association which worked was brought about late in 1872. What happened? Prices for refined oil were run up from 23 cents a gallon in June to 27 cents a gallon in November, and the margin increased from 13.6 cents to 17.7 cents. From a profit of about 1½ cents a gallon they rose to one of over 4 cents. Unfortunately, however, the refiners of that period were not educated to the self-restraint necessary to carry out this scheme. They very soon failed to keep down their output of oil and overstocked the market, and the whole machine went to pieces. Mr. Rockefeller had been able to make oil dear for a short time, but only for a short time. Worse than that, what he had been able to do brought severe public condemnation. It had, indeed, produced exactly the result the economists tell us too high prices must produce—limitation of the market and stimulation of competition in rival goods. Mr. Rockefeller's second scheme to work out the good of the oil business by making oil dear resulted in decreasing oil exports for the first time since the discovery of oil.* It also increased one of the chief grievances of the American refinery—that was, the exporting of the crude oil to be refined in Europe. Where the exports of crude had been something over eleven million gallons in 1871, they were now over sixteen millions. And it set the shale-oil factories of Scotland to work merrily. It was cheaper for Great Britain to use oil from Scottish shales than to buy oil sold under Mr. Rockefeller's great plan for benefiting the oil business. So for the time the scheme fell down.

* In 1871 there was something over 132,000,000 gallons of illuminating oil exported. In 1872 it fell to about 118,000,000 gallons.

As the diagram shows, the margin dropped rapidly back after this brief success from eighteen to thirteen cents, nor did it stay there. With the return of competition, in the fall of 1873, it continued to drop rapidly. By the end of the year

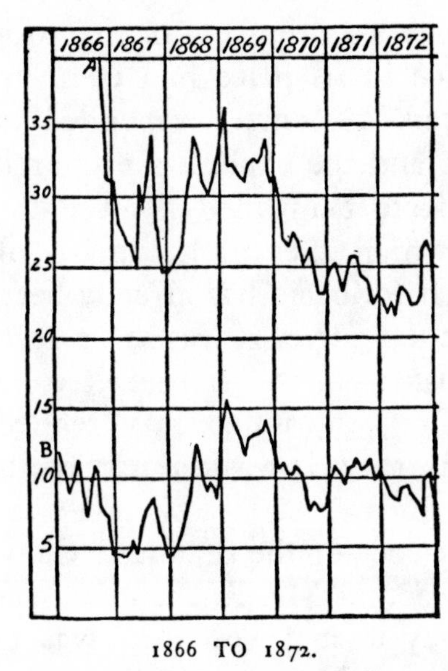

1866 TO 1872.

Fragment of oil chart, showing decline of margin between crude and refined oil in the first seven years after the pipe-line was proved practical. Notice sudden rise in refined oil in 1872 caused by the first Refiners' Association.

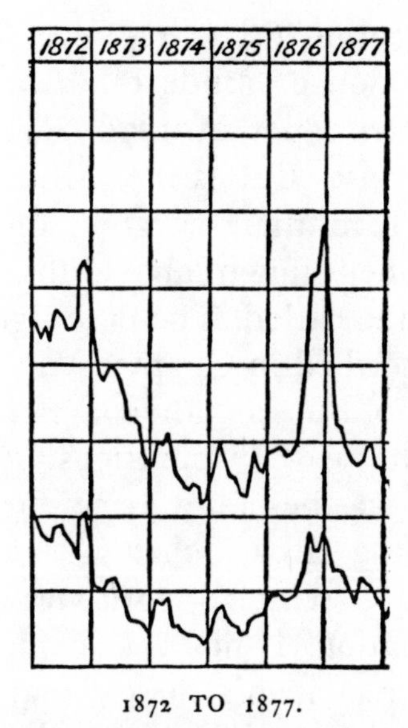

1872 TO 1877.

Fragment of oil chart, showing decline in margin after the failure of the Refiners' Association in 1872, and the abnormal increase in the margin in 1876, when the next combination was perfected.

it was down to eleven cents; by the end of 1874 to nine. What had done it? A decline in expenses, coming from the multiplication of pipe-lines, reduction in freight charges, and free competition in the markets. Nothing else.

In spite of the obvious economic effects of his scheme in 1872 Mr. Rockefeller did not give up his theory that to make oil dear was for the good of the business. He went steadily ahead, developing quietly his plan of a union of all refiners, pledged to limit their output of oil to an allotment he should assign, to accept the freight rates he should arrange for, to buy and sell at the prices he set. It was a year before the alliance was nearly enough complete to make its power felt. By the summer of 1876 it claimed to have nine-tenths of the refiners in the country in line. At that time a situation rose in the crude oil market well calculated to help it in its intention to raise prices. This was a falling off in the production of crude oil. An advance in its price had come in the summer of 1876. Refined had, of course, responded to the rise. But as the fall came on and the exporters prepared to load their cargoes, the syndicate demanded a price for refined much above that for which the market price of crude called. The embargo which followed has already been described in Chapter VII of this narrative. It was as straight a hold-up as our commercial history offers, rich as it is in that sort of operations. From October to February refined oil was held at a price purely arbitrary. It was the first fruits of the Great Scheme.

The winter's work was a great one for the Standard Combination. It not only demonstrated that Mr. Rockefeller was correct in his theory that the way to make oil dear was to refuse to sell it cheap, but not since the coup of 1872, with the South Improvement Company, had Mr. Rockefeller reaped such rewards. The profits were staggering. One of the leading gentlemen in this pretty affair told the writer once that he had sold one cargo at thirty-five cents a gallon, oil which cost him on board the ship a trifle under ten cents. To-day one-fourth of a cent profit a gallon is considered large on export oil. The Standard Oil Company of Ohio had always paid a good dividend,* but the year of this raid,

* According to the statement of the Standard Oil Company, made in a suit for taxes brought by the state of Pennsylvania in 1881, it declared dividends as follows: In 1873, year ending the first Monday in November, $347,610; in 1874, $358,605; in 1875 (the capital stock was raised from $2,500,000 to $3,500,000 in 1875), $514,230; in 1876, $501,285; in 1877, $3,248,650.01; in 1878, $875,000; in 1879, $3,150,000; in 1880, $1,050,000.

1877, it surpassed all bounds. On a capitalisation of $3,500,-000 it paid $3,248,650.01, only a fraction less than 100 per cent. One of its stockholders, the late Samuel Andrews, when on the witness-stand in 1879, said they might have paid the dividend twice over and had money to spare.

The profits were great, but notice the forces set in motion by this coup. The exporters were angry. The buyers in Europe were angry. If the Americans are going to force up prices in this way, they said, we will not buy their refined oil. We will import their crude and refine it ourselves. We will go back to shale oil. A first result, then, of this attempt to hold prices up to a point conspicuously out of proportion to the raw product was that the exports of illuminating oil fell off—they were less by a million gallons in 1878 than in 1877. In the United States the market was threatened in the same way. There had been much trouble in the years just preceding these events with extortionate prices for gas—particularly in New York and Brooklyn. Illuminating oil was so much cheaper that it had been largely substituted, but this artificial forcing of the oil market in 1876–1877 caused a threat to return the next year to gas.

The effect on the refiners who were operating with Mr. Rockefeller in running arrangements was decidedly bad. Each refiner was under bonds to use only a certain percentage of his capacity, and to shut down entirely if Mr. Rockefeller said so. Scofield, Shurmer and Teagle, independents of Cleveland, who had yielded to the attractiveness of Mr. Rockefeller's scheme, and had gone into a running arrangement with him to limit their output, made $2.52 a barrel on their oil from July, 1876, to July, 1877! They had been satisfied with thirty-four cents profit a barrel the year before. Since making oil paid so well, why not make more? Why keep their allotment down to exactly 85,000 barrels, as they had agreed, when they were prepared to make 180,000?

They did not. They put out a few extra thousand barrels each year. Others did the same. It was, of course, fatal to the "good of the oil business." Not only did these profits tempt many refiners to overrun their allotment; the few independents left profited by the prices and increased their plants; the great Empire Transportation Company combined refineries with its pipe-lines as Mr. Rockefeller was adding pipe-lines to his refineries. Thus competition was stimulated.

The effect on the men who produced oil was, of course, bad. They had found it impossible at any time, while the refined was kept so high, to force crude up to a corresponding point, though every effort was made. The producers threatened to combine and refine their own oil. When the Empire Transportation Company went into refining the producers heartily favoured the movement, and throughout the next year a severe competition kept prices down. The Empire was finally wiped out; the producers, aroused by this failure, combined against the Standard in one of the greatest associations they ever had. From 1878 to 1880 they fought continuously to restore competition. They secured the introduction into Congress of a bill to regulate interstate commerce; they fought for more drastic laws against railroad discrimination in the state of Pennsylvania; they persuaded the state to prosecute the Pennsylvania Railroad for discrimination; they indicted Mr. Rockefeller and eight of his colleagues for criminal conspiracy; and they supported by money and influence a scheme for a seaboard pipe-line connected with the independent refineries.*

If one will look at the chart he will see graphically the effect on Mr. Rockefeller's ambition of this fundamentally sound independent movement. The margin between crude and refined, thrust up to over twenty cents by the combination of 1878, fell rapidly under the combined efforts of the

* See Chapter VII.

independents through 1877, 1878 and 1879. In the latter year it touched five cents for the first time in the history of the business. Competition resulting in economies, in a revolutionising transportation invention—the seaboard pipe-line

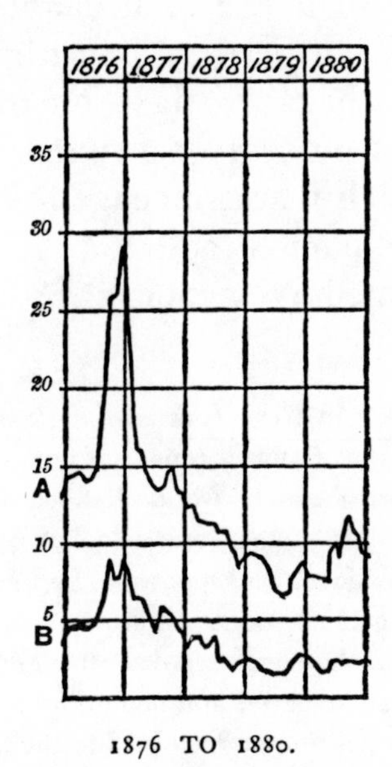

1876 TO 1880.

Fragment of chart, showing decline in margin after the coup of 1876–1877, caused by alliance of independent oil men and the success of the first seaboard pipe-line.

—in a greatly extended foreign market, brought down this margin in 1879. Nothing else.

Those who have read this history know what became of the competitive movement of these years of 1878–1879. They remember how the Producers' Union compromised its suits and abandoned its efforts for interstate commerce regulation. They remember, too, how, just before the great seaboard pipe-line project was proved to be a success, all but one of the independent refineries were, by one means or another, persuaded to sell or to combine with the Standard, leaving the Tidewater without an outlet for its oil. Before the end of 1879 the Standard claimed ninety-five per cent. of the refining business. Now examine the chart for the effect on the price of oil in 1880, of this doing away with competition —another sudden uplift of the price of refined, this time without the excuse of a rise or probable rise in crude. For three years oil had not been sold so high as it was in 1880, when the exporters began to take on their winter's supply. An interesting contemporary account of this coup of 1880, and the way in which it was managed, is found in the excellent monthly Petroleum Trade Report, published by John C. Welch. It is dated November, 1880, and headed "Very Sharp Practice":

"There is made each day in New York what is known as an official quotation for refined oil, this official quotation being made as a matter of convenience in cabling the price of refined oil throughout the world. Refined oil not being sold at an open board, it is sometimes difficult to quote it accurately, but by having an 'official quotation' this can be quoted, and the difficulty is supposed to be, in a measure at least, remedied. The 'official quotation' is made by three petroleum brokers appointed by the Produce Exchange for that purpose, who meet each day after exchange hours for the purpose of establishing it. There is one party, and one party only, that have very large lots to sell, and so important a position do they hold in the business that their prices are ordinarily the market. Of course, to make transactions, their prices and buyers' prices have to come together, and transactions establish a market much better than prices offered to buy or sell at, but without transactions. At many times, if the Standard do not sell, there are no transactions, and, consequently, the Standard's asking price is leaned upon to establish an official quotation. During September, the official quotation went up from 9⅜ cents to 11⅞ cents, with comparatively little demand, as the foreign stocks were large, and very little oil was required to supply the world's wants. The upward movement was, consequently, purely arbitrary. Arbitrary prices are, however, a part of the Standard's every-day life, and I am not taking at this time any exception to them. All through October and up to November 13, the official quotation was 12 cents, or sometimes a little over and sometimes a little under, and as this price did not meet the views of buyers to but slight extent, the Standard were supposed to be exercising a Roman virtue in not selling. Twelve cents continued as the official quotation to November 13, without any wavering, but from the 13th to the 18th, while '12 cents asked by refiners' continued in the quotation, such sentences as these were included at different dates: 'Other lots obtainable at 11 cents.' 'Sales at 10½ cents, offered at that.' 'Other lots obtainable at irregular prices, from 10 to 10½ cents.' On November 18, the quotation was '10 to 12 cents.' I give the following quotation of the New York refined market as published in my Oil City daily report of November 11: 'The New York market yesterday closed, secretly offered and unsalable at 11½ cents, and probably at 11¼ cents by resales and outside refiners, and likely by Standard, though they openly ask 12.'

"The point that seems apparent is that the official quotation of 12 cents ceased to be an honest quotation a considerable time before it was abandoned. The committee making the quotation can probably justify their position by the custom of the trade of regarding the prices the Standard openly ask as the market, nevertheless they, and the Produce Exchange whom they represent, were the bulwark from behind which the Standard were able to get off their hot shot against the consuming trade in the United States and the consuming trade in Europe, who all this time were buying Standard oil on the basis of 12 cents at New York, the supplies at the time being drawn from their stock in Europe and from their various depots in the United States."

But the performance of 1876 and 1877 was not forgotten in Europe. In 1879 the exporters and buyers from all the great foreign markets had met in Bremen in an indignation meeting over the way the Standard was handling the oil business. Remonstrances came from the consuls at Antwerp and Bremen to our State Department concerning even the quality of oil which had been sent to Europe by the Standard. John C. Welch, who was abroad in 1879, was told by a prominent Antwerp merchant: "I am of the opinion that if the petroleum business continues to be conducted as it has been in the past in Europe, it will go to smash." * The attempt to repeat in 1880 what had been done in 1876 failed. The exports of illuminating oil that year fell much below what they had been the year before. In 1879, 365,000,000 gallons of refined oil were exported; in 1880, only 286,000,000 gallons. Exports of crude, on the contrary, rose from about 28,000,000 gallons to nearly 37,000,000 gallons. The foreigners could export and refine their own oil cheaper than they could buy from Mr. Rockefeller. Competition was after him,

* Report of the Special Committee on Railroads, New York Assembly, 1879. Volume IV, page 3680

too, for the Tidewater, whose refineries he had cut off, had stored their oil, built new plants, and were again ready to compete in the market.

This third corner of the oil market seems to have convinced Mr. Rockefeller and his colleagues at last that, however great the fun and profits of making oil very dear, in the long run it does not pay; that it weakens markets and stimulates competition. They learned a lesson in these years they have never forgotten—that when you make a scoop it must not be so big that you will never have a chance to make another one; that if you want to keep your power to manipulate the market you must use that power so modestly that the public in general will not realise you have it. Again and again the effect of the experiences of 1872, 1876 and 1880 crops out in the testimony of Standard officials. Benjamin Brewster once said to a Federal Investigating Committee, which had asked if the Standard could not fix the price of oil as it wished: "At the moment many things may be done, but the reaction is like a relapse of typhoid fever. The Standard Oil Company can never afford to sell goods dear. The people would go to dipping tallow candles in the old-fashioned way if we got the price too high." The after-effects of the first great raids, then, were salutary. The Standard learned the limitations set on monopolies by certain great economic laws.

But if the Standard Oil Company learned in its first attempts to raise the price of oil that they could not in the long run afford to make from 100 to 350 per cent., they by no means gave up their attempt to keep their control, and to hold up profits as high as they could without injuring the market or inviting too strong competition. If one will look at the chart showing the fluctuations from 1879, when control was achieved, to the beginning of 1889, one will find that for ten years the margin between refined oil and crude never fell below the point reached by competitive influences in the former year, though frequently it rose considerably above. Yet it is in this period that the Standard did all its great work in extending markets, in developing by-products, and in introducing the small and varied economies on which it rests its claim to be a great

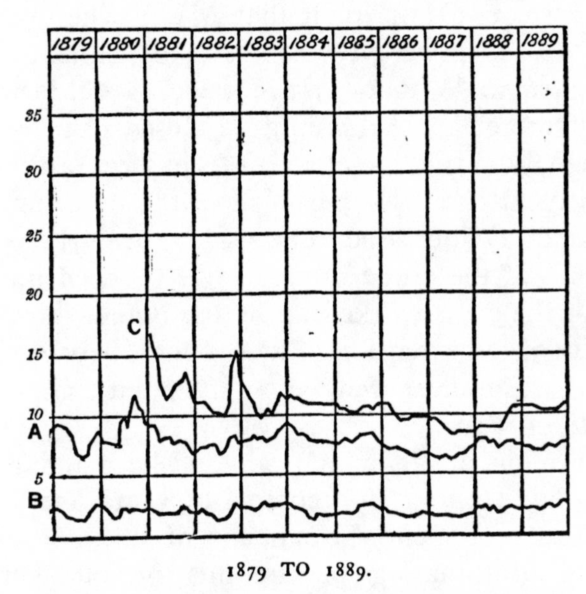

1879 TO 1889.

Fragment of chart, showing how margin reached in 1879 by competition was raised and sustained for ten years under the monopoly achieved by the Standard Oil Company in 1880. The sudden rise in refined in the fall of 1880 was a purely arbitrary price. Notice that crude was stationary at the time.

public benefactor. The first eight years of its existence had been spent in bold and relentless warfare on its competitors. Competition practically out of the way, it set all its great energies to developing what it had secured. In this period it brought into line the foreign markets and aided in increasing the exports of illuminating oil from 365,000,000 gallons in 1879 to 455,000,000 in 1888; of lubricating, from 3,000,000 to 24,000,000, and yet this great extension of the volume of business profited the consumer nothing. In this period it laid hands on the idea of the Tidewater, the long-distance pipe-lines for transporting crude oil, and so rid itself practically of the railroads, and yet this immense economy profited the public nothing. In spite of the immense development of this system and the enormous economies it brought about—a system so important that Mr. Rockefeller himself has said: "The entire oil business is dependent upon this pipe-line system. Without it every well would shut down, and every foreign market would be closed to us"—the margins never fell the fraction of a cent from 1879 to 1889, though it frequently rose. In this period, too, the by-products of oil were enormously increased. The waste, formerly as much as ten per cent. of the crude product, was reduced until practically all of the oil is worked up by the Standard people, and yet, in spite of the extension of by-products between 1879 and 1889, the margin never went below the point competition had forced it to in 1879.

The enormous profits which came to the Standard in these ten years by keeping out competition are evident if we consider for a moment the amount of business done. The exports of illuminating oil in this period were nearly 5,000,000,000 gallons; of this the Standard handled well toward ninety per cent. Consider what sums lay in the ability to hold up the price on such an amount even an eighth of a cent a gallon. Combine this control of the price of refined oil with the control over the crude product, the ability to depress the market for purchasing, an ability used most carefully, but most constantly; add to this the economies and development Mr. Rockefeller's able and energetic machine was making, and the great profits of the Standard Oil Trust between 1879 and 1889 are easily explained. In 1879, on a capital of $3,500,000, the Standard Oil Company paid $3,150,000 dividends; in 1880 it paid $1,050,000. In 1882 it capitalised itself at $70,000,000. In 1885, three years later, its net earnings were over $8,000,000; in 1886, over $15,-000,000; in 1888, over $16,000,000; in 1889, nearly $15,000,-000. In the meantime the net value of its holdings had increased from $72,000,000; in 1883, to over $101,000,000. While the Standard was making these great sums, the men who produced the oil saw their property depreciating, and the value of their oil actually eaten up every two years by the prices the Standard charged for gathering and storing it.

But to return to the chart. With the beginning of 1889 the margin begins to fall. This is so in spite of a rising crude line. It would look as if the Standard Oil Company had suddenly had a change of heart. In the report of that year's business made to the trustees of the Standard Oil Trust, the following elaborate and interesting calculation was presented:

"The quantity of crude oil consumed by the Standard manufacturing interests in 1889 was 896,250,325 gallons, or 20,339,293 barrels, an increase over the previous year of 119,073,589 gallons, or 2,835,085 barrels, an increase of 15.3 per cent.

"The sales of crude oil by our interests for purposes other than their own manufacture were 135,788,959 gallons, or 3,232,832 barrels, an increase of 43½ per cent. over the previous year, making the total consumption of crude oil through our interests 1,032,029,284 gallons, or 24,572,126 barrels, an increase over 1888 of 3,809,-917 barrels, or 18.35 per cent., and exceeding the consumption of 1887, which was the largest of any previous year, by 12.7 per cent.

"The quantity of refined oil produced was 666,742,547 gallons, or 13,334,851 barrels of 50 gallons each; of lubricating paraffine and compounded oils 43,862,795 gallons, or 877,256 barrels, and of other products 160,712,183 gallons, or 3,214,243 barrels, making a total of all products of 871,371,525 gallons, or 17,426,350 barrels, valued at over $46,000,000.

"The average cost of the crude consumed in refining was .211 of a cent more than in 1888, while the average price realised per gallon of crude was .090 of a cent less, showing a decrease in the margin between the crude and finished product of .301 of a cent. This represents a saving to the consumer over what the finished products would have cost him if the same margin had been maintained on the increased price of crude of $2,697,000. This has been done without a corresponding loss to our interests by a decrease in cost of manufacturing and marketing, and by the increased quantity handled .204 of a cent, effecting a saving of $1,860,000, and the difference has been more than made up by further reductions of cost of marketing by our distributing interests, as well as in the increased quantity handled. Although the average price of crude has been the highest this year of any of the last five years, the increase over the price of 1887 (when the price on both crude and refined was the lowest for that period) being about 22½ per cent., the average price of products has increased but 12½ per cent., showing a saving to the consumer of 10 per cent. We have therefore continued to make good the claim that the Standard has heretofore maintained of cheapening the cost of the products to the consumers by giving them the benefits of the saving in costs effected by consolidation of interests." *

This certainly sounds just—even philanthropic. It is exactly what the consumer claims is his due—to have a share of the economies which undoubtedly may be effected by such complete and intelligent consolidation as Mr. Rockefeller has effected. But was it combination that caused this falling of the margin? As a matter of fact this lowering of the margin was the direct result of competition. In 1888 a German firm, located in New York City, erected large oil plants in Rotterdam and Bremerhaven. They put up storage tanks at each place of 90,000 barrels' capacity. They also established a storage depot of 30,000 barrels at Mannheim, and took steps to extend their supply stations in Germany and Switzerland. They built tank steamers in order to ship their oil in bulk. These oil importers allied themselves with certain independent refiners, and interested themselves also in the co-operative movement which the producers of Pennsylvania were striving to get into operation at this time. The extent of the undertaking threatened serious competition. In the same year imports of Russian oil into the markets of Western Europe began for the first time to assume serious proportions. Russian oil had, from the beginning, been a possible menace to American petroleum, for the wonderful fields on the Caspian were known long before oil was "struck" in

* Plaintiff's Exhibit, Number 51, in the case of James Corrigan *vs.* John D. Rockefeller in the Court of Common Pleas, Cuyahoga County, Ohio, 1897.

Pennsylvania. They did not begin to be exploited in a way to threaten competition until late in the eighties. In 1885 consuls at European ports began to report its appearance—fifty barrels were landed at Bremen that year as against 180,855 of American oil. In this year, too, the first Russian oil went to Asia Minor, where "Pratt" oil had long held sway. The first cargo reported at Antwerp was in March, 1886. In April, 1890, the consul at Rotterdam, in calling attention to the independent American competition, said of Russian oil: "It is no longer a serious competitor for the petroleum trade of Western Continental Europe." The consul said that while the American oil shipments to the five principal continental ports were fully 4,000,000 barrels per year, those of Russian were less than a tenth of that number. However, a growth of 400,000 barrels in five years was something, and the Standard Oil Trust was the last to underestimate such a growth. Prices of export oil immediately fell. There was nothing in the world that gave oil consumers the benefit of the Standard's savings by economies in 1889 but the competition threatened by Russia and the American and German independent alliance. The Standard, to offset it, not only lowered its price, but it followed the German company to Rotterdam in order to put up an oil plant similar to the one which had been erected by those independents. They also purchased at this time the great oil establishments at Bremen and Hamburg which had hitherto been owned and operated by Germans. A full account of this new development in the oil trade was reported by the American consul at Rotterdam in April of 1890, and is to be found in the consular reports of that year.

Follow the lines a little farther. Notice how, in 1892, the price of refined oil begins to fall, although crude is stationary. Notice how the refined line remains steady throughout 1893 and 1894, although the crude line steadily rises. This went on for nearly three years, until there was a margin of only three cents between crude and refined oil. The barrel, which is always reckoned in the official quotations of export refined oil, costs two and a half cents per gallon, and the price of manufacturing is usually put at one-half a cent. The cost of transporting the oil was not covered by the margin the

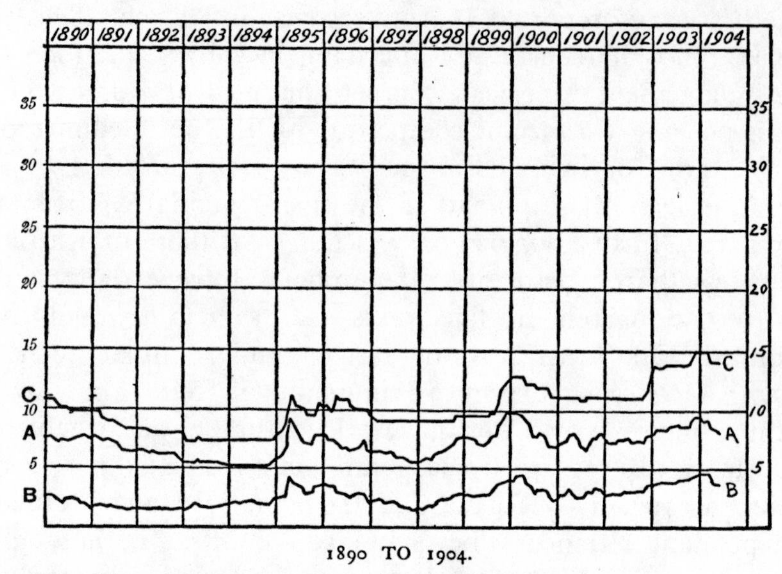

1890 TO 1904.

Fragment of chart, showing relation between crude and refined oil in the last fourteen years. Notice effect on margin from 1890 to 1894 of rise of strong competitive forces. Notice also how margin between price of crude and of domestic oil increased in the winter of 1903–1904, during the coal famine.

greater part of the year 1894. Now, the Standard Oil Company were not selling oil at a loss at this time out of love for the consumers, although they made enough money in 1894 on by-products and domestic oil to have done so—their net earnings were over $15,000,000 in 1894, and they reckoned an increase in net value of property of over $4,000,000—they were fighting Russian oil and the independent combination started in 1889. By 1892 this combination was in active operation. The extent of this movement was described in the last chapter of this narrative. At the same time certain large producers in the McDonald oil field built a pipe-line from Pittsburg to Baltimore, the Crescent Line, and began to ship crude oil to France in great quantities. It looked as if both combinations meant to do business, and the Standard set out to get them out of the way. One method they took was to prevent the refiners in the combination making any money on export oil.

The extent to which cutting was carried on for two years, beginning with the fall of 1892, has been referred to in the last chapter, but is perhaps worth repeating in this connection. In January of 1892 crude oil was selling at 53½ cents a barrel at the wells, and refined oil for export at 5.33 cents a gallon in barrels. Throughout the year the price of crude advanced, until in December it was 78⅜ cents. Refined, on the contrary, fell, and it was actually 18 points lower in December than it had been twelve months before. Throughout 1894 the Standard kept refined oil down; the average price of the year was 5.19 cents a gallon, in face of an average crude market of 83¾ cents, lower than in January, 1893, with crude at 53½ cents a barrel.

After two years they gave it up. It was too expensive. The Crescent Line sold to them, but the other independents were too plucky. They had lost money for two years, but they were still hanging on like grim death, and the Standard concluded to concentrate their attacks on other points of the combination rather than on this export market where it was costing them so much.

About the end of 1894 the depression of export oil was abandoned, as the chart shows. Notice that from 1895 to 1898 the margin remained at about four cents, that in 1900 it rose to six cents, and from that time until June, 1904, it swung between four and a half and five. The increasing competition in Western Europe of independent American oils, and the rapid rise since 1895, particularly of Russian oil, are what has kept this margin down. It is doubtful, such is the growing strength of these various competitive forces, if the Standard Oil Trust will ever be able to put up the margin on export oils. If there were only the American independents to reckon with, a compromise might be possible, but Russia, Burmah and Sumatra are all in the game. By 1896 Russia was exporting 210,000,000 gallons of petroleum products (America in that year exported over 931,000,000 gallons), and these products were going to nearly every part of Europe and Asia. They began to cut heavily into the trade of the Standard in China, India, Great Britain and France. By 1899 the exports of Russian oil were over 347,000,000 gallons; in 1901, over 428,000,000 gallons. In China, India, and Great Britain particularly, has the Russian competition increased. While at one time the Standard Oil Company had almost the entire oil trade at the port of Calcutta, last year, 1903, out of 91,500,000 gallons imported, only about 6,500,000 gallons were of American oil. In China, Sumatra oil is now ahead of American, the report for 1903 being: American, 31,060,527 gallons; Sumatra, 39,859,508.

For the Standard there is good profit in this margin of four and a half cents for export oil. The expenses the margin must cover are the transportation of the crude from the wells to New York, the cost of manufacture, the barrel and the loading. For twenty-five years the published charge of the Standard Oil Company for gathering oil from the wells has been twenty cents a barrel. The charge for bringing it to New York has been forty cents, a little less than one and a half cents a gallon. It costs, by rough calculation, one-half a cent to make the oil and load it. The barrel is usually reckoned at two and a half cents. Here are four and a half cents for expenses—the entire margin. Where the Standard has the advantage is in its ownership of oil transportation. A common carrier gathering and transporting in 1902 all but perhaps 10,000 barrels of the 150,000 barrels' daily production of Eastern oil, the service for which the outsider pays sixty cents, costs it from ten to twelve cents at the most liberal estimate. Here is over a cent saved on a gallon, and a cent saved, where millions of gallons are in question, makes not only great profits, but keeps down competition. The refiner who to-day must pay the Standard rates for transportation cannot compete in export oil with them. In January of 1904, when the chart shows the margin to have been about four and three-quarter cents, an independent refiner in the state of Ohio, dependent on the Standard for oil, gave the writer a detailed statement of costs and selling prices of products in his refinery. According to his statement he lost one and three-fifth cents on his export oil. He was forced, of course, to pay Standard transportation prices for crude and railroad charges for refined from Ohio to New York harbour.*

That there would have been such a transportation situation to-day had it not been for the discrimination by the railways, which threw the pipes into the Standard's hands in the first place, and the long story of aggression by which the Standard has kept out rival pipes, and so been able for twenty-five years to sustain the price for transportation, is of course evident. To-day, as thirty years ago, it is transportation advantages, unfairly won, which give the Standard Oil Company its hold. It is not only on transportation that the Standard to-day has great advantages over the independent refiner in the export market. As said at the beginning of this chapter, the Standard Oil Company "makes the price of refined oil" —within strict limits. Of course, making the market, it has all the advantages of the "inside track." Its transactions can

* It costs the Cleveland refiner .64 of a cent a gallon to bring oil in bulk from the Oil Regions to his refinery, and 1.44 cents per gallon to send it refined in bulk to New York.

be carried on in anticipation of the rise or fall. For instance, in January of 1904, when there were strong fluctuations in the water-white (150 degrees test) prices, the agent of an independent refiner, who was in Wall Street trying to keep track of markets for out-of-town competitors, reported the price as 9.20 cents a gallon. The refiners' goods were refused on the ground that this was above the market. The Standard Oil export man and a broker who worked with the company were consulted. The market was 9.20. Further investigation, however, showed that at headquarters the figure given out privately was 8.70 cents. The disadvantage of the outsider in disposing of his goods is obvious. The Standard makes the official market, and undersells it. The situation seems to be the same in practice as that described by Mr. Welch, in 1880, though now the fiction of a committee of brokers has been done away with. Of course there is nothing else to be expected when one body of men control a market.

Thus far the illustrations of Mr. Rockefeller's use of his power over the oil market have been drawn from export oil. It is the only market for which "official" figures can be obtained for the entire period, and it is the market usually quoted in studying the movement of prices. It is of this grade of oil that the largest percentage of product is obtained in distilling petroleum. For instance, in distilling Pennsylvania crude, fifty-two per cent. is standard-white or export oil, twenty-two per cent. water-white—the higher grade commonly used in this country—thirteen per cent. naphtha, ten per cent. tar, three per cent. loss. The runs vary with different oils, and different refiners turn out different products. The water-white oils, while they cost the same to produce, sell from two to three cents higher. The naphtha costs the same to make as export oil, but sells at a higher price, and many refiners have pet brands, for which, through some

A TYPICAL OIL FARM OF THE EARLY DAYS

marketing trick, they get a fancy price. The Standard Oil Company has a great number of fancy brands of both illuminating and lubricating oils, for which they get large prices —although often the oil itself comes from the same barrels as the ordinary grade. Now it is from the extra price obtained from naphtha, water-white, fancy brands, and by-products that the independent refiner makes up for his loss on export oil, and the Standard Oil Trust raises its dividends to forty-eight per cent. The independent refiner quoted above, who in January of 1904 lost 1⅜ cents on export oil, made enough on other products to clear 8.3 cents a barrel on his output—eighty-three dollars a day clear on a refinery of 1,000 barrels capacity, which represents an investment of $150,000.

Turn now to the price of domestic oil, and examine the chart to see if we have fared as well as the exporters. The line C on the chart represents the price per gallon in New York City of 150° water-white oil in barrels from the beginning of 1881 to June, 1904.* The figures used are those of the Oil, Paint and Drug Reporter. A glance at the chart is enough to show that the home market has suffered more violent, if less frequent, fluctuations than the export market. A suggestive observation for the consumer is the effect of a rise in crude on the price of domestic oil. The refined line usually rises two or three points to every one of the crude line. It is interesting to note, too, how frequently high domestic prices are made to offset low export prices; thus, in 1889, when the Standard was holding export oil low to fight competition in Europe, it kept up domestic oil. The same thing is happening to-day. We are helping pay for the Standard's fight with Russian, Roumanian and Asiatic oils. But this line, while it shows what the New York trade has paid, is a poor guide for the country as a whole. Domestic oil, indeed, has no regular price. Go back as far as anything

* Trustworthy and regular quotations are not to be obtained earlier than 1881.

like trustworthy documents exist, and we find the most astonishing vagaries, even in the same state. For instance, in a table presented to a Congressional Committee in 1888, and compiled from answers to letters sent out by George Rice, the price of 110° oil in barrels in Texas ranged from 10 to 20 cents; in Arkansas, of 150° oil in barrels, from 8 to 18; in Tennessee, the same oil, from 8 to 16; in Mississippi, the same, from 11 to 17. In the eighties, prime white oil sold in barrels, wholesale, in Arkansas, all the way from 8 to 14 cents; in Illinois, from 7½ to 10; in Mississippi, from 7¼ to 13½; in Nebraska, 7½ to 18; in South Carolina, 8 to 12½; and in Utah, 13 to 23. Freight and handling might, of course, account for one to two cents of the difference, but not more.

A table of the wide variation in the price of oil, compiled in 1892, showed the range of price of prime white oil in the United States to be as follows:

In barrels	6 to 25 cents
In cases	14 to 34½ cents
In bulk	3½ to 25 cents

The same wide range was found in water-white oil:

In barrels	6½ to 30 cents per gallon
In cases	16 to 35 cents per gallon
In bulk	3½ to 29 cents per gallon

In 1896 an investigation of prices of oil sold from tank-wagons in the different towns of Ohio, in the same week, was made, and was afterward offered as sworn testimony in a trust investigation in that state. The price per gallon ranged from 4¾ cents to 8¾ cents.

The most elaborate investigation of oil prices ever made was that instigated by the recent Industrial Commission. In February, 1901, the commission sent out inquiries to 5,000 retail dealers, scattered from the Atlantic to the Pacific and from the Lakes to the Gulf, asking the prices of certain commodities, among them illuminating oils; 1,578 replies were received. The tables prepared offered striking examples of the variability of prices—thus:

In Colorado the wholesale price of illuminating oil (150° test) varied from 13 to 20 cents; in Delaware, 8 to 10; in Illinois, 6 to 10; in Alabama, 10.50 to 16; in Michigan, 5.50 to 12.25; in Missouri, 7.50 to 12.50; in Kentucky, 7 to 11.50; in Ohio, 5.50 to 9.75; in California, 12.50 to 20; in Utah, 20 to 22; in Maine, 8.25 to 12.75 (freight included in all these prices).

The difference between the highest and the lowest wholesale prices in the same states varies from 8 cents in Oregon (12.50 to 20.50) to 1.50 in Rhode Island (8.50 to 10). Of course, in the former case, two or even three cents of the difference may be due to freight, but hardly more. Take adjoining states, for instance. In Vermont there is a difference of 4.50 cents between the highest and lowest price of oil; in New Hampshire, only 1.75. In Delaware there is a difference of 2 cents; in Virginia, of 6.

Compare, now, the lowest price in different states. In Ohio and Pennsylvania oil was sold as low as 5.50; 6.50 is the lowest in New York State, 8.50 the lowest in Rhode Island, and 7 the lowest in New Jersey. In Indiana oil sells as low as 5.50, but in Kansas nothing below 8.50 is reported (the freight rate to Atchison, Kansas, from Whiting, Indiana, which supplies both of these states, is 1.7 per gallon. The freight rate from Whiting to Indianapolis is .5 per gallon).

Not long ago there fell into the writer's hands a sheet from one of the ledgers forming a part of the Standard Oil Company's remarkable system of bookkeeping. This sheet gave the cost and selling price per gallon of different grades of refined oil at over a dozen stations in the same state in October, 1901. In the account of cost of oil were included net cost, freight, inspection, cost of barrels and cost of marketing. The selling price was given and the margin of profit computed. The selling price of water-white from tank-wagons (it is customary for Standard tank-wagons to deliver oil from their stations to local dealers) ranged from 8½ to 11½ cents, and the profit on the oil sold from the wagons varied from about one-half cent to over three cents.

Now, in considering these differences, liberal allowance for freight rates must be made. Something of what these allowances should be can be judged from the table of oil freights which the Industrial Commission published with its schedule of prices. From this table many interesting comparisons can be made. For instance, it cost the Standard Oil Company (if they paid the open rate their rivals did) 1.5 cents to send a gallon of oil from Whiting, Indiana, their supply station, to Mobile, Alabama. They sold their oil in Alabama at wholesale from 11½ to 16 cents. The net cost of this oil was under five cents in February, 1901. It cost them the same 1.5 cents to send a gallon of oil to Des Moines, Iowa (if they paid the open rate), but in Iowa they sold it from 7 to 11. The freight from Whiting to New Orleans was the same 1.5 cents, but prices in Louisiana ranged from 9 to 14 cents. According to the investigation the average wholesale price of oil, including freight, ranged from 8.27 in Pennsylvania to 25.78 in Nevada.

Freights and handling considered, there is, it is evident, nothing like a settled price or profit for illuminating oil in the United States. Now, there is no one who will not admit that it is for the good of the consumer that the normal market price of any commodity should be such as will give a fair and even profit all over the country. That is, that freights and expense of handling being considered, oil should sell at the same profit in Texas as in Ohio. That such must be the case where there is free and general competition is evident.

But from the beginning of its power over the market the Standard Oil Company has sold domestic oil at prices varying from less than the cost of the crude oil it took to make it up to a profit of 100 per cent. or more. Wherever there has been a loss, or merely what is called a reasonable profit of, say, ten per cent., an examination of the tables quoted above shows conclusively it has been due to competition. The competition is not, and has not been since 1879, very great. In that year the Standard Oil Company claimed ninety-five per cent. of the refining interests of the country. In 1888 they claimed about eighty per cent.; in 1898, eighty-three per cent. This five to seventeen per cent. of independent interest is too small to come into active competition, of course, at all points. So long as one interest handles eighty-three per cent. of a product it is clear that it has the trade as a whole in its hands. The competition it encounters will be local only. But it is this local competition, unquestionably, that has brought down the price of oil at various points and caused the striking variation in prices recorded in the charts of the Industrial Commission and other investigations. The writer has before her a pile of a hundred or more letters written in the eighties by dealers in twelve different states. These letters tell the effect on the prices of the introduction of an independent oil into a territory formerly occupied exclusively by the Standard:

Calvert, Tenn.—The Waters-Pierce Oil Company (Standard) so reduced the price of their oil here when mine arrived that I will have some trouble to dispose of mine.

Chattanooga, Tenn.— . . . Cut the price of oil that had been selling at 21 cents to 17 cents.

Pine Bluff, Ark.—While the merchants here would like to buy from some other than the Standard they cannot afford to take the risks of loss. We have just had an example of one hundred barrels opposition oil which was brought here, which had the effect of bringing Waters-Pierce Oil Company's oil down from 18 to 13 cents—one cent less than cost of opposition, with refusal on their part to sell to anyone that bought from other than their company.

Vicksburg, Miss.—The Chess Carley Company (Standard) is now offering 110° oil at nine cents to any and every one. Shall we meet their prices? All they want is to get us out of the market, then they would at once advance price of oil.

These are but illustrations of the entire set of letters; prices dropped at once by Standard agents on the introduction of an independent oil. A table offered to Congress in 1888, giving the extent of their cutting in the Southwest, shows that it ranged from 14 to 220 per cent.

Every investigation made since shows that it is the touch of the competitor which brings down the price. For instance, in the cost and profit sheet from a Standard ledger referred to above, there was one station on the list at which oil was selling at a loss. On investigation the writer found it to be a point at which an independent jobber had been trying to get a market. If one examines the tables of prices in the recent report of the Industrial Commission, he finds that wherever there is a low price there is competition. Thus, at Indianapolis, the only town in the state of Indiana reporting competition, the wholesale price of oil was 5½ cents, although forty out of the fifty-three Indiana towns reporting gave from 8 cents to 10½ cents as the wholesale price per gallon. (These prices included freight. Taking Indianapolis as a centre, the local freight on oil to any point in Indiana is in no case over a cent.) In April, 1904, inquiry showed the same striking difference between prices in Indianapolis, where six independent companies are now established, and neighbouring towns to which competition has not as yet reached.

The advent of an independent concern in Morristown, New Jersey, brought down the price to grocers to 7½ cents and to housewives to 10, but in the neighbouring towns of Elizabeth and Plainfield, where only the Standard is reported, the grocers pay 9 cents and the housewives 12 and 11, respectively. In Akron, Ohio, where an independent company was operating at the time the investigation was made, oil was sold at wholesale at 5¾ cents; at Painesville, nearer Cleveland, the shipping point, at 9¼ cents. In Richmond, Virginia, one dealer reported to the commission a wholesale price of 5 cents, and added: "A cut rate between oil companies; has been selling at 9 and 10 cents."

In the month of April of 1904 150° oil was selling from tank-wagons in Baltimore, where there is competition, at 9 cents. In Washington, where there is no competition, it sold at 10½ cents, and in Annapolis (no competition) at 11 cents. In Seaford, Delaware, the same oil sold at 8 cents under competition. The freight rates are practically the same to all these points. And so one might go on indefinitely, showing how the introduction of an independent oil has always reduced the price. As a rule, the appearance of the oil has led to a sharp contest or "Oil War," at which, not infrequently, both sides have sold at a loss. The Standard, being able to stand a loss indefinitely, usually won out.

An interesting local "Oil War," which occurred in 1896 and 1897 in New York and Philadelphia, figured in the reports of the Industrial Commission, and illustrates very well the usual influence on Standard prices of the incoming of competition. On March 20, 1896, the Pure Oil Company put three tank-wagons into New York City. The Standard's price of water-white oil from tank-wagons that day was 9½ cents, and the Pure Oil Company followed it. In less than a week the Standard had cut to 8 cents * *along the route of the Pure Oil Company wagons.* In April the price was cut to 7 cents. By December, 1896, it had fallen to 6 cents; by December, 1897, to 5.4. It is true that crude oil was falling at this time, but the fall in water-white was out of all proportion. For, while between the price of refined on March 20 and the average price of refined in April along the Pure

* Report of the Industrial Commission, 1900. Volume I, page 365.

Oil Company route, there was a fall of 2½ cents, in crude there was a fall of but four-tenths of a cent. Refined fell from 7 cents in April to 6 cents in May, and crude fell one-tenth of a cent. John D. Archbold, in answering the figures given by the Pure Oil Company to the Industrial Commission, accused them of "carelessness," and gave the average monthly price of crude and refined to show that no such glaring discrepancy had taken place. Mr. Archbold gives the average price in March, for instance, as 7.98 and in April as 7.31 cents. However, his price is the average to "all the trade of Greater New York and its vicinity," whereas the prices of the Pure Oil Company are those they met in their limited competition. As Professor Jenks remarked at the examination: "It might easily be, therefore, that your" (Standard) "average price would be what you had given, and that to a good many special customers with whom the Pure Oil Company was trying to deal it could be five and a half cents." That this was the fact seems to be proved by the quotations for water-white oil from tank-wagons, which were published from week to week in trade journals like the Oil, Paint and Drug Reporter. These prices show 9⅞ cents for water-white on March 21, and an average of 9.4 cents in April. Evidently only a part of the trade of "all Greater New York and vicinity" got the benefit of averages quoted to the Industrial Commission by Mr. Archbold.

If competition persists the result usually has been permanently lower prices than in territory where competition has been run out or has never entered. For instance, why should oil be sold to a dealer at nearly four cents more on an average in Kansas than in Kentucky, when the freight from Whiting to Kansas is only a cent more? For no reason except that in Kentucky there has been persistent competition for twenty-five years, and in Kansas none has ever secured a solid foothold. Why should Colorado pay an average of 16.90 cents for oil per gallon and California 14.60 cents, when the freight from Whiting differs but one-tenth of one cent? For no reason except that a few years ago competition was driven from Colorado, and in California it still exists.

Indeed, any consecutive study of the Standard Oil Company's use of its power over the price of either export or domestic oil must lead to the conclusion that it has always been used to the fullest extent possible without jeopardising it; that we have always paid more for our refined oil than we would have done if there had been free competition. But why should we expect anything else? This is the chief object of combinations. Certainly the candid members of the Standard Oil Company would be the last men to argue that they give the public any more of the profits they may get by combination than they can help. One of the ablest and frankest of them, H. H. Rogers, when before the Industrial Commission in 1899, was asked how it happened that in twenty years the Standard Oil Company had never cheapened the cost of gathering and transporting oil in pipe-lines by the least fraction of a cent; that it cost the oil producer just as much now as it did twenty years ago to get his oil taken away from the wells and to transport it to New York. And Mr. Rogers answered, with delightful candour: "We are not in business for our health, but are out for the dollars."

John D. Archbold was asked at the same time if it were not true that, by virtue of its great power, the Standard Oil Company was enabled to secure prices that, on the whole, were above those under competition, and Mr. Archbold said: "Well, I hope so." *

But these are frank answers, perhaps surprised out of the gentlemen. The able and wary president of the great con-

* See Appendix, Number 58. John D. Archbold's statement on the prices the Standard receives for refined oil.

cern, John D. Rockefeller, is more cautious in his admissions. On the witness-stand in 1888 he was forced to admit, after some skilful evasion, that the control the Standard Oil Company had of prices was such that they could raise or lower them at will. "But," added Mr. Rockefeller, "we would not do it." The whole colloquy between the examiner and Mr. Rockefeller is interesting:

Q. Isn't it a fact that the nine trustees controlling the large amount of capital which the Standard Oil Trust does could very easily advance or depress the market price of oil if they saw fit? . . .

A. I don't think they would.

Q. I don't ask whether they would; could they do it?

A. I suppose it would be possible for these gentlemen; if they should buy enough oil, it would make the price go up.

There was considerable sparring, Mr. Rockefeller trying to explain away his answer.

Q. I can't get you down to my question . . . that is a very great power to wield.

A. Certainly; an individual or a combination of men can advance the price or more or less depress the price of any commodity.

Q. But if you desire to increase—to put up the price of the refined oil, or to put down the price of the crude oil, is it within your power to do it, in the way I have indicated, by staying out of the market or going into the market to purchase, controlling 75 per cent. of the demand for the crude oil?

A. It would be a temporary effect, but that is all. . . .

Q. By stopping the manufacture of refined oil your refineries representing so large a proportion would tend to raise the price?

A. That is something we never do; our business is to increase all the time, not to decrease.

. . . .

Q. Really your notion is that the Standard Oil Trust is a beneficial organisation to the public?

A. I beg with all respect to present the record which shows that it is.*

For many of the world it is a matter of little moment, no doubt, whether oil sells for eight or twelve cents a gallon.

* Report on Investigation Relative to Trusts, New York Senate, 1888, pages 434-435 and 396-398.

It becomes a tragic matter sometimes, however, as in 1902-1903 when, in the coal famine, the poor, deprived of coal, depended on oil for heat. In January, 1903, oil was sold to dealers from tank-wagons in New York City at eleven cents a gallon. That oil cost the independent refiner, who paid full transportation charges and marketed at the cost of a cent a gallon, not over 6.4 cents. It cost the Standard Oil Company probably a cent less. That such a price could prevail under free competition is, of course, impossible. Throughout the hard winter of 1902–1903 the price of refined oil advanced. It was claimed that this was due to the advance in crude, but in every case it was considerably more than that of crude. Indeed, a careful comparative study of oil prices shows that the Standard almost always advances the refined market a good many more points than it does the crude market. The chart shows this. While this has been the rule, there are exceptions, of course, as when a rate war is on. Thus, in the spring of 1904, the severe competition in England of the Shell Transportation Company and of Russian oil caused the Standard to drop export refined considerably more than crude. But, as the chart shows, domestic oil has been kept up.

As a result of the Standard's power over prices, not only does the consumer pay more for oil where competition has not reached or has been killed, but this power is used steadily and with consummate skill to make it hard for men to compete in any branch of the oil business. This history has been but a rehearsal of the operations practised by the Standard Oil Company to get rid of competition. It was to get rid of competition that the South Improvement Company was formed. It was to get rid of competition that the oil-carrying railroads were bullied or persuaded or bribed into unjust discriminations. It was to get rid of competition that the Empire Transportation Company, one of the finest transportation companies ever built up in this country, was wrested from the hands of the men who had developed it. It was to get rid of competition that war was made on the Tidewater Pipe Line, the Crescent Pipe Line, the United States Pipe Line, not to mention a number of similar smaller enterprises. It was to get rid of competition that the Standard's spy system was built up, its oil wars instituted, all its perfect methods for making it hard for rivals to do business developed.

The most curious feature perhaps of this question of the Standard Oil Company and the price of oil is that there are still people who believe that the Standard has made oil cheap! Men look at this chart and recall that back in the late sixties and seventies they paid fifty and sixty cents a gallon for oil, which now they pay twelve and fifteen cents for. This, then, they say, is the result of the combination. Mr. Rockefeller himself pointed out this great difference in prices. "In 1861," he told the New York Senate Committee, "oil sold for sixty-four cents a gallon, and now it is six and a quarter cents." The comparison is as misleading as it was meant to be. In 1861 there was not a railway into the Oil Regions. It cost from three to ten dollars to get a barrel of oil to a shipping point. None of the appliances of transportation or storage had been devised. The process of refining was still crude, and there was great waste in the oil. Besides, the markets were undeveloped. Mr. Rockefeller should have noted that oil fell from 61½ in 1861 to 25⅝ in the year he first took hold of it, and that by his first successful manipulation it went up to 30! He should point out what the successive declines in prices since that day are due to—to the seaboard pipe-lines, to the development of by-products, to bulk instead of barrel transportation, to innumerable small economies. People who point to the differences in price, and call it combination, have never studied the price-line history in hand. They do not know the meaning of the variation of the line; that it was forced down from 1866 to 1876, when Mr. Rockefeller's first effective combination was secured by competition, and driven up in 1876 and 1877 by the stopping of competition; that it was driven down from 1877 to 1879 by the union of all sorts of competitive forces—producers, independent refiners, the developing of an independent seaboard pipe-line—to a point lower than it had ever been before. They forget that when these opposing forces were overcome, and the Standard Oil Company was at last supreme, for ten years oil never fell a point below the margin reached by competition in 1879, though frequently it rose above that margin. They forget that in 1889, when for the first time in ten years the margin between crude and refined oil began to fall, it was the competition coming from the rise of American independent interests and the development of foreign oil fields that did it.

To believe that the Standard Oil Combination, or any other similar aggregation, would lower prices except under the pressure of the competition they were trying to kill, argues an amazing gullibility. Human experience long ago taught us that if we allowed a man or a group of men autocratic powers in government or church, they used that power to oppress and defraud the public. For centuries the struggle of the nations has been to obtain stable government, with fair play to the masses. To obtain this we have hedged our kings and emperors and presidents about with a thousand constitutional restrictions. It has not been possible for us to allow even the church, inspired by religious ideals, to have the full power it has demanded in society. And yet we have here in the United States allowed men practically autocratic powers in commerce. We have allowed them special privileges in transportation, bound in no great length of time to kill their competitors, though the spirit of our laws and of the charters of the transportation lines forbade these privileges.

We have allowed them to combine in great interstate aggregations, for which we have provided no form of charter or of publicity, although human experience long ago decided that men united in partnerships, companies, or corporations for business purposes must have their powers defined and be subject to a reasonable inspection and publicity. As a natural result of these extraordinary powers, we see, as in the case of the Standard Oil Company, the price of a necessity of life within the control of a group of nine men, as able, as energetic, and as ruthless in business operations as any nine men the world has ever seen combined. They have exercised their power over prices with almost preternatural skill. It has been their most cruel weapon in stifling competition, a sure means of reaping usurious dividends, and, at the same time, a most persuasive argument in hoodwinking the public.

CHAPTER SEVENTEEN

THE LEGITIMATE GREATNESS OF THE STANDARD OIL COMPANY

CENTRALISATION OF AUTHORITY—ROCKEFELLER AND EIGHT OTHER TRUSTEES MANAGING THINGS LIKE PARTNERS IN A BUSINESS—NEWS-GATHERING ORGANIZATION FOR COLLECTING ALL INFORMATION OF VALUE TO THE TRUSTEES—ROCKEFELLER GETS PICKED MEN FOR EVERY POST AND CONTRIVES TO MAKE THEM COMPETE WITH EACH OTHER—PLANTS WISELY LOCATED—THE SMALLEST DETAILS IN EXPENSE LOOKED OUT FOR —QUICK ADAPTABILITY TO NEW CONDITIONS AS THEY ARISE—ECONOMY INTRODUCED BY THE MANUFACTURE OF SUPPLIES—A PROFIT PAID TO NOBODY—PROFITABLE EXTENSION OF PRODUCTS AND BY-PRODUCTS—A GENERAL CAPACITY FOR SEEING BIG THINGS AND ENOUGH DARING TO LAY HOLD OF THEM.

WHILE there can be no doubt that the determining factor in the success of the Standard Oil Company in securing a practical monopoly of the oil industry has been the special privileges it has enjoyed since the beginning of its career, it is equally true that those privileges alone will not account for its success. Something besides illegal advantages has gone into the making of the Standard Oil Trust. Had it possessed only the qualities which the general public has always attributed to it, its overthrow would have come before this. But this huge bulk, blackened by commercial sin, has always been strong in all great business qualities—in energy, in intelligence, in dauntlessness. It has always been rich in youth as well as greed, in brains as well as unscrupulousness. If it has played its great game with contemptuous indifference to fair play, and to nice legal points of view, it has played it with consummate ability, daring and address. The silent, patient, all-seeing man who has led it in its transportation raids has led it no less successfully in what may be called its legitimate work. Nobody has appreciated more fully than he those qualities which alone make for permanent stability and growth in commercial ventures. He has insisted on these qualities, and it is because of this insistence that the Standard Oil Trust has always been something besides a fine piece of brigandage, with the fate of brigandage before it, that it has been a thing with life and future.

If one attempts to analyse what may be called the legitimate greatness of Mr. Rockefeller's creation in distinction to its illegitimate greatness, he will find at the foundation the fact that it is as perfectly centralised as the Catholic church or the Napoleonic government. As was pointed out in a former chapter, the entire business was placed in 1882 in the hands of nine trustees, of whom Mr. Rockefeller was president. These trustees have always acted exactly as if they were nine partners in a business, and the only persons concerned in it. They met daily, giving their whole time to the management and development of the concern, as the partners in a dry-goods house would. Anything in the oil world might come under their ken, from a smoking wick in Oshkosh to the competition of Russian oil in China. Everything; but nothing came unless it was necessary; for below them, and sifting things for their eyes, were committees which dealt with the various departments of the business. There was a Crude Committee which considered the subject of crude oil, the world over; a Manufacturing Committee which studied the making of refined, the utilisation of waste, the development of new products; a Marketing Committee which considered the markets. Before each of these committees was laid daily all the information to be found on earth concerning its particular field; not only were there reports made to it of what was doing in its line in the Standard Oil Trust, but information came of everything

S. C. T. DODD

Chief counsel of the Standard Oil Company. Framer of the Trust agreement of 1882.

JABEZ A. BOSTWICK

From 1872 to 1892 the chief oil buyer of the Standard Oil Company.

JOSEPH SEEP

Head of the "Seep Agency," through which all oil transported by the Standard Oil Company goes.

DANIEL O'DAY IN 1872

Vice-president of the National Transit Company, the pipe-line company owned by the Standard Oil Company.

connected with such work everywhere by everybody. These committees not only knew all about their own business, they knew all about everybody else's. The Manufacturing Committee knew just what each of the feeble independent refiners still existing was doing—what its resources and advantages were; the Transportation Committee knew what rates it got; the Marketing Committee knew its market. Thus the fullest information about new developments of crude, new openings for refined, new processes of manufacture, was always at the command of the nine trustees of the trust.

How did they get this information? As the press does—by a wide-spreading system of reporters. In 1882 the Standard had correspondents in every town in the oil fields, and to-day it has them not only there but in every capital of the globe. It is a common enough thing, indeed, in European capitals to run across high-class newspaper correspondents, consuls, or business men who add to their incomes by private reporting to the Standard Oil Company. The people in their employ naturally report all they learn. There are also outsiders who report what they pick up—"occasional contributions." There is more than one man in the Oil Regions who has made his livelihood for years by picking up information for the Standard. "Spies," they are called there. They may deserve the name sometimes, but the service may be perfectly legitimate.

These trustees then "know everything" about the oil business and they have used their information. Nobody ever used information more profitably. What was learned was applied, and affected the whole great structure, for by a marvellous genius in organisation Mr. Rockefeller had devised a machine with a head whose thinking was felt from the seat of power in New York City to the humblest pipe-line patrol on Oil Creek. This head controlled each one of the scattered plants with absolute precision. Take the refineries; they were individual plants, having a manager and a board of directors like any outside plant, but these plants were not free agents. According to J. J. Vandergrift's testimony in 1879, the Imperial Refinery, of which he was president, had no control of its oil after it was made. The Standard Oil Company of Cleveland took charge of it at Oil City, and arranged for transportation and for marketing. The managers of the Central Association, into which the allied refiners went in 1875 under Mr. Rockefeller's presidency, had "irrevocable authority to make all purchases of crude oil and sales of refined oil," as well as to "negotiate for all railroad and pipe-line freights and transportation expenses" for each of the refineries. Each plant, of course, was limited as to the amount of oil it could make. Thus, in 1876, when the Cleveland firm of Scofield, Shurmer and Teagle went into a running arrangement with Mr. Rockefeller on condition that he get for them the same rebates he enjoyed, it was agreed that the firm should manufacture only 85,000 barrels a year, though they had a capacity of 180,000 barrels.

One of Mr. Rockefeller's greatest achievements has been to bring men who had built up their own factories and managed them to suit themselves to work harmoniously under such limitations. As this history has shown, the first attempt to harness the refiners failed because they would not obey the rules. No doubt the chief reason why they finally consented to them was that only by so doing could they get transportation rates equally advantageous to those of the Standard Oil Company; but, having consented and finding it profitable, they were kept in line by an ingenious system of competition which must have done much to satisfy their need of individual effort and their pride in independent work. In the investigation of 1879, when the producers were trying to find out the real nature of the Standard alliance, they were much puzzled by the sworn testimony of certain Standard men that the factories they controlled were competing, and competing hard, with the Standard Oil Company of Cleveland. How could this be? Being bitter in heart and reckless in tongue, the oil men denounced the statements as perjury, but they were the literal truth. Each refinery in the alliance was required to make to Mr. Rockefeller each month a detailed statement of its operations. These statements were compared and the results made known. If the Acme at Titusville had refined cheaper that month than any other member of the alliance, the fact was made known. If this cheapness continued to show, the others were sent to study the Acme methods. Whenever an improvement showed, that improvement received credit, and the others were sent to find the secret. The keenest rivalry resulted—every factory was on its mettle.

This supervision took account of the least detail. There is a story often told in the Oil Regions to illustrate the minuteness of the supervision. In commenting as usual on the monthly "competitive statements," as they are called, Mr. Rockefeller called the attention of a certain refiner to a discrepancy in his reports. It referred to *bungs*—articles worth about as much in a refinery as pins are in a household. "Last month," the comment ran, "you reported on hand 1,119 bungs. Ten thousand were sent you at the beginning of this month. You have used 9,527 this month. You report 1,012 on hand. What has become of the other five hundred and eighty?" The writer has it on high authority that the current version of this story is not true, but it reflects very well the impression the Oil Regions have of the thoroughness of Mr. Rockefeller's supervision. The Oil Regions, which were notoriously extravagant in their business methods, resented this care and called it meanness, but the Oil Regions were wrong and Mr. Rockefeller was right. Take care of the bungs and the barrels will take care of themselves, is as good a policy in a refinery as the old saw it paraphrases is in financiering.

There were other features of this revolutionary management

which caused deep resentment in the oil world. Chief among them was the dismantling or abandoning of plants which the Standard had "acquired," and which it claimed were so badly placed or so equipped that it did not pay to run them. There was reason enough in many cases for dissatisfaction with the process of acquisition, but having acquired the refineries, the Standard showed its wisdom in abandoning many of them. Take Pittsburg, for instance. When Mr. Lockhart began to absorb his neighbours, in 1874, there were some twenty-five plants in and around the town. They were of varying capacity, from little ten-barrel stills of antiquated design and out-of-the-way location, to complete plants like the Citizens', which Mr. Tack described in Chapter V. But how could Mr. Lockhart manage these as they stood to good advantage? It might pay the owner of the little refinery to run it, for he was his own stillman, his own pipe-fitter, his own foreman, and did not expect large returns; but it would have been absurd for Mr. Lockhart to try to run it. He simply carted away any available machinery, sold what he could for junk, and left the *débris*. Now, one of the most melancholy sights on earth is an abandoned oil refinery; and it was the desolation of the picture, combined, as it always was in the Oil Regions, with the history of the former owners, that caused much of the outcry. It was a thing that the oil men could not get over, largely because it was a sight always before their eyes.

Bitter as this policy was for those who had suffered by the Standard's campaigns, it was, of course, the only thing for the trust to do—indeed, that was what it had been waging war on the independents for: that it might shut them down and dismantle them, that there might be less oil made and higher prices for what it made. This wisdom in locating factories has continued to characterise the Standard operations. It works only plants which pay, and it places its plants where they can be operated to the best advantage. Many fine examples of the relation of location in manufacturing to crude supply and to markets are to be seen in the Standard Oil Company plants to-day. For example, refined for foreign shipments is made at the seaboard, and the vessels which carry it are loaded at docks, as at the works at Bayonne, New Jersey. The cost of transportation from factory to ship, a large item in the old days, is eliminated entirely. The Middle West market is now supplied almost entirely from the Standard factories at Whiting, Indiana, a town built by the Standard Oil Company for refining Ohio oil. Here 25,000 barrels of oil are refined daily, and from this central point distributed to the Mississippi Valley.

All of the industries which have been grafted on to the refineries have always been run with the same exact regard to minute economies. These industries were numerous because of Mr. Rockefeller's great principle, "pay a profit to nobody." From his earliest ventures in combination he had applied this principle. Mr. Blanchard's explanation to the Hepburn Commission in 1879 of why the Standard had controlled the Erie's yards at Weehawken since 1874, shows exactly Mr. Rockefeller's point of view.* This policy of paying nobody a profit took Mr. Rockefeller into the barrel business. In 1872, when Mr. Rockefeller became master of the Cleveland oil business, the purchase of barrels was one of a refiner's heaviest expenses. In an estimate of the cost of producing a gallon of refined oil in 1873, made in the Oil City Derrick and accepted as correct by that paper, the cost of the barrel is put at four cents a gallon, which was more than the crude oil cost at that date. Even at four cents a gallon barrels were hard to get, so great was the demand. If a refiner could get his barrels back, of course there was a saving (a returned barrel was estimated to be worth 2¾ cents), but the return could not be counted on; empty barrels coming from Europe particularly, and con-

* See Chapter V.

signed to Western shippers, were frequently seized in New York by Eastern refiners. The need was held to justify the deed, like thieving in famine time. Fortunes were made in barrels, and dealers hearing of a big supply in Europe have been known to charter a vessel and go for them, and reap rich profits. In fact, a whole volume of commercial tragedy and comedy hangs around the oil barrel. Now it was to the barrel —the "holy blue barrel"—that Mr. Rockefeller gave early attention. He determined to make it himself. One of the earliest outside ventures of the Standard Oil Company in Cleveland was barrel works, and Mr. Rockefeller was soon getting for two and a half cents what his rivals paid four for, though he was by no means the only refiner who manufactured barrels in the early days—each factory aimed to add barrel works as soon as able. The amount the Standard Oil Company saved on this one item is evident when the extent of its business is considered. The year before the trust was formed (1881) they manufactured 4,500,000 barrels, an average of about 15,000 a day. Since that time the barrel has been gradually going out of the oil business, bulk transportation taking its place very largely. Nevertheless, in 1901 the Standard Oil Company manufactured about 3,000,000 new barrels. In the period since they began the manufacture of barrels their factories have introduced some small savings which in the aggregate amount to large sums. For instance, they have improved the lap of the hoop—a small thing, but one which amounted in 1901 to something like $15,000. Some $50,000 a year was saved by a slight increase in the size of the tankage. The Standard claims that these economies are so small in themselves that it only pays to practise them where there is a large aggregate business.

More important than the barrel to-day, however, is the tin can—for it is in tin cans that all the enormous quantities of refined sent to tropical and Oriental countries must go to prevent deterioration—and nowhere does the policy of economy which Mr. Rockefeller has worked out show better than in one of the Standard canning works. In 1902 the writer visited the largest of the Standard can factories, the Devoe, on the East River, Long Island City. It has a capacity of 70,000 five-gallon cans a day, and is probably the largest can factory in the world. At the entrance of the place a man was sweeping up carefully the dirt on the floor and wheeling it away—not to be dumped in the river, however. The dirt was to be sifted for tin filings and solder dust. At every step something was saved. The Standard buys the tin for its cans in Wales, because it is cheaper. It would not be cheaper if it were not for a vagary in administering the tariff by which the duty on tin plate is refunded if the tin is made into receptacles to be exported. This clause was probably made for the benefit of the Standard, it being the largest single consumer of tin plate in the United States. In 1901 the Standard Oil Company imported over 60,000 tons of tin with a value of over $1,000,000. This tin comes in sheets packed in flat boxes, which are opened by throwing—it is quicker than opening by a hammer, and time is considered as valuable as tin filings. The empty boxes are sold by the hundred to the Long Island gardens for growing plants in, and the broken covers are sold for kindling. The trimmings which result from shaping the tin sheets for a can are gathered into bundles and sold to chemical works or foundries. There is the same care taken with solder as with tin, the amount each workman uses being carefully gauged. The canning plants, like the refineries, compare their results monthly, and the laurels go to the manager who has saved the most ounces of solder, the most hours, the most footsteps.

The five-gallon can turned out at the Devoe is a marvel of evolution. The present methods of manufacture are almost entirely the work of Herman Miller, known in Standard circles as the "father of the five-gallon can"; and a fine type of the German inventor he is. The machinery for making the can has been so developed that while, in 1865, when Mr. Miller began his work under Charles Pratt, one man and a boy soldered 850 cans in a day, in 1880 three men made 8,000, and since 1893 three men have made 24,000. It is an actual fact that a tin can is made by Miller in just about the time it takes to walk from the point in the factory where the sheets of tin are unloaded to the point where the finished article is filled with oil.

And here is a nice point in combination. Not far away from the canning works, on Newtown Creek, is an oil refinery. This oil runs to the canning works, and, as the new-made cans come down by a chute from the works above, where they have just been finished, they are filled, twelve at a time, with the oil made a few miles away. The filling apparatus is admirable. As the new-made cans come down the chute they are distributed, twelve in a row, along one side of a turn-table. The turn-table is revolved, and the cans come directly under twelve measures, each holding five gallons of oil—a turn of a valve, and the cans are full. The table is turned a quarter, and while twelve more cans are filled and twelve fresh ones are distributed, four men with soldering coppers put the caps on the first set. Another quarter turn, and men stand ready to take the cans from the filler, and while they do this, twelve more are having caps put on, twelve are filling, and twelve are coming to their place from the chute. The cans are placed at once in wooden boxes standing ready, and, after a twenty-four-hour wait for discovering leaks, are nailed up and carted to a near-by door. This door opens on the river, and there at anchor by the side of the factory is a vessel chartered for South America or China or where not—waiting to receive the cans which a little more than twenty-four hours before were tin sheets lying in flat boxes. It is a marvellous example of economy not only in materials, but in time and in footsteps.

With Mr. Rockefeller's genius for detail, there went a sense of the big and vital factors in the oil business, and a daring in laying hold of them which was very like military genius. He saw strategic points like a Napoleon, and he swooped on them with the suddenness of a Napoleon. This master ability has been fully illustrated already in this work. Mr. Rockefeller's capture of the Cleveland refineries in 1872 was as dazzling an achievement as it was a hateful one. The campaign by which the Empire Transportation Company was wrested from the Pennsylvania Railroad, viewed simply as a piece of brigandage, was admirable. The man saw what was necessary to his purpose, and he never hesitated before it. His courage was steady—and his faith in his ideas unwavering. He simply knew that was the thing to do, and he went ahead with the serenity of the man who knows.

After the formation of the trust the demand for these qualities was constant. For instance, the contract which the Standard signed with the producers in February, 1880, pledged them to take care of a production of 65,000 barrels a day. When they signed this agreement there was above ground nearly nine and one-half million barrels of oil. The production increased at a frightful rate for four years. At the end of 1880 there were stocks of over 17,000,000 above ground; in 1881, over 25,000,000; 1882, over 34,000,000; 1883, over 35,000,000; and 1884, over 36,000,000, and the United Pipe Lines took care of this production—with the aid of the producers, who built tanks neck and neck with them. In 1880 the Standard people averaged over one iron tank a day, the tanks holding from 25,000 to 35,000 barrels. There were not tank-builders enough in the United States to do the work, and crews were brought from Canada and England. This, of course, called for an enormous expenditure of money, for tanks cost from $7,000 to $10,000 apiece. Rich as the United Pipe Lines were they were forced to borrow money in these years of excessive production, for they had to lay lines as well as build tanks. There were nearly 4,000 miles of pipe-line laid in the Bradford region alone from 1878 to 1884, and these lines connected with upward of 20,000 wells.

From the time it completed its pipe-line monopoly the Standard has followed oil wherever found. It has had to do it to keep its hold on the business, and its courage never yet has faltered, though it has demanded some extraordinary efforts. In 1891 a great deposit of oil was tapped in the McDonald field of Southwestern Pennsylvania. The monthly production increased from 50,000 barrels in June to 1,600,000 in December. It is an actual fact that in the McDonald field the United Pipe Lines increased the daily capacity of 3,500 barrels, which they had at the beginning of July, to one of 26,000 barrels by the first of September, and by the first of December they could handle 90,000 barrels a day. If one considers what this means one sees that it compares favourably with the great ordnance and mobilising feats of the Civil War. To accomplish it, rolling mills and boiler shops in various cities worked night and day to turn out the pipe, the pumps, the engines, the boilers which were needed. Transportation had to be arranged, crews of men obtained, a wild country prepared, sawmills to cut the quantities of timber needed built, and this vast amount of material placed and set to work.

The same audacity and effectiveness are shown by the Standard in attacking situations created by new developments in handling business. The seaboard pipe-line is a notable example. When the Standard completed its pipe-line monopoly at the end of 1877, the pipe-line was still regarded as the feeder of the railroad. Naturally the railroads were seriously opposed to its becoming anything more. In Pennsylvania particularly the laws had been so manipulated by the Pennsylvania Railroad as to prevent the pipe-line carrying oil even for short distances in competition with them. Now, for many years it had been believed that the pipe-line could carry oil long distances—many claimed to the seaboard—and as soon as the independents found that the oil-bearing roads were acting solely in the interest of the Standard they began an agitation for a seaboard line which finally terminated in the Tidewater Line, one hundred and four miles long, carrying oil from the Bradford field to Williamsport on the Reading Railroad, and it was certain that the Tidewater eventually would get to the seaboard. That the day of the railroad as a carrier of crude oil was over when the Tidewater began to pump oil was obvious both to Mr. Rockefeller and to the railroad presidents, and without hesitation he seized the idea. By 1883 the Standard was pumping oil to New York, and the railroads that had served so effectively in building up the trust were practically out of the crude business. It was this audacious and splendid stroke, practically freeing him from the railroads which had made him, which made the passage of the Interstate Commerce Bill a matter of comparatively small importance to Mr. Rockefeller. To be sure, he still needed the railroads for refined, but he could so place his refineries that this service would be greatly minimised. The legislation which the Oil Regions of Pennsylvania demanded for fifteen years in hope of securing an equal chance in transportation came too late. By the time the bill was passed the pipe had replaced the rail as the great oil carrier, and the pipes were not merely under Mr. Rockefeller's control, as the rails had been; they belonged to him. It was little wonder, then, that the passage of the great bill did not ruffle his serenity. Little wonder that the Oil Regions, realising the situation, so tragic in its irony, as fully as Mr. Rockefeller did, felt an exasperation almost uncontrolled over it. Yet the seaboard pipe-line was no development of the Standard Oil Company. The idea had been conceived and the practicability demonstrated by others, but it was seized by the Standard as soon as it proved possible. This quick sense of the real value of new developments, and this alertness in seizing them, have been among the strongest elements in the Standard's success.

And every new line of action was developed to its utmost. Take the work the Standard began in 1879 on the foreign market. Before the Standard Oil Company was known, save as one of several prosperous Cleveland refineries, the foreign trade had been developed until petroleum was *fourth* in our list of exports, and it went literally to every civilised country on the globe. In 1874 Colonel Forney made a trip through the Orient, and he wrote in one of his letters that he found both Babylon and Nineveh to be lighted with American petroleum, and that while he was in Damascus a census was taken to ascertain how much petroleum was needed for each house in the place, and a proposition was made for its entire use. "At present," said the Derrick, in commenting on this letter, "petroleum is the chief commercial representative of the United States in the Levant and the Orient."

The same dithyrambic paragraphs were written by oil men then, as by the Standard now, concerning foreign trade. For instance, compare the two paragraphs below—the one found in 1874 in the Derrick, the second in a defence of the Oil Trust published in 1900:

1874—"It lights the dwellings, the temples, and the mosques amid the ruins of ancient Babylon and Nineveh; it is the light of Bagdad, the city of the Thousand and One Nights; of Orfa, birthplace of Abraham; of Mardeen, the ancient *Macius* of the Romans, and of Damascus, gem of the Orient. It burns in the grotto of the Nativity at Bethlehem; in the Church of the Holy Sepulchre in Jerusalem; amidst the Pyramids of Egypt; on the Acropolis of Athens; on the plains of Troy; and in cottage and palace on the banks of the Bosporus and the Golden Horn."

1900—"Petroleum to-day is the light of the world. It is carried wherever a wheel can roll or a camel's hoof be planted. The caravans on the desert of Sahara go laden with Pratt's Astral, and elephants in India carry cases of 'Standard-white,' while ships are constantly loading at our wharves for Japan, Java and the most distant isles of the sea."

Exports grew rapidly through the same machinery which had created the foreign market. In 1870 there were something over one hundred and forty million gallons of petroleum products going abroad, in 1873 nearly two and one-half hundred million, in 1878 three and one-half hundred million. In 1870 the Standard began its work on the foreign trade by sending a representative abroad. Country after country seems to have been taken up, the idea being that the daily Standard Oil meeting should have the same full information before it concerning every place of foreign trade as it had of the American trade, and that gradually the company should control the foreign trade as it did the American industry, doing away with middlemen, "paying nobody a profit." This work, begun in 1879, has been carried on steadily ever since. Through it the Standard soon became largely its own exporter. It established stations of its own in one port after another of Europe, Asia, South America, and has built up a large oil fleet. It carried on an aggressive campaign for developing markets; it looked after hostile legislation; it studied the possible competition of native oils; it met every difficulty—prejudice, ignorance, poverty. Little by little it has done in foreign countries what it has done in the United States. To-day it even carts oil from door to door in Germany and Portugal and other countries, as it does in America, thus realising Mr. Rockefeller's vision of controlling the petroleum of America from the time it leaves the ground until it is put into the lamp of the consumer.

The same economy and alertness were applied to the matter of making oils. In laying hands on the refineries of the country, Rockefeller had acquired by 1882 about all the processes of manufacturing known, both patented and free. These processes, including all the essential ones of to-day, had been developed entirely outside of the Standard Oil Company. As early as 1865, the year Mr. Rockefeller went into the business, William Wright wrote an exhaustive book on the Oil Regions of Pennsylvania. Among other things, he reported

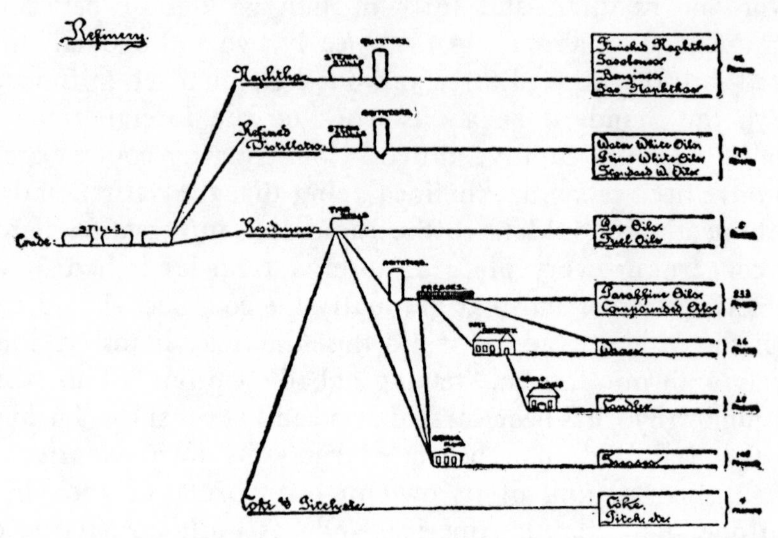

PRODUCTS OBTAINED FROM THE DISTILLATION OF CRUDE OIL IN A REFINERY.

quite fully what was being done in the refining of petroleum. He found that in several factories they were making naphtha, gasoline and benzine; that three grades of illuminating oils —"prime white," "standard white" and "straw colour"— were made everywhere; that paraffine, refined to a pure white article like that of to-day, was manufactured in quantities by the Downer works; and that lubricating oils were beginning to be made.

In 1872, the year that Mr. Rockefeller took things in hand, all of these original products had been greatly extended, as we have seen. Joshua Merrill had succeeded in deodorising lubricating oil, making it possible to put the petroleum lubricants on the foreign market, and in 1871 Mr. Merrill's factory sold 50,000 gallons in England alone. By 1872 paraffine wax was being made in many factories, and one maker of chewing gum in Maine used 70,000 pounds that year. The foreign trade in all the products of petroleum outside of illuminating oil was already considerable.* Many of the factories in making their oils gave them names; thus, Pratt's

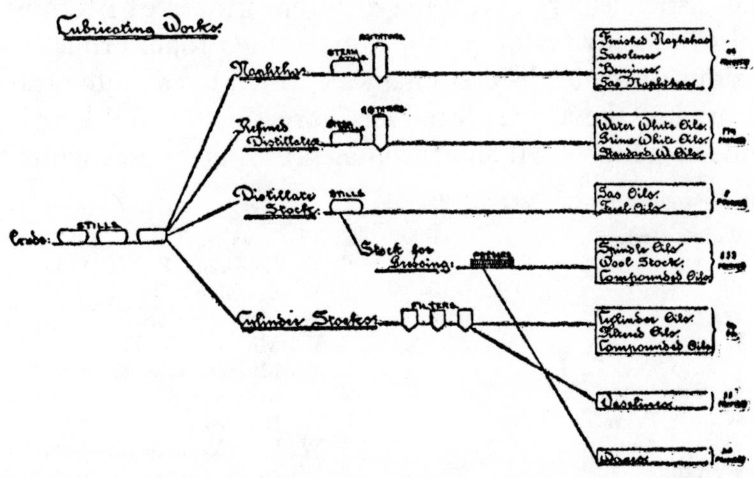

PRODUCTS OBTAINED FROM THE DISTILLATION OF CRUDE OIL IN LUBRICATING WORKS.

Astral was a name for a water-white oil made by the Pratt works of Brooklyn. It was a high-grade oil, made exactly as the oil made by many other refineries, but it had a name—a valuable one.

* In 1872 there were exported as follows:

Crude	16,363,975	gallons.
Naphtha, benzine, gasoline, etc	8,688,257	"
Lubricating, heavy paraffine, etc	438,425	"
Residuum, pitch and tar	568,218	"
Illuminating	118,259,832	"

—*Derrick Handbook.*

The tables (pages 246–247) analysing the products of crude oil obtained to-day at the Standard factories show the results tabulated. Now all of the products in these groups could be made in 1872, but certainly there were not forty-six distinct products under the naphthas as the table shows—nor were there 174 refined distillates. In fact, these are not really products; they are rather brands. Thus, though the table shows twenty-nine different kinds of odorised or deodorised naphthas, the main difference between them is their name. The 174 refined distillates are really the different grades of illuminating oil which any factory can get, given the proper crude base, with a multitude of different names applied to catch the trade. Thus among these 174 "products" are thirty-three kinds of "Standard-white" * oil and forty-one kinds of "water-white" †

* The "Standard-whites" are as follows:

S. W. 100 (fl).
S. W. 110.
S. W. 112.
S. W. 115.
S. W. 120.
S. W. 130 Dia. H. L.
S. W. 130.
S. W. 130 P. W. H. L.
S. W. 73 Abel.
S. W. 150.
S. W. 160.
S. W. Canadian Legal Test.
S. W. Georgia P. W. H. L.
S. W. Georgia Dia. H. L.
S. W. Indiana P. W. H. L.
S. W. Indiana S. T.
S. W. Indiana Dia. H. L.
S. W. Iowa S. T.
S. W. Louisiana P. W. H. L.
S. W. Louisiana Dia. H. L.
S. W. Massachusetts S. T.
S. W. Michigan S. T.
S. W. Minnesota S. T.
S. W. Montana S. T.
S. W. Nebraska S. T.
S. W. New York S. T.
S. W. North Dakota S. T.
S. W. Ohio S. T.
S. W. South Dakota S. T.
S. W. Tennessee Dia. H. L.
S. W. Tennessee P. W. H. L.
S. W. Tennessee S. T.
S. W. Wisconsin S. T.

† The "water-whites" are as follows:

W. W. 110.
W. W. 112.
W. W. 115.
W. W. 120.
W. W. 120 Eupion.
W. W. 130 Sunlight.
W. W. 130.
W. W. 130 Eupion.
W. W. 130 Fireproof.
W. W. 150.
W. W. 150 Headlight.
W. W. 150 for extra Star.
W. W. 150 forty-nine grav.
W. W. 160.

—the principal difference between them being the different fire tests at which they are put out. The real service of the Standard has been not this multiplication of so-called products, but in finding processes by which a poor oil like the famous Lima oil could be refined. In the case of the Lima oil the Standard claims it spent millions of dollars before it solved the problem of its usefulness. The amount of sulphur in the Lima or Ohio oil prevented its use as an illuminating oil, for the odour was intolerable, there was a disagreeable smoke, and the wick charred rapidly. The problem of deodorising it was attacked by many experimenters, and was finally practically solved by the Frasch process, which the Standard acquired after spending a large amount of money in testing its efficacy. Probably sixty per cent. of the illuminating oil used in the United States now is manufactured from an Ohio oil base.

This multiplication of varieties is, of course, a perfectly legitimate merchandising device, but it is not a development of products, properly speaking. Nor indeed was it for discoveries and inventions that the Standard Oil Trust was great in 1882, or that it is now—it is in the way it adapts and handles the discoveries and inventions it acquires. Take the matter of lubricating oils. After a long struggle it gathered to itself

W. W. 165.
W. W. Canadian Legal Test.
W. W. Electric.
W. W. Georgia Sunlight.
W. W. Georgia S. T.
W. W. Indiana Perfection.
W. W. Indiana S. T.
W. W. Iowa Perfection.
W. W. Iowa S. T.
W. W. Kansas Perfection.
W. W. Kansas S. T.
W. W. Louisiana S. T.
W. W. Louisiana Sunlight.
W. W. Massachusetts S. T.
W. W. Michigan S. T.
W. W. Minnesota S. T.
W. W. Nebraska S. T.
W. W. Nebraska Perfection.
W. W. New York S. T.
W. W. North Dakota S. T.
W. W. Ohio Perfection.
W. W. Ohio S. T.
W. W. South Dakota S. T.
W. W. South Dakota Perfection.
W. W. Tennessee S. T.
W. W. Tennessee Sunlight.
W. W. Wisconsin S. T.

the factories and the patents of lubricating oils, and it has developed the trade amazingly; for, while in 1872 less than a half million gallons of petroleum lubricants were going abroad, in 1897 over 50,000,000 gallons went. The extension of the lubricating trade was made possible largely by the discovery of Mr. Merrill referred to above. In 1869 Mr. Merrill discovered a process by which a deodorised lubricating oil could be made. He had both the apparatus for producing the oil and for the oil itself patented. The oil was so favourably received that the market sale was several hundred per cent. greater in a single year than the firm had ever sold before. Naturally, an attempt was made by other lubricating works to imitate Mr. Merrill's new product. The most successful imitation was made by Dr. S. D. Tweedle of Pittsburg. The oil he put upon the market was considered an infringement by Mr. Merrill, who commenced suit against the agents handling it. The case was before the courts for some six years, and Mr. Merrill spent over $100,000 in maintaining the patent. The case was finally decided in his favour by the Supreme Court in Washington. During this suit the Standard Oil Company stood behind Dr. Tweedle, furnishing the money to defend the suit. When finally they were defeated they took a license under the new patent which Mr. Merrill was obliged to get out, and paid him a royalty on the oil until within about a year and a half before the end of the life of the patent, when they bought it outright for a large sum, Mr. Merrill reserving the right to manufacture and sell the oil without a royalty. Most lubricating oils from petroleum are now made after Mr. Merrill's process.

Having obtained control of the lubricating oils, the Standard showed the greatest intelligence in studying the markets and in developing the products. It makes lubricants for every machine that works. It offers scores of cylinder oils, scores of spindle lubricants, of valve lubricants, of gas-engine lubricants, special brands for sewing machines, for looms, for sole leather, for dynamos, for marine engines, for everything that runs and works by steam power, by air, by electricity, by gas, by man, or by beast power. Now any lubricating factory can produce the six or eight primary lubricants. Given these, the varieties to be produced by skilful compounding are infinite. They can be made more or less viscous, flowing, heavy, light, according to the needs of the machines and the idiosyncrasies of individuals who run them. The man who runs a machine soon knows what oil suits him, and if his trade is big enough an oil is put up especially for him with a name to tickle his vanity. It may be exactly like a dozen other oils on the market, but having its own name it is reckoned a new product. Skilful compounders insist that they can duplicate any of the 833 lubricating oils of the Standard if they can have samples. Of course this close study of the needs of a market, and this adaptation of one's goods to the requirements, are the highest sort of merchandising.

Unquestionably the great strength of the Standard Trust in 1882, when it was founded as it is to-day, was the men who formed it. However sweeping Mr. Rockefeller's commercial vision, however steady his purpose, however remarkable his insight into what was essential to the realisation of his ambition, he would have never gone far had he not drawn men into his concern who understood what he was after and knew how to work for it. His principle concerning men was laid down early. "We want only the big ones, those who have already proved they can do a big business. As for the others, unfortunately they will have to die." The scheme had no provision for mediocrity—nor for those who could not stomach his methods. The men who in 1882 formed the Standard alliance were all from the foremost rank in the petroleum trade, men who without question would be among those at the top to-day if there had never been a Standard Oil Company. In Pittsburg it was Charles Lockhart, a man interested in petroleum before the Drake well was struck, who had begun oil operations on Oil Creek in March, 1860, who had carried samples of crude and refined to Europe as early as May, 1860, who had built one of the first refineries in Pittsburg, and who was easily the largest refiner there in 1874 when Mr. Rockefeller bought him up. In Philadelphia, the largest refiner in 1874 was W. G. Warden of the Atlantic Refining Company, and it was he whom Mr. Rockefeller wanted. In New York it was the concern of Charles Pratt and Company, one of the three largest concerns around Manhattan—the concern to which H. H. Rogers belonged. Charles Pratt had been in the oil and paint business since 1850, and he had become a refiner of petroleum at Greenpoint, Long Island, in 1867. Before Standard Oil was known outside of New York the fame of Pratt's Astral Oil had gone around the world. Mr. Pratt's concern was rated at the same daily capacity as Mr. Rockefeller's (1,500 barrels) in the spring of 1872, when the latter wiped up the Cleveland refineries and grew in a night to 10,000 barrels. Mr. Vandergrift, who united his interests with Mr. Rockefeller's in 1874 and 1875, had been a far better known man in the oil business and controlled much greater and more varied interests up to South Improvement times. When he went into the Standard he controlled the largest refinery on Oil Creek, the Imperial, of about 1,400 barrels. He was president of a large system of pipe-lines, and he was a member of one of the largest oil-producing concerns of the time—the H. L. Taylor Company.

There is no doubt but that Mr. Rockefeller had plenty of brains in his great trust. It was those who had done business with him who were the first to point this out when critics declared that the concern could not—or must not—live. "There is no question about it," W. H. Vanderbilt told the Hepburn Commission in 1879, "but these men are smarter than I am a great deal. They are very enterprising and smart men. I never came in contact with any class of men as smart and able as they are in their business. They would never have got into the position they now are without a great deal of ability —and one man would hardly have been able to do it; it is a combination of men."

It was not only that first-rate ability was demanded at the top; it was required throughout the organisation. The very day-labourers were picked men. It was the custom to offer a little better day wages for labourers than was current and then to choose from these the most promising specimens; those men were advanced as they showed ability. To-day the very errand boys at 26 Broadway are chosen for the promise of development they show, and if they do not develop they are discharged. No dead wood is taken into the concern unless it is through the supposed necessities of family or business relations, as probably occurs to a degree in every human organisation.

The efficiency of the working force of the Standard was greatly increased when the trust was formed by the opportunity given to the employees of taking stock. They were urged to do it, and where they had no savings money was lent them on easy terms by the company. The result is that a great number of the employees of the Standard Oil Company are owners of stock which they bought at eighty, and on which for several years they have received from thirty to forty-eight per cent. dividends. It is only natural that under such circumstances the company has always a remarkably loyal and interested working force.

Mr. Rockefeller's great creation has really been strong, then, in many admirable qualities. The force of the combination has been greater because of the business habits of the independent body which has opposed it. To the Standard's caution the Oil Regions opposed recklessness; to its economy, extravagance; to its secretiveness, almost blatant frankness; to its far-sightedness, little thought of the morrow; to its close-fistedness, a spendthrift generosity; to its selfish unscrupulousness, an almost quixotic love of fair play. The Oil Regions had, besides, one fatal weakness—its passion for speculation. Now, Mr. Rockefeller never speculates. He deals only in those things which other people have proved sure!

It is when one examines the inside of the Standard Oil Trust that one sees how much reason there is for the opinion of those people who declare that Mr. Rockefeller can always sustain the monopoly of the oil business he has achieved. One begins to see what Mr. Vanderbilt meant in 1879 when he said: "I don't believe that by any legislative enactment or anything else, through any of the states or all of the states, you can keep such men down. You can't do it! They will be on top all the time, you see if they are not." * It is not surprising that those who realise the compactness and harmony of the Standard organisation, the ability of its members, the solidity of the qualities governing its operations, are willing to forget its history. Such is the blinding quality of success! "It has achieved this," they say; "no matter what helped to rear this structure, it is here, it is admirably managed. We might as well accept it. We must do business." They are weary of contention, too—who so unwelcome as an agitator?—and they began to accept the Standard's explanation that the critics are indeed "people with a private grievance," "moss-backs left behind in the march of progress." Again and again in the history of the oil business it has looked to the outsider as if henceforth Mr. Rockefeller would have to have things his own way, for who was there to interfere with him, to dispute his position? No one, save that back in Northwestern Pennsylvania, in scrubby little oil towns, around greasy der-

* See Appendix, Number 59. W. H. Vanderbilt's characterisation of Standard Oil men.

ricks, in dingy shanties, by rusty, deserted oil stills, men have always talked of the iniquity of the railroad rebate, the injustice of restraint of trade, the dangers of monopoly, the right to do an independent business; have always rehearsed with tiresome persistency the evidence by which it has been proved that the Standard Oil Company is a revival of the South Improvement Company. It has all seemed futile enough with the public listening in wonder and awe to the splendid rehearsal of figures, and the unctuous logic of the Mother of Trusts, and yet one can never tell. It was the squawking of geese that saved the Capitol.

Certain it is that many and great as are his business qualities, John D. Rockefeller has never been allowed to enjoy the fruits of his victory in that atmosphere of leisure and adulation which the victor naturally craves. Certain it is that the incessant agitation of men with a "private grievance" has ruined some of his fairest schemes, has hauled him again and again before investigating committees, and has contributed greatly to securing a federal law authorising so fundamental and obvious a right as equal rates on common carriers. Certain it is that the incessant efforts of those who believed they had a right to do an independent business have resulted in the most important advances made in the oil business since the beginning of Mr. Rockefeller's combination, namely, the seaboard pipe-line, for transporting crude oil, due to the Tidewater Pipe Line, and later the use of the seaboard pipe-line for transporting refined oil, due to the United States Pipe Line. Certain it is, too, that all of competition which we have, with its consequent lowering of prices, is due to independent efforts.

CHAPTER EIGHTEEN

CONCLUSION

CONTEMPT PROCEEDINGS BEGUN AGAINST THE STANDARD IN OHIO IN 1897 FOR NOT OBEYING THE COURT'S ORDER OF 1892 TO DISSOLVE THE TRUST —SUITS BEGUN TO OUST FOUR OF THE STANDARD'S CONSTITUENT COMPANIES FOR VIOLATION OF OHIO ANTI-TRUST LAWS—ALL SUITS DROPPED BECAUSE OF EXPIRATION OF ATTORNEY-GENERAL MONNETT'S TERM—STANDARD PERSUADED THAT ITS ONLY CORPORATE REFUGE IS NEW JERSEY—CAPITAL OF THE STANDARD OIL COMPANY OF NEW JERSEY INCREASED, AND ALL STANDARD OIL BUSINESS TAKEN INTO NEW ORGANISATION—RESTRICTION OF NEW JERSEY LAW SMALL—PROFITS ARE GREAT AND STANDARD'S CONTROL OF OIL BUSINESS IS ALMOST ABSOLUTE—STANDARD OIL COMPANY ESSENTIALLY A REALISATION OF THE SOUTH IMPROVEMENT COMPANY'S PLANS—THE CRUCIAL QUESTION NOW, AS ALWAYS, IS A TRANSPORTATION QUESTION—THE TRUST QUESTION WILL GO UNSOLVED SO LONG AS THE TRANSPORTATION QUESTION GOES UNSOLVED—THE ETHICAL QUESTIONS INVOLVED.

FEW men in either the political or industrial life of this country can point to an achievement carried out in more exact accord with its first conception than John D. Rockefeller, for both in purpose and methods the Standard Oil Company is and always has been a form of the South Improvement Company, by which Mr. Rockefeller first attracted general attention in the oil industry. The original scheme has suffered many modifications. Its most offensive feature, the drawback on other people's shipments, has been cut off. Nevertheless, to-day, as at the start, the purpose of the Standard Oil Company is the purpose of the South Improvement Company—the regulation of the price of crude and refined oil by the control of the output; and the chief

JOHN D. ROCKEFELLER

From a photograph by Allen Ayrault Green, taken about 1892.

means for sustaining this purpose is still that of the original scheme—a control of oil transportation giving special privileges in rates.

It is now thirty-two years since Mr. Rockefeller applied the fruitful idea of the South Improvement Company to the Standard Oil Company of Ohio, a prosperous oil refinery of Cleveland, with a capital of $1,000,000 and a daily capacity for handling 1,500 barrels of crude oil. And what have we as a result? What is the Standard Oil Company to-day? First, what is its organisation? It is no longer a trust. As we have seen, the trust was obliged to liquidate in 1892. It became a "trust in liquidation," and there it remained for some five years. It seemed to have come into a state of stationary liquidation, for at the end of 1892 477,881 shares were uncancelled; at the end of 1896 the same number were out. The situation of the great corporation was indeed curious. There began to be comments on it, for complications arose—one over taxes. In 1893 an auditor in Ohio tried to collect taxes on 225 shares of the Standard Oil Trust. The owner refused to pay and took the case into court. He won it. The Standard Oil Trust is an unlawful organisation, said the court. Its certificates have no validity. It would seem strange that a certificate which was void to all purpose would still be valid as to taxable purposes.* Here was an anomaly indeed. The certificates were drawing big quarterly dividends, had a big market value, but were illegal. Owners of small certificates naturally refused to exchange. In 1897 it took 194½ shares in the Standard Oil Trust to bring back one share in each of the twenty companies. Thus one share in the Standard Oil Company of Ohio was worth twenty-seven shares in the Standard Oil Trust. If a man owned twenty-five shares he got only fractional parts of a share in each company. On these fractional parts he received no dividends, it not being considered practical

* Ohio Circuit Court Reports, Volume VII, 1893, page 508.

to consider such small sums. To raise his twenty-five shares to 194, and so secure dividends, took a good sum of money, since Standard Oil Trust shares were worth at least 340 then. But why should he trouble? He received his quarterly dividends promptly, and they were large! He paid no taxes, for his stock was illegal! The trustees were not pushing him to liquidate. Besides, it was doubtful if they could do anything. Joseph Choate said they could not. On May 3, 1894, before the attorney-general of New York, in an application for the forfeiture of the charter of the Standard Oil Company of New York, Mr. Choate said:

"I happen to own 100 shares in the Standard Oil Trust, and I have never gone forward and claimed my aliquot share. Why not? Because I would get ten in one company, and ten in another company, and two and three-fifths in another company.

"There is no power that this company can exercise to compel me and other indifferent certificate holders, if you please, to come forward and convert our trust certificates."

If there was a way, the trustees were indifferent to it. They evidently were contented to let things alone. It is quite possible that they would have been holding to-day 477,881 uncancelled shares of Standard Oil Trust if it had not been for the irrepressible George Rice. Since October, 1892, Mr. Rice had held a Standard Oil Trust certificate for six shares. He had never cancelled it. He had received no invitation to do so. He received his dividends regularly on it. Later, he purchased one share, called "assignment of legal title"—the new form given the trust certificate—and on this he received dividends, exactly as on the original trust certificate. Finally Mr. Rice made up his mind, without knowing any of the facts of the liquidation outlined above, that there was no intention to carry out the dissolution, that some means of evasion had been devised, and he proposed to find out what it was.

To do this he transferred his assignment of legal title to an agent with the order to liquidate it. A long correspondence followed between Mr. Kemper, Mr. Rice's agent, and Mr. Dodd, who objected to making the transfer on the ground that it cut the share into a "multitude of almost infinitesimal fractions of corporate shares." They were obviating this difficulty, Mr. Dodd said, by purchasing certificates calling for one or a few shares and uniting them until sufficient were had by one party to call for the issue of full corporate shares. Mr. Kemper insisted, however, and finally received scrip for his share. "Infinitesimal" it was, indeed, $\frac{5,000}{972,500}$ of one share in one company, $\frac{10,000}{972,500}$ of one share in another, and so on through nineteen constituent companies.*

Arguing from these experiences and what else he could gather, Mr. Rice decided that the trust was not dissolved and had no intention of doing so. Furthermore, he argued that the scheme was one to entice the small shareholders to sell their shares and thus enable the trustees to increase their holdings! And he sought legal counsel in Ohio as to the possibility of bringing suit against the Standard Oil Company of Ohio for failing to obey the court's orders in March, 1892. The attorneys, one of whom was Mr. Watson, advised Mr. Rice to lay his facts before the attorney-general of the state, Frank S. Monnett. Like Mr. Watson, when he brought his suit, Mr. Monnett was young and held firmly to the belief that the business of an attorney-general is to enforce the laws. The facts Mr. Rice and his counsel laid before him seemed to him to indicate that the Standard Oil Company of Ohio had taken advantage of the leniency of the court in allowing it time to disentangle itself from the trust, and had devised a skilful plan to evade the judgment pronounced against it five years before. He asked Mr. Rice and his attorneys to go with him and lay the case before the judges of the Supreme Court

* See Appendix, Number 60. Facsimile of one of Mr. Kemper's shares.

in chambers, and ask if it did not justify proceedings against the company. The judges agreed with the attorney-general and ordered him to bring the company before the court for contempt. Information was filed in November, 1897. The suit which followed proved one of the most sensational ever instituted against the Standard Oil Combination.

The first substantial point gained by the attorney-general in the proceedings was securing answers to a long series of questions concerning the history of the operations of the Standard Oil Company of Ohio, both within and without the trust. These answers were made by the president of that company, who was at the same time the president of the trust, John D. Rockefeller. They furnish a mass of facts of value and interest, and they include the minutes of the meeting at which the trust was dissolved on March 11, 1892, as well as the minutes of all the quarterly meetings the liquidating trustees held from 1892 to October, 1897. It was from the information obtained from this set of questions that Mr. Monnett secured proof that the liquidation scheme had been held up, as Mr. Rice claimed. The minutes showed, as related in Chapter XIV, that from November, 1892, to March, 1896, 477,881 shares were reported every three months to the trustees as uncancelled. In July, 1896, the number fell suddenly to 477,880. George Rice had succeeded in having his assignment of legal title liquidated! Mr. Monnett learned from the result of this inquiry another suggestive fact, that while only one share was cancelled in the five years *before* the contempt proceedings were brought, in the first three months *after,* 100,583 shares were cancelled! *

It took Mr. Monnett some six months to secure the answers from Mr. Rockefeller, but his information was still incomplete, and he asked the court to appoint a master commis-

* History of Standard Oil Case in Supreme Court of Ohio, 1897–1898. Part II, page 39.

sioner, with power to examine the officers, affairs and books of the Standard, to take testimony within or without the state, and to report. This was done, the commissioner holding his first court at the New Amsterdam Hotel, in New York, on October 11 and 12, 1898. Mr. Rockefeller was the only witness examined at the sessions, and his deliberation and self-control, his almost detached attitude as a witness, were the subject of remark by more than one observer. He answered no question promptly. He had the air of reflecting always before he spoke. He consulted frequently with his counsel. His counsel, his colleagues who were present, the counsel of the prosecution, were sometimes irate, never Mr. Rockefeller. From beginning to end he was the soul of self-possession. His only sign of impatience—if it was impatience—was an incessant slight tapping of the arm of his chair with his white fingers.

The outcome of this examination of Mr. Rockefeller was that Mr. Monnett and his colleagues called for those books of the trust which would show exactly how the original trust certificates had been liquidated. It was then that the copies of the transfers of Mr. Rockefeller's trust certificates and of his assignments of legal title printed in the Appendix, Number 54, were obtained. Although Mr. Monnett had added to his knowledge of the Standard's operations between 1892 and 1898, he was not yet convinced that the Standard Oil Company of Ohio was conducting its own business. He had found that, in spite of the order of the court in 1892, 13,593 shares of that company's stock were still outstanding in trust certificates. He knew these certificates drew dividends. Was the company paying money directly or indirectly to the liquidating trustees? They said no, that they had been paying no dividends since 1892, that the money paid the holders of trust certificates came from the other nineteen companies, that all their earnings had been used in improving their plant, or were invested in government bonds. Besides, said they, we are not the thrifty concern we used to be. Mr. Monnett demanded proof from their books. The secretary of the company, on advice of his counsel, Virgil P. Kline, refused to produce the books asked for, on the ground that they would incriminate the company. The court supported Mr. Monnett, and ordered the company to produce those of their records showing the gross earnings since 1892, and what had been done with them. The order met with a second refusal.

Such was the status of the proceedings when Mr. Monnett received an anonymous communication stating that, about the time the company was ordered by the court to produce its records, a great quantity of books had been taken from the Standard's office in Cleveland and burned. An investigation was at once made by the attorney-general, and a number of witnesses examined. The fact of the burning of sixteen boxes of books from the Standard offices in Cleveland was established, but these books, the officers of the company contended, were not the ones wanted by Mr. Monnett. "Then produce the ones we want," ordered the court. But, on the ground that such records might incriminate them, the officers still refused.

The fact was, the Standard Oil Company of Ohio was in a very tight place, and it is difficult to see how an examination of their books could have failed to incriminate not only it, but three other of the constituent companies of the trust which held charters from the same state. These three companies were the Ohio Oil Company, which produced oil; the Buckeye Pipe Line, which transported it; and the Solar Refining Company, which refined it. Mr. Monnett had learned enough about these organisations in the course of his investigations since November, 1897, to convince him that these companies—all of them enormously profitable—were, for all practical purposes, one and the same combination, and that they were all working with the Standard Oil Company of Ohio, and that their operations were in direct violation of a state anti-trust law recently passed. As soon as he had sufficient evidence he had filed petitions against all four of them. Now, these petitions were filed about the time he demanded the books showing the earnings of the Standard Oil Company of Ohio, for use in his contempt case. It was the old story of one suit being used as a shield in another. A witness cannot be made to incriminate himself.

The reasons F. B. Squire, the secretary of the Standard Oil Company of Ohio, gave for refusing to produce the books as ordered by the court were as follows:

1st. Because they are demanded in an action instituted against the Standard Oil Company for contempt of court, and for the purpose of proving said company guilty of contempt in order that the penalties for contempt may be inflicted upon it and its officers; and I am informed that, to enforce their production in such a case and for such a purpose, is an unreasonable search and seizure.

2nd. Because the books disclose facts and circumstances which may be used against the Standard Oil Company, tending to prove it guilty of offences made criminal by an act of the Legislature of Ohio, passed April 19, 1898, entitled "An Act to define trusts and to provide for criminal penalties, civil damages, and the punishment of corporations," etc.

3rd. Because they disclose facts and circumstances which may be used against myself personally as an officer of said company, tending to prove me guilty of offences made criminal by the act aforesaid.*

All through the winter of 1898 and 1899, up to the end of March, when the commission declared the taking of testimony closed, the wrangle over the production of the books went on. Depositions had begun to be taken at the same time in the cases against the constituent companies for violation of the anti-trust laws, and by the time the contempt case was closed in March, 1899, the exasperation of both sides had reached fever pitch. Nor did the judgment of the court quiet it, for three judges voted for finding the company guilty of contempt, and three for clearing it.

Unsatisfactory as this was, Mr. Monnett still had his anti-

* History of Standard Oil Case in Supreme Court of Ohio, 1897–1898. Part II, page 248.

trust suits, through which he expected and through which he did secure much further evidence that the four Standard companies in Ohio were practically one concern so shrewdly and secretly handled that they were evading not only the laws of the state, but that policy of all states which decrees that it is unsafe to allow men to work together in industrial combinations without charters defining their privileges, and subjecting them to reasonable examinations and publicity. Mr. Monnett's work on these suits came to an end with the expiration of his term in January, 1900, and the suits were suppressed by his successor, John M. Sheets! Unfinished as they were, they were of the greatest value in dragging into the light information concerning the methods and operations of the Standard Oil Combination to which the public has the right, and which it must digest if it is to succeed in working out a legal harness for combinations which, like the Standard, demand freedom to do what they like and do it secretly.

The only refuge offered in the United States for the Standard Oil Trust in 1898, when the possibility arose by these suits of the state of Ohio taking away the charters of four of its important constituent companies for contempt of court and violation of the anti-trust laws of the state, lay in the corporation law of the state of New Jersey, which had just been amended, and here it settled. Among the twenty companies which formed the trust was the Standard Oil Company of New Jersey, a corporation for manufacturing and marketing petroleum products. Its capital was $10,000,000. In June, 1899, this capital of $10,000,000 was increased to one of $110,000,000, and into this new organisation was dumped the entire Standard aggregation. The old trust certificates outstanding and the assignments of legal title which had succeeded them were called in, and for them were given common stock of the new Standard Oil Company. The amount of this stock which had been issued, in January, 1904, when the last report was made, was $97,448,800. Its market value at that date was $643,162,080. How it is divided is of course a matter of private concern. The number of stockholders in 1899 was about 3,500, according to Mr. Archbold's testimony to the Interstate Commerce Commission, but over one-half of the stock was owned by the directors, and probably nearly one-third was owned by Mr. Rockefeller himself.

The companies which this new Standard Oil Company has bought up with its stock are numerous and scattered. They consist of oil-producing companies like the South Penn Oil Company, the Ohio Oil Company, and the Forest Oil Company; of transporting companies like the National Transit Company, the Buckeye Pipe Line Company, the Indiana Pipe Line Company, and the Eureka Pipe Line Company; of manufacturing and marketing companies like the Atlantic Refining Company of Pennsylvania, and the Standard Oil Companies of many states—New York, Indiana, Kentucky, Ohio, Iowa; of foreign marketing concerns like the Anglo-American Company. In 1892 there were twenty of these constituent companies. There have been many added since, in whole or part, like gas companies; new producing concerns, made necessary by developments in California, Kansas and Texas; new marketing concerns for handling oil directly in Germany, Italy, Scandinavia and Portugal. What the total value of the companies owned by the present Standard Oil Company is it is impossible to say. In 1892, when the trust was on trial in Ohio, it reported the aggregate capital of its twenty companies as $102,233,700, and the appraised value was given as $121,631,312.63; that is, there was an excess of about $19,000,000.

In 1898, when Attorney-General Monnett of Ohio had the Standard Oil Company of the state on trial for contempt of court, he tried to find out from Mr. Rockefeller what the surplus of each of the various companies in the trust was at that date. Mr. Rockefeller answered: "I have not in my possession or power data showing . . . the amount of such surplus money in their hands after the payment of the last dividends." Then Mr. Rockefeller proceeded to repeat as the last he knew of the value of the holdings of the trust the list of values given six years before.* This list has continued to be cited ever since as authoritative. There is a later one, whether Mr. Rockefeller had it in his "possession or power," or not, in 1898. It is the last trustworthy valuation of which the writer knows, and is found in testimony taken in 1899, in a private suit to which Mr. Rockefeller was party. It is for the year 1896. This shows the "total capital and surplus" of the twenty companies to have been, on December 31 of that year, something over one hundred and forty-seven million dollars, nearly forty-nine millions of which was scheduled as "undivided profits." † Of course there has been a constant increase in value since 1896.

The new Standard Oil Company is managed by a board of fourteen directors.‡ They probably collect the dividends of the constituent companies and divide them among stockholders in exactly the same way the trustees of 1882 and the liquidating trustees of 1892 did. As for the charter under which they are operating, never since the days of the South Improvement Company has Mr. Rockefeller held privileges so in harmony with his ambition. By it he can do all kinds of mining, manufacturing, and trading business; transport goods and merchandise by land and water in any manner; buy, sell, lease, and improve lands; build houses, structures,

* See Appendix, Number 53.

† See Appendix, Number 61. General balance sheet, Standard Oil interests, December 31, 1896.

‡ The present directors are John D. Rockefeller, William Rockefeller, Henry M. Flagler, John D. Archbold, Henry H. Rogers, W. H. Tilford, Frank Q. Barstow, Charles M. Pratt, E. T. Bedford, Walter Jennings, James A. Moffett, C. W. Harkness, John D. Rockefeller, Jr., Oliver H. Payne.

vessels, cars, wharves, docks, and piers; lay and operate pipe-lines; erect and operate telegraph and telephone lines, and lines for conducting electricity; enter into and carry out contracts of every kind pertaining to his business; acquire, use, sell, and grant licenses under patent rights; purchase, or otherwise acquire, hold, sell, assign, and transfer shares of capital stock and bonds or other evidences of indebtedness of corporations, and exercise all the privileges of ownership, including voting upon the stocks so held; carry on its business and have offices and agencies therefor in all parts of the world, and hold, purchase, mortgage, and convey real estate and personal property outside the state of New Jersey. These privileges are, of course, subject to the laws of the state or country in which the company operates. If it is contrary to the laws of a state for a foreign corporation to hold real estate in its boundaries, a company must be chartered in the state. Its stock, of course, is sold to the New Jersey corporation, so that it amounts to the same thing as far as the ability to do business is concerned. It will be seen that this really amounts to a special charter allowing the holder not only to do all that is specified, but to create whatever other power it desires, except banking.* A comparison of this summary of powers with those granted by the South Improvement Company shows that in sweep of charter, at least, the Standard Oil Company of to-day has as great power as its famous progenitor.†

The profits of the present Standard Oil Company are enormous. For five years the dividends have been averaging about forty-five million dollars a year, or nearly fifty per cent. on its capitalisation, a sum which capitalised at five per cent. would give $900,000,000. Of course this is not all that the combination makes in a year. It allows an annual average of

* See Appendix, Number 62. Amended certificate of incorporation of the Standard Oil Company of New Jersey.

† See Appendix, Number 9.

5.77 per cent. for deficit, and it carries always an ample reserve fund. When we remember that probably one-third of this immense annual revenue goes into the hands of John D. Rockefeller, that probably ninety per cent. of it goes to the few men who make up the "Standard Oil family," and that it must every year be invested, the Standard Oil Company becomes a much more serious public matter than it was in 1872, when it stamped itself as willing to enter into a conspiracy to raid the oil business—as a much more serious concern than in the years when it openly made warfare of business, and drove from the oil industry by any means it could invent all who had the hardihood to enter it. For, consider what must be done with the greater part of this $45,000,000. It must be invested. The oil business does not demand it. There is plenty of reserve for all of its ventures. It must go into other industries. Naturally, the interests sought will be allied to oil. They will be gas, and we have the Standard Oil crowd steadily acquiring the gas interests of the country. They will be railroads, for on transportation all industries depend, and, besides, railroads are one of the great consumers of oil products and must be kept in line as buyers. And we have the directors of the Standard Oil Company acting as directors on nearly all of the great railways of the country, the New York Central, New York, New Haven and Hartford, Chicago, Milwaukee and St. Paul, Union Pacific, Northern Pacific, Delaware, Lackawanna and Western, Missouri Pacific, Missouri, Kansas and Texas, Boston and Maine, and other lesser roads. They will go into copper, and we have the Amalgamated scheme. They will go into steel, and we have Mr. Rockefeller's enormous holdings in the Steel Trust. They will go into banking, and we have the National City Bank and its allied institutions in New York City and Boston, as well as a long chain running over the country. No one who has followed this history can expect these holdings will be acquired on a rising market. Buy cheap and sell high is a rule of business, and when you control enough money and enough banks you can always manage that a stock you want shall be temporarily cheap. No value is destroyed for you—only for the original owner. This has been one of Mr. Rockefeller's most successful manœuvres in doing business from the day he scared his twenty Cleveland competitors until they sold to him at half price. You can also sell high, if you have a reputation of a great financier, and control of money and banks. Amalgamated Copper is an excellent example. The names of certain Standard Oil officials would float the most worthless property on earth a few years ago. It might be a little difficult for them to do so to-day with Amalgamated so fresh in mind. Indeed, Amalgamated seems to-day to be the worst "break," as it certainly was one of the most outrageous performances of the Standard Oil crowd. But that will soon be forgotten! The result is that the Standard Oil Company is probably in the strongest financial position of any aggregation in the world. And every year its position grows stronger, for every year there is pouring in another $45,000,000 to be used in wiping up the property most essential to preserving and broadening its power.

And now what does the law of New Jersey require the concern which it has chartered, and which is so rapidly adding to its control of oil the control of iron, steel, copper, banks, and railroads, to make known of itself? It must each year report its name, the location of its registration office, with name of agent, the character of its business, the amount of capital stock issued, and the names and addresses of its officers and directors!

So much for present organisation, and now as to how far through this organisation the Standard Oil Company is able to realise the purpose for which it was organised—the control of the output, and, through that, the price, of refined oil. That is, what per cent. of the whole oil business does Mr.

Rockefeller's concern control. First as to oil production. In 1898 the Standard Oil Company reported to the Industrial Commission that it produced 35.58 per cent. of Eastern crude —the production that year was about 52,000,000 barrels.* (It should be remembered that it is always to the Eastern oil fields—Pennsylvania, Ohio, Indiana, West Virginia—that this narrative refers. Texas, Kansas, Colorado and California are newer developments. These fields have not as yet been determining factors in the business, though Texas particularly has been a distributing factor.) But while Mr. Rockefeller produces only about a third of the entire production, he controls all but about ten per cent. of it; that is, all but about ten per cent. goes immediately into his custody on coming from the wells. It passes entirely out of the hands of the producers when the Standard pipe-line takes it. The oil is in Mr. Rockefeller's hands, and he, not the producer, can decide who is to have it. The greater portion of it he takes himself, of course, for he is the chief refiner of the country. In 1898 there were about twenty-four million barrels of petroleum products made in this country.† Of this amount about twenty million were made by the Standard Oil Company; fully a third of the balance was produced by the Tidewater Company, of which the Standard holds a large minority stock, and which for twenty years has had a running arrangement with the Standard. Reckoning out the Tidewater's probable output, and we have an independent output of about 2,500,000 in twenty-four million. It is obvious that this great percentage of the business gives the Standard the control of prices. This control can be kept in the domestic markets so long as the Standard can keep under competition as successfully as it has in the past. It can be kept

* See Appendix, Number 63. Production of Pennsylvania and Lima crude oil by Standard Oil Company, 1890–1898.

† See Appendix, Number 64. Business of Standard Oil Company and other refiners, 1894–1898.

in the foreign market as long as American oils can be made and sold in quantity cheaper than foreign oils. Until a decade ago the foreign market of American oils was not seriously threatened. Since 1895, however, Russia, whose annual output of petroleum had been for a number of years about equal in volume to the American output, learned to make a fairly decent product; more dangerous, she had learned to market. She first appeared in Europe in 1885. It took ten years to make her a formidable rival, but she is so to-day, and, in spite of temporary alliances and combinations, it is very doubtful whether the Standard will ever permanently control Russian oil.

In 1899 Mr. Archbold presented to the Industrial Commission a most interesting list of foreign corporations and individuals doing an oil business in various countries. According to this there were more than a score of large concerns in Russia, and many small ones. The aggregate capitalisation shown by Mr. Archbold's list was over forty-six and a half millions, and the capitalisation of a number of the concerns named was not given. In Galicia, four companies, with an aggregate capital of $3,775,100, and in Roumania six large companies, with an aggregate capital of $12,500,000, were reported. Borneo was shown to have nearly three millions invested in the oil fields; Sumatra and Java each over twelve millions. Since this report was made these companies have grown, particularly in marketing ability. In the East the oil market belonged practically to the Standard Oil Company until recently. Last year (1903), however, Sumatra imported more oil into China than America, and Russia imported nearly half as much.* About 91,500,000 gallons of kerosene went

* America imported into China, 1893...... 31,060,527 gallons
Borneo " " " " ...... 574,615 "
Russia " " " " ...... 13,503,685 "
Sumatra " " " " ...... 39,859,508 "

into Calcutta last year, and of this only about six million gallons came from America. In Singapore representatives of Sumatra oil claim that they have two-thirds of the trade.

Combinations for offensive and defensive trade campaigns have also gone on energetically among these various companies in the last few years. One of the largest and most powerful of these aggregations now at work is in connection with an English shipping concern, the Shell Transport and Trading Company, the head of which is Sir Marcus Samuel, formerly Lord Mayor of London. This company, which formerly traded almost entirely in Russian oil, undertook a few years ago to develop the oil fields in Borneo, and they built up a large Oriental trade. They soon came into hot competition with the Royal Dutch Company, handling Sumatra oil, and a war of prices ensued which lasted nearly two years. In 1903, however, the two competitors, in connection with four other strong Sumatra and European companies, drew up an agreement in regard to markets which has put an end to their war. The "Shell" people have not only these allies, but they have a contract with the Guffey Petroleum Company, the largest Texas producing concern, to handle its output, and they have gone into a German oil company, the Petroleum Produkten Aktien Gesellschaft. Having thus provided themselves with a supply they have begun developing a European trade on the same lines as their Oriental trade, and they are making serious inroads on the Standard's market.

The naphthas made from the Borneo oil have largely taken the place of American naphtha in many parts of Europe. One load of Borneo benzine even made its appearance in the American market in 1904. It is a sign of what well may happen in the future with an intelligent development of these Russian and Oriental oils—the Standard's domestic market invaded. It will be interesting to see to what further extent the American government will protect the Standard

Oil Company by tariff on foreign oils if such a time does come. It has done very well already. The aggressive marketing of the "Shell" and its allies in Europe has led to a recent Oil War of great magnitude. For several months in 1904 American export oil was sold at a lower price in New York than the crude oil it takes to make it costs there. For instance, on August 13, 1904, the New York export price was 4.80 cents per gallon for Standard-white in bulk. Crude sold at the well for $1.50 a barrel of forty-two gallons, and it costs sixty cents to get it to seaboard by pipe-line; that is, forty-two gallons of crude oil costs $2.10, or five cents a gallon in New York—twenty points loss on a gallon of the raw material! But this low price for export affects the local market little or none. The tank-wagon price keeps up to ten and eleven cents in New York. Of course crude is depressed as much as possible to help carry this competition. For many months now there has been the abnormal situation of a declining crude price in face of declining stocks. The truth is the Standard Oil Company is trying to meet the competition of the low-grade Oriental and Russian oils with high-grade American oil—the crude being kept as low as possible, and the domestic market being made to pay for the foreign cutting. It seems a lack of foresight surprising in the Standard to have allowed itself to be found in such a dilemma. Certainly, for over two years the company has been making every effort to escape by getting hold of a supply of low-grade oil which would enable it to meet the competition of the foreigner. There have been more or less short-lived arrangements in Russia. An oil territory in Galicia was secured not long ago by them, and an expert refiner with a full refining plant was sent over. Various hindrances have been met in the undertaking, and the works are not yet in operation. Two years ago the Standard attempted to get hold of the rich Burma oil fields. The press of India fought them out of the country, and their weapon was the

Standard Oil Company's own record for hard dealings! The Burma fields are in the hands of a monopoly of the closest sort which has never properly developed the territory, but the people and government prefer their own monopoly to one of the American type!

Altogether the most important question concerning the Standard Oil Company to-day is how far it is sustaining its power by the employment of the peculiar methods of the South Improvement Company. It should never be forgotten that Mr. Rockefeller never depended on these methods alone for securing power in the oil trade. From the beginning the Standard Oil Company has studied thoroughly everything connected with the oil business. It has known, not guessed at conditions. It has had a keen authoritative sight. It has applied itself to its tasks with indefatigable zeal. It has been as courageous as it has been cautious. Nothing has been too big to undertake, as nothing has been too small to neglect. These facts have been repeatedly pointed out in this narrative. But these are the American industrial qualities. They are common enough in all sorts of business. They have made our railroads, built up our great department stores, opened our mines. The Standard Oil Company has no monopoly in business ability. It is the thing for which American men are distinguished to-day in the world.

These qualities alone would have made a great business, and unquestionably it would have been along the line of combination, for when Mr. Rockefeller undertook to work out the good of the oil business the tendency to combination was marked throughout the industry, but it would not have been the combination whose history we have traced. To the help of these qualities Mr. Rockefeller proposed to bring the peculiar aids of the South Improvement Company. He secured an alliance with the railroads to drive out rivals. For fifteen years he received rebates of varying amounts on at least the greater part of his shipments, and for at least a portion of that time he collected drawbacks of the oil other people shipped; at the same time he worked with the railroads to prevent other people getting oil to manufacture, or if they got it he worked with the railroads to prevent the shipment of the product. If it reached a dealer, he did his utmost to bully or wheedle him to countermand his order. If he failed in that, he undersold until the dealer, losing on his purchase, was glad enough to buy thereafter of Mr. Rockefeller. How much of this system remains in force to-day? The spying on independent shipments, the effort to have orders countermanded, the predatory competition prevailing, are well enough known. Contemporaneous documents, showing how these practices have been worked into a very perfect and practically universal system, have already been printed in this work.* As for the rebates and drawbacks, if they do not exist in the forms practised up to 1887, as the Standard officials have repeatedly declared, it is not saying that the Standard enjoys no special transportation privileges. As has been pointed out, it controls the great pipe-line handling all but perhaps ten per cent. of the oil produced in the Eastern fields. This system is fully 35,000 miles long. It goes to the wells of every producer, gathers his oil into its storage tanks, and from there transports it to Philadelphia, Baltimore, New York, Chicago, Buffalo, Cleveland, or any other refining point where it is needed. This pipe-line is a common carrier by virtue of its use of the right of eminent domain, and, as a common carrier, is theoretically obliged to carry and deliver the oil of all comers, but in practice this does not always work. It has happened more than once in the history of the Standard pipes that they have refused to gather or deliver oil. Pipes have been taken up from wells belonging to individuals running or working with independent refiners. Oil has been refused delivery at points practical for inde-

* See Chapter X

pendent refiners. For many years the supply of oil has been so great that the Standard could not refuse oil to the independent refiner on the ground of scarcity. However, a shortage in Pennsylvania oil occurred in 1903. A very interesting situation arose as a result. There are in Ohio and Pennsylvania several independent refiners who, for a number of years, have depended on the Standard lines (the National Transit Company) for their supply of crude. In the fall of 1903 these refiners were informed that thereafter the Standard could furnish them with only fifty per cent. of their refining capacity. It was a serious matter to the independents, who had their own markets, and some of whom were increasing their plants. Supposing we buy oil directly from the producers, they asked one another, must not the Standard as a common carrier gather and deliver it? The experienced in the business said: "Yes. But what will happen? The producer rash enough to sell you oil may be cut off by the National Transit Company. Of course, if he wants to fight in the courts he may eventually force the Standard to reconnect, but they could delay the suit until he was ruined. Also, if you go over Mr. Seep's head"—Mr. Seep is the Standard Oil buyer, and all oil going into the National Transit system goes through his hands—"you will antagonise him." Now, "antagonise" in Standard circles may mean a variety of things. The independent refiners decided to compromise, and an agreement terminable by either party at short notice was made between them and the Standard, by which the members of the former were each to have eighty per cent. of their capacity of crude oil, and were to give to the Standard all of their export oil to market. As a matter of fact, the Standard's ability to cut off crude supplies from the outside refiners is much greater than in the days before the Interstate Commerce Bill, when it depended on its alliance with the railroads to prevent its rival getting oil. It goes without saying that this is an absurd power to allow in the hands of any manufacturer of a great necessity of life. It is exactly as if one corporation aiming at manufacturing all the flour of the country owned all but ten per cent. of the entire railroad system collecting and transporting wheat. They could, of course, in time of shortage, prevent any would-be competitor from getting grain to grind, and they could and would make it difficult and expensive at all times for him to get it.

It is not only in the power of the Standard to cut off outsiders from it, it is able to keep up transportation prices. Mr. Rockefeller owns the pipe system—a common carrier—and the refineries of the Standard Oil Company pay in the final accounting cost for transporting their oil, while outsiders pay just what they paid twenty-five years ago. There are lawyers who believe that if this condition were tested in the courts, the National Transit Company would be obliged to give the same rates to others as the Standard refineries ultimately pay. It would be interesting to see the attempt made.

Not only are outside refiners at just as great disadvantage in securing crude supply to-day as before the Interstate Commerce Commission was formed; they still suffer severe discrimination on the railroads in marketing their product. There are many ways of doing things. What but discrimination is the situation which exists in the comparative rates for oil freight between Chicago and New Orleans, and Cleveland and New Orleans? All, or nearly all, of the refined oil sold by the Standard Oil Company through the Mississippi Valley and the West is manufactured at Whiting, Indiana, close to Chicago, and is shipped on Chicago rates. There are no important independent oil works at Chicago. Now at Cleveland, Ohio, there are independent refiners and jobbers contending for the market of the Mississippi Valley. See how prettily it is managed. The rates between the two Northern cities and New Orleans in the case of nearly all commodities is about two cents per hundred pounds in favour of Chicago. For example, the rate on flour from Chicago is 23 cents per 100 pounds; from Cleveland, 25 cents per 100 pounds; on canned goods the rates are 33 and 35; on lumber, 31 and 33; on meats, 51 and 54; on all sorts of iron and steel, 26 and 29; but on petroleum and its products they are 23 and 33!

In the case of Atlanta, Georgia, a similar vagary of rates exists. Thus Cleveland has, as a rule, about two cents advantage per 100 pounds over Chicago. Flour is shipped from Chicago to Atlanta at 34 cents, and from Cleveland at 32½; lumber at 32 and 28½; but Cleveland refiners actually pay 48 cents to Atlanta, while the Standard only pays 45 from Whiting.

There is a curious rule in the Boston and Maine Railroad in regard to petroleum shipments. On all commodities except petroleum, what is known as the Boston rate applies, but oil does not get this. For instance, the Boston rate applies to Salem, Massachusetts, on all traffic except petroleum, and that pays four cents more per 100 pounds to Salem than to Boston.

The New York, New Haven and Hartford Railroad gives no through rates on petroleum from Western points, although it gives them on every other commodity. It does not refuse to take oil, but it charges the Boston rate plus the local rates. Thus, to use an illustration given by Mr. Prouty, of the Interstate Commerce Commission, in a recent article, if a Cleveland refiner sends into the New Haven territory, say to New Haven, a car-load of oil, he pays 24 cents per 100 pounds to Boston and the local rate of 12 cents from Boston to New Haven. On any other commodity he would pay the Boston rate. Besides, the rates on petroleum have been materially advanced over what they were when the Interstate Commerce Bill was passed in 1887, although on other commodities they have fallen. In 1887 grain was shipped from Cleveland to Boston for 22 cents, iron for 22, petroleum for 22. In 1889 the rate on grain was 15 cents, on iron 20 cents, and on petroleum 24. Of course it may be merely a coincidence that the New Haven territory can be supplied by the Standard Oil Company from its New York refineries by barge, and that William Rockefeller is a director of the New York, New Haven and Hartford Railroad.

An independent refiner of Titusville, Pennsylvania, T. B. Westgate, told the Industrial Commission in 1898 that his concern was barred from shipping their products to nearly all New England and Canadian points by the refusal of the roads to give the same advantages in tariff which other freight was allowed. Mr. Westgate made the suggestive comment that very few railroads ever solicited oil trade. He pointed out that when the United States Pipe Line was building, agents of various roads were after the oil men soliciting shipments of the pipe, etc., to be used. "We could ship iron, but the oil —we must not handle. That is probably the password that goes over."

Examples of this manipulation might be multiplied. There is no independent refiner or jobber who tries to ship oil freight that does not meet incessant discouragement and discrimination. Not only are rates made to favour the Standard refining points and to protect their markets, but switching charges and dock charges are multiplied. Loading and unloading facilities are refused, payment of freights on small quantities are demanded in advance, a score of different ways are found to make hard the way of the outsider. "If I get a barrel of oil out of Buffalo," an independent dealer told the writer not long ago, "I have to *sneak* it out. There are no public docks; the railroads control most of them, and they won't let me out if they can help it. If I want to ship a car-load they won't take it if they can help it. They are all afraid of offending the Standard Oil Company."

This may be a rather sweeping statement, but there is too much truth in it. There is no doubt that to-day, as before the Interstate Commerce Commission, a community of interests exists between railroads and the Standard Oil Company sufficiently strong for the latter to get any help it wants in making it hard for rivals to do business. The Standard owns stock in most of the great systems. It is represented on the board of directors of nearly all the great systems, and it has an immense freight not only in oil products, but in timber, iron, acids, and all of the necessities of its factories. It is allied with many other industries, iron, steel, and copper, and can swing freight away from a road which does not oblige it. It has great influence in the money market and can help or hinder a road in securing money. It has great influence in the stock market and can depress or inflate a stock if it sets about it. Little wonder that the railroads, being what they are, are afraid to "disturb their relations with the Standard Oil Company," or that they keep alive a system of discriminations the same in effect as those which existed before 1887.

Of course such cases as those cited above are fit for the Interstate Commerce Commission, but the oil men as a body have no faith in the effectiveness of an appeal to the Commission, and in this feeling they do not reflect on the Commission, but rather on the ignorance and timidity of the Congress which, after creating a body which the people demanded, made it helpless. The case on which the Oil Regions rests its reason for its opinion has already been referred to in the chapter on the co-operative independent movement which finally resulted in the Pure Oil Company. The case first came before the Commission in 1888. At that time there was a small group of independent refiners in Oil City and Titusville, who were the direct outgrowth of the compromise of 1880 between the Producers' Protective Association and the Pennsylvania Railroad. The railroad, having promised open rates to all, urged the men to go into business. Soon after came the great fight between the railroads and the seaboard pipe-

A 25,000-BARREL TANK OF OIL IN FLAMES

line, with the consequent low rates. This warfare finally ended in 1884, after the Standard had brought the Tidewater into line, in a pooling arrangement between the Standard, now controlling all seaboard pipe-lines, and the Pennsylvania Railroad, by which the latter was guaranteed twenty-six per cent. of all Eastern oil shipments on condition that they keep up the rate to the seaboard to fifty-two cents a barrel.

Now, most of the independents shipped by barrels loaded on rack cars. The Standard shipped almost entirely by tank-cars. The custom had always been in the Oil Regions to charge the same for shipments whether by tank or barrel. Suddenly, in 1888, the rate of fifty-two cents on oil in barrels was raised to one of sixty-six cents. The independents believed that the raise was a manipulation of the Standard intended to kill their export trade, and they appealed to the Commission. They pointed out that the railroads and the pipe-lines had been keeping up rates for a long time by a pooling arrangement, and that now the roads made an unreasonable tariff on oil in barrels, at the same time refusing them tank cars. The hearing took place in Titusville in May, 1889. The railroads argued that they had advanced the rate on barrelled oil because of a decision of the Commission itself—a case of very evident discrimination in favour of barrels. The Commission, however, argued that each case brought before it must stand on its own merits, so different were conditions and practices, and in December, 1892, it gave its decision. The pooling arrangement it did not touch, on the ground that the Commission had authority only over railroads in competition, not over railroads and pipe-lines in competition. The chief complaint, that the new rate of sixty-six cents on oil in barrels and not on oil in tanks was an injurious discrimination, the Commission found justified. It ordered that the railroads make the rates the same on oil in both tanks and barrels, and that they furnish shippers tanks whenever reasonable notice was given. As the amounts wrongfully collected by the railroads from the refiners could not be ascertained from the evidence already taken, the Commission decided to hold another hearing and fix the amounts. This was not done until May, 1894, five years after the first hearing. Reparation was ordered to at least eleven different firms, some of the sums amounting to several thousand dollars; the entire award ordered amounted to nearly $100,000.

In case the railroads failed to adjust the claims the refiners were ordered to proceed to enforce them in the courts. The Commission found at this hearing that none of their orders of 1892 had been followed by the roads and they were all repeated. As was to be expected, the roads refused to recognise the claims allowed by the Commission, and the case was taken by the refiners into court. It has been heard three times. Twice they have won, but each time an appeal of the roads has forced them to appear again. The case was last heard at Philadelphia in February, 1904, in the United States Circuit Court of Appeals. No decision had been rendered at this writing.

It would be impossible to offer direct and conclusive proof that the Standard Oil Company persuaded or forced the roads to the change of policy complained of in this case, but the presence of their leading officials and counsel at the hearings, the number of witnesses furnished from their employ, the statement of President Roberts of the Pennsylvania Railroad that the raise on barrelled oil was insisted on by the seaboard refiners (the Standard was then practically the only seaboard refiner), as well as the perfectly well-known relations of the railroad and the Standard, left no doubt in the minds of those who knew the situation that the order originated with them, and that its sole purpose was harassing their competitors. The Commission seems to have had no doubt of this. But see the helplessness of the Commission. It takes full testimony in 1889, digests it carefully, gives its orders in 1892, and they are not obeyed. More hearings follow, and in 1895 the orders are repeated and reparation is allowed to the injured refiners. From that time to this the case passes from court to court, the railroad seeking to escape the Commission's orders. The Interstate Commerce Commission was instituted to facilitate justice in this matter of transportation, and yet here we have still unsettled a case on which they gave their judgment twelve years ago. The lawyer who took the first appeal to the Commission, that of Rice, Robinson and Winthrop, of Titusville, M. J. Heywang, of Titusville, has been continually engaged in the case for sixteen years!

In spite of the Interstate Commerce Commission, the crucial question is still a transportation question. Until the people of the United States have solved the question of free and equal transportation it is idle to suppose that they will not have a trust question. So long as it is possible for a company to own the exclusive carrier on which a great natural product depends for transportation, and to use this carrier to limit a competitor's supply or to cut off that supply entirely if the rival is offensive, and always to make him pay a higher rate than it costs the owner, it is ignorance and folly to talk about constitutional amendments limiting trusts. So long as the great manufacturing centres of a monopolistic trust can get better rates than the centres of independent effort, it is idle to talk about laws making it a crime to undersell for the purpose of driving a competitor from a market. You must get into markets before you can compete. So long as railroads can be persuaded to interfere with independent pipe-lines, to refuse oil freight, to refuse loading facilities, lest they disturb their relations with the Standard Oil Company, it is idle to talk about investigations or anti-trust legislation or application of the Sherman law. So long as the Standard Oil Company can control transportation as it does to-day, it will remain master of the oil industry, and the people of the United States will pay for their indifference and folly in regard to transportation a good sound tax on oil, and they will yearly see an increasing concentration of natural resources and transportation systems in the Standard Oil crowd.

If all the country had suffered from these raids on competition, had been the limiting of the business opportunity of a few hundred men and a constant higher price for refined oil, the case would be serious enough, but there is a more serious side to it. The ethical cost of all this is the deep concern. We are a commercial people. We cannot boast of our arts, our crafts, our cultivation; our boast is in the wealth we produce. As a consequence business success is sanctified, and, practically, any methods which achieve it are justified by a larger and larger class. All sorts of subterfuges and sophistries and slurring over of facts are employed to explain aggregations of capital whose determining factor has been like that of the Standard Oil Company, special privileges obtained by persistent secret effort in opposition to the spirit of the law, the efforts of legislators, and the most outspoken public opinion. How often does one hear it argued, the Standard Oil Company is simply an inevitable result of economic conditions; that is, given the practices of the oil-bearing railroads in 1872 and the elements of speculation and the over-refining in the oil business, there was nothing for Mr. Rockefeller to do but secure special privileges if he wished to save his business.

Now in 1872 Mr. Rockefeller owned a successful refinery in Cleveland. He had the advantage of water transportation a part of the year, access to two great trunk lines the year around. Under such able management as he could give it his concern was bound to go on, given the demand for refined oil. It was bound to draw other firms to it. When he went into the South Improvement Company it was not to save his own business, but to destroy others. When he worked so persistently to secure rebates after the breaking up of the

South Improvement Company, it was in the face of an industry united against them. It was not to save his business that he compelled the Empire Transportation Company to go out of the oil business in 1877. Nothing but grave mismanagement could have destroyed his business at that moment; it was to get every refinery in the country but his own out of the way. It was not the necessity to save his business which compelled Mr. Rockefeller to make war on the Tidewater. He and the Tidewater could both have lived. It was to prevent prices of transportation and of refined oil going down under competition. What necessity was there for Mr. Rockefeller trying to prevent the United States Pipe Line doing business?—only the greed of power and money. Every great campaign against rival interests which the Standard Oil Company has carried on has been inaugurated, not to save its life, but to build up and sustain a monopoly in the oil industry. These are not mere affirmations of a hostile critic; they are facts proved by documents and figures.

Certain defenders go further and say that if some such combination had not been formed the oil industry would have failed for lack of brains and capital. Such a statement is puerile. Here was an industry for whose output the whole world was crying. Petroleum came at the moment when the value and necessity of a new, cheap light was recognised everywhere. Before Mr. Rockefeller had ventured outside of Cleveland kerosene was going in quantities to every civilised country. Nothing could stop it, nothing check it, but the discovery of some cheaper light or the putting up of its price. The real "good of the oil business" in 1872 lay in making oil cheaper. It would flow all over the world on its own merit if cheap enough.

The claim that only by some such aggregation as Mr. Rockefeller formed could enough capital have been obtained to develop the business falls utterly in face of fact. Look at the enormous amounts of capital, a large amount of it speculative, to be sure, which the oil men claim went into their business in the first ten years. It was estimated that Philadelphia alone put over $168,000,000 into the development of the Oil Regions, and New York $134,000,000, in their first decade of the business. How this estimate was reached the authority for it does not say.* It may have been the total capitalisation of the various oil companies launched in the two cities in that period. It shows very well, however, in what sort of figures the oil men were dealing. When the South Improvement Company trouble came in 1872, the producers launched a statement in regard to the condition of their business in which they claimed that they were using a capital of $200,000,000. Figures based on the number of oil wells in operation or drilling at that time of course represent only a portion of the capital in use. Wildcatting and speculation have always demanded a large amount of the money that the oil men handled. The almost conservative figures in regard to the capital invested in the Oil Regions in the early years were those of H. E. Wrigley, of the Geological Survey of Pennsylvania. Mr. Wrigley estimates that in the first twelve years of the business $235,000,000 was received from wells. This includes the cost of the land, of putting down and operating the well, also the profit on the product. This estimate, however, makes no allowance for the sums used in speculation—an estimate, indeed, which it was impossible for one to make with any accuracy. The figures, unsatisfactory as they are, are ample proof, however, that there was plenty of money in the early days to carry on the oil business. Indeed, there has always been plenty of money for oil investment. It did not require Mr. Rockefeller's capital to develop the Bradford oil fields, build the first seaboard pipe-line, open West Virginia, Texas, or Kansas. The oil business would no more have suf-

* The Petroleum Age, Volume I, page 35.

fered for lack of capital without the Standard combination than the iron or wheat or railroad or cotton business. The claim is idle, given the wealth and energy of the country in the forty-five years since the discovery of oil.

Equally well does both the history and the present condition of the oil business show that it has not needed any such aggregation to give us cheap oil. The margin between crude and refined was made low by competition. It has rarely been as low as it would have been had there been free competition. For five years even the small independent refineries outside of the Pure Oil Company have been able to make a profit on the prices set by the Standard, and this in spite of the higher transportation they have paid on both crude and refined, and the wall of seclusion the railroads build around domestic markets.

Very often people who admit the facts, who are willing to see that Mr. Rockefeller has employed force and fraud to secure his ends, justify him by declaring, " It's business." That is, "it's business" has to come to be a legitimate excuse for hard dealing, sly tricks, special privileges. It is a common enough thing to hear men arguing that the ordinary laws of morality do not apply in business. Now, if the Standard Oil Company were the only concern in the country guilty of the practices which have given it monopolistic power, this story never would have been written. Were it alone in these methods, public scorn would long ago have made short work of the Standard Oil Company. But it is simply the most conspicuous type of what can be done by these practices. The methods it employs with such acumen, persistency, and secrecy are employed by all sorts of business men, from corner grocers up to bankers. If exposed, they are excused on the ground that this is business. If the point is pushed, frequently the defender of the practice falls back on the Christian doctrine of charity, and points that we are erring mortals and must allow for each other's weaknesses!—an excuse which, if carried to its legitimate conclusion, would leave our business men weeping on one another's shoulders over human frailty, while they picked one another's pockets.

One of the most depressing features of the ethical side of the matter is that instead of such methods arousing contempt they are more or less openly admired. And this is logical. Canonise "business success," and men who make a success like that of the Standard Oil Trust become national heroes! The history of its organisation is studied as a practical lesson in money-making. It is the most startling feature of the case to one who would like to feel that it is possible to be a commercial people and yet a race of gentlemen. Of course such practices exclude men by all the codes from the rank of gentlemen, just as such practices would exclude men from the sporting world or athletic field. There is no gaming table in the world where loaded dice are tolerated, no athletic field where men must not start fair. Yet Mr. Rockefeller has systematically played with loaded dice, and it is doubtful if there has ever been a time since 1872 when he has run a race with a competitor and started fair. Business played in this way loses all its sportsmanlike qualities. It is fit only for tricksters.

The effects on the very men who fight these methods on the ground that they are ethically wrong are deplorable. Brought into competition with the trust, badgered, foiled, spied upon, they come to feel as if anything is fair when the Standard is the opponent. The bitterness against the Standard Oil Company in many parts of Pennsylvania and Ohio is such that a verdict from a jury on the merits of the evidence is almost impossible! A case in point occurred a few years ago in the Bradford field. An oil producer was discovered stealing oil from the National Transit Company. He had tapped the main line and for at least two years had run a small but steady stream of Standard oil into his private tank. Finally the thieving pipe was discovered, and the owner of it, after acknowledging his guilt, was brought to trial. The jury gave a verdict of Not guilty! They seemed to feel that though the guilt was acknowledged, there probably was a Standard trick concealed somewhere. Anyway it was the Standard Oil Company and it deserved to be stolen from! The writer has frequently heard men, whose own business was conducted with scrupulous fairness, say in cases of similar stealing that they would never condemn a man who stole from the Standard! Of course such a state of feeling undermines the whole moral nature of a community.

The blackmailing cases of which the Standard Oil Company complain are a natural result of its own practices. Men going into an independent refining business have for years been accustomed to say: "Well, if they won't let us alone, we'll make them pay a good price." The Standard complains that such men build simply to sell out. There may be cases of this. Probably there are, though the writer has no absolute proof of any such. Certainly there is no satisfactory proof that the refinery in the famous Buffalo case was built to sell, though that it was offered for sale when the opposition of the Everests, the managers of the Standard concern, had become so serious as later to be stamped as criminal by judge and jury, there is no doubt. Certainly nothing was shown to have been done or said by Mr. Matthews, the owner of the concern which the Standard was fighting, which might not have been expected from a man who had met the kind of opposition he had from the time he went into business.

The truth is, blackmail and every other business vice is the natural result of the peculiar business practices of the Standard. If business is to be treated as warfare and not as a peaceful pursuit, as they have persisted in treating it, they cannot expect the men they are fighting to lie down and die without a struggle. If they get special privileges they must expect their competitors to struggle to get them. If they will find it more profitable to buy out a refinery than to let it live, they must expect the owner to get an extortionate price if he can. And when they complain of these practices and call them blackmail, they show thin sporting blood. They must not expect to monopolise hard dealings, if they do oil.

These are considerations of the ethical effect of such business practices on those outside and in competition. As for those within the organisation there is one obvious effect worth noting. The Standard men as a body have nothing to do with public affairs, except as it is necessary to manipulate them for the "good of the oil business." The notion that the business man must not appear in politics and religion save as a "stand-patter"—not even as a thinking, aggressive force—is demoralising, intellectually and morally. Ever since 1872 the organisation has appeared in politics only to oppose legislation obviously for the public good. At that time the oil industry was young, only twelve years old, and it was suffering from too rapid growth, from speculation, from rapacity of railroads, but it was struggling manfully with all these questions. The question of railroad discriminations and extortions was one of the "live questions" of the country. The oil men as a mass were allied against it. The theory that the railroad was a public servant bound by the spirit of its charter to treat all shippers alike, that fair play demanded open equal rates to all, was generally held in the oil country at the time Mr. Rockefeller and his friends sprung the South Improvement Company. One has only to read the oil journals at the time of the Oil War of 1872 to see how seriously all phases of the transportation question were considered. The country was a unit against the rebate system. Agreements were signed with the railroads that all rates henceforth should be equal. The signatures were not on before Mr. Rockefeller had a rebate, and gradually others got them until the Standard had won the advantages it expected the South Improvement Company to give it. From that time to this Mr. Rockefeller has had to fight the best sentiment of the oil country and of the country at large as to what is for the public good. He and his colleagues kept a strong alliance in Washington fighting the Interstate Commerce Bill from the time the first one was introduced in 1876 until the final passage in 1887. Every measure looking to the freedom and equalisation of transportation has met his opposition, as have bills for giving greater publicity to the operations of corporations. In many of the great state Legislatures one of the first persons to be pointed out to a visitor is the Standard Oil lobbyist. Now, no one can dispute the right of the Standard Oil Company to express its opinions on proposed legislation. It has the same right to do this as all the rest of the world. It is only the character of its opposition which is open to criticism, the fact that it is always fighting measures which equalise privileges and which make it more necessary for men to start fair and play fair in doing business.

Of course the effect of directly practising many of their methods is obvious. For example, take the whole system of keeping track of independent business. There are practices required which corrupt every man who has a hand in them. One of the most deplorable things about it is that most of the work is done by youngsters. The freight clerk who reports the independent oil shipments for a fee of five or ten dollars a month is probably a young man, learning his first lessons in corporate morality. If he happens to sit in Mr. Rockefeller's church on Sundays, through what sort of a haze will he receive the teachings? There is something alarming to those who believe that commerce should be a peaceful pursuit, and who believe that the moral law holds good throughout the entire range of human relations, in knowing that so large a body of young men in this country are consciously or unconsciously growing up with the idea that business is war and that morals have nothing to do with its practice.

And what are we going to do about it? for it is *our* business. We, the people of the United States, and nobody else, must cure whatever is wrong in the industrial situation, typified by this narrative of the growth of the Standard Oil Company. That our first task is to secure free and equal transportation privileges by rail, pipe and waterway is evident. It is not an easy matter. It is one which may require operations which will seem severe; but the whole system of discrimination has been nothing but violence, and those who have profited by it cannot complain if the curing of the evils they have wrought bring hardship in turn on them. At all events, until the transportation matter is settled, and settled right, the monopolistic trust will be with us, a leech on our pockets, a barrier to our free efforts.

As for the ethical side, there is no cure but in an increasing scorn of unfair play—an increasing sense that a thing won by breaking the rules of the game is not worth the winning. When the business man who fights to secure special privileges, to crowd his competitor off the track by other than fair competitive methods, receives the same summary disdainful ostracism by his fellows that the doctor or lawyer who is "unprofessional," the athlete who abuses the rules, receives, we shall have gone a long way toward making commerce a fit pursuit for our young men.

THE END

APPENDIX

NUMBER 37 (See page 4)

ARTICLES OF INCORPORATION OF THE TIDEWATER PIPE LINE

Incorporation Tidewater Pipe Company, Limited, of Titusville, Pennsylvania. Recorded November 22, 1878. William F. Dickson, Recorder.

The undersigned persons, to wit: Byron David Benson, Robert Emmet Hopkins, Andrew Worton Perrin, Alanson Ashford Sumner, David Boyd Stewart, David McKelvy, Samuel Queen Brown, Adam Clark Hawkins, Willis Booth Benedict, Marcus Brownson, William Henry Nicholson, Calvin Nathaniel Payne, John Hahn Dilks, Hascal Ledger Taylor, William Henry Conley, Thomas Benton Riter, Clark Isaac Hayes, Gershom Hyde, James Henry Caldwell, George Lawrence Benton, George Hill Graham, Elisha Gilbert Patterson, Benjamin Bakewell Campbell, Delos Olcott Wickham, Joseph Henry Simmonds, Lewis Henry Smith, desire to form a partnership association, pursuant to the provisions of an act of the General Assembly of the Commonwealth of Pennsylvania, entitled, "An Act, authorising the formation of partnership association in which the capital subscribed shall alone be responsible for the debts of the association except under certain circumstances," approved the second day of June, A.D. 1874, and the several supplements thereto for the purpose of conducting a legal business or occupation, within the United States or elsewhere, whose principal office or place of business shall be established and maintained within the state of Pennsylvania, by subscribing and contributing capital thereto, which capital shall alone be liable for the debts of such association, and to that end sign and acknowledge the following statement:

Full names of the persons desiring to form such association are: Byron David Benson, Robert Emmet Hopkins, Andrew Worton Perrin, Alanson Ashford Sumner, David Boyd Stewart, David McKelvy, Samuel Queen Brown, Adam Clark Hawkins, Willis Booth Benedict, Marcus Brownson, William Henry Nicholson, Calvin Nathaniel Payne, John Hahn Dilks, Hascal Ledger Taylor, William Henry Conley, Thomas Benton Riter, Clark Isaac Hayes, Gershom Clark Hyde, James Henry Caldwell, George Lawrence Benton, George Hill Graham, Elisha Gilbert Patterson, Benjamin Bakewell Campbell, Delos Olcott Wickham, Joseph Henry Simmonds, Lewis Henry Smith.

The amount of capital of said association subscribed for by each is as follows, to wit:

Said Byron David Benson has subscribed for $100,300 of the capital of said association; the said Robert Emmet Hopkins has subscribed for $72,400 of the capital of said association; said Andrew Worton Perrin has subscribed for $24,700 of the capital of said association; said David Boyd Stewart has subscribed for $16,800 of the capital of said association; said David McKelvy has subscribed for $72,500 of the capital of said association; said Samuel Queen Brown has subscribed for $25,000 of the capital of said association; said Adam Clark Hawkins has subscribed for $6,000 of the capital of said association; said Willis Booth Benedict has subscribed for $5,000 of the capital of said association; said Marcus Brownson has subscribed for $10,000 of the capital of said association; said William Henry Nicholson has subscribed for $5,000 of the capital of said association; said Calvin Nathaniel Payne has subscribed for $5,000 of the capital of said association; said John Hahn Dilks has subscribed $82,300 of the capital of said association; said Hascal Ledger Taylor has subscribed for $50,000 of the capital of said association; said William Henry Conley has subscribed for $2,500 of the capital of said association; said Thomas Benton Riter has subscribed for $2,500 of the capital of said association; said Clark Isaac Hayes has subscribed for $10,000 of the capital of said association; said Gershom Clark Hyde has subscribed for $1,000 of the capital of said association; said James Henry Caldwell has subscribed for $2,500 of the capital of said association; said George Lawrence Benton has subscribed for $1,000 of the capital of said association; said George Hill Graham has subscribed for $1,000 of the capital of said association; said Elisha Gilbert Patterson has subscribed for $5,000 of the capital of said association; said Benjamin Bakewell Campbell has subscribed for $10,000 of the capital of said association; said Delos Olcott Wickham has subscribed for $2,500 of the capital of said association; said Joseph Henry Simmonds has subscribed for $1,000 of the capital of said association; said Lewis Henry Smith has subscribed for $1,000 of the capital of said association.

Second.—The total amount of the capital of the said association is $625,000, and said capital shall be paid at the times and in the manner following, to wit: Twenty-five per cent. thereof on the second day of December, A.D. 1878; twenty-five per cent. thereof on the second day of January, A.D. 1879; twenty-five per cent. thereof on the first day of February, A.D. 1879, and the balance of twenty-five per cent. thereof the third day of March, A.D. 1879. The whole of said capital shall be paid in lawful money to the treasurer of said association at the principal office or place of business of said association at Titusville, Pennsylvania.

Third.—The character of the business to be conducted by said association is the production, shipping, refining, storing, insuring, buying and selling of petroleum and its products, and the acquisitions, manufacture and management of such property, real, personal and mixed, as may be deemed necessary or advisable to use in such business or in connection therewith. The location of the business to be conducted by said association is at the city of Titusville, in the county of Crawford, and state of Pennsylvania, where the principal office or place of business of said association is established and shall be maintained.

Fourth.—The name of the said association is the Tidewater Pipe Company (Limited).

Fifth.—The contemplated duration of said association is twenty years from the date of this statement.

Sixth.—The names of the officers of said association selected in conformity with the provisions of said act are as follows:

The managers of said association so elected are: Byron David Benson, Hascal Ledger Taylor, Alanson Ashford Sumner, Robert Emmet Hopkins, and John Hahn Dilks, of whom said Byron David Benson is so selected chairman of said association; said Robert Emmet Hopkins is so selected treasurer of said association; and said Alanson Ashford Sumner is so selected secretary of said association.

In Witness Whereof, the persons named in this statement have hereunto severally signed their names, this thirteenth day of November, *Anno Domini* one thousand eight hundred and seventy-eight:

ELISHA GILBERT PATTERSON, BYRON DAVID BENSON, MARCUS BROWNSON, HASCAL LEDGER TAYLOR, GEORGE LAWRENCE BENTON, ALANSON ASHFORD SUMNER, DELOS OLCOTT WICKHAM, DAVID MCKELVY, ADAM CLARK HAWKINS, DAVID BOYD STEWART, JOHN HAHN DILKS, GEORGE HILL GRAHAM, WILLIAM HENRY NICHOLSON, JOSEPH HENRY SIMMONDS, GERSHOM CLARK HYDE, LEWIS HENRY SMITH, WILLIS BOOTH BENEDICT, BENJAMIN BAKEWELL CAMPBELL, WILLIAM HENRY CONLEY, CALVIN NATHANIEL PAYNE, THOMAS BENTON RITER, JAMES HENRY CALDWELL, CLARK ISAAC HAYES, ANDREW NORTON PERRIN, SAMUEL QUEEN BROWN, ROBERT EMMET HOPKINS.

NUMBER 38 (See page 15)

TESTIMONY OF HENRY M. FLAGLER IN REGARD TO THE TIDEWATER CONTEST

[Proceedings in Relation to Trusts, House of Representatives, 1888. Report Number 3,112, page 783.]

Q. Now you can make your statement.

A. I want to say this: The Tidewater Pipe Line was the first line built to the seaboard, and it had a connection with the Reading Railroad, by which the railroad and the line jointly undertook to do business. We had several discussions of pipe-lines of the future with the representatives of the Tidewater Pipe Line, and would have had no difficulty whatever in making satisfactory arrangements with them, which would have removed all unnecessary competition, but the New York Central, the Erie road, and the Pennsylvania Central said to us: "Gentlemen, we don't want you to make any alliance of any formal nature with the Tidewater Pipe Line." They added: "We will protect you in the matter of rates as against any competition furnished by the Reading and Tidewater Pipe Line." I replied to that: "I have never seen a contest begun of this kind but what there was an end to it. Now, we can make a satisfactory arrangement with the Tidewater Pipe Line and avoid all this contest. It is not necessary for you to throw away any money. We are not seekers after low rates. We have done our business by you, and are willing to continue, but only upon one single, solitary condition: we would prefer not to have this contest; it is better that the Tidewater and Reading Railroad should be recognised." The reply was: "We never will recognise them as carriers of oil."

Q. That was the reply of these three trunk lines?

A. Yes, sir. I said: "Gentlemen, the other thing is of a great deal more importance than the rates. The rates are short-lived affairs." Now, I will make this explanation in justice to ourselves, in reply to the remark you made of our contest with the Tidewater Line. We had no contest. It was simply a contest of the transportation lines, and we, like fools, allowed ourselves, instead of making arrangements with the Tidewater Line, to say to the trunk lines: "Very well, then, we will stick to you and leave you to fight out this battle." They fought it for a year or two, and you know how it ended.

Q. Three or four years, was it not?

A. I thought it was two years.

Q. Then I understand you to say that all that struggle, and the low rate that the trunk line charged at the time the competition with the Tidewater and Reading came into existence, was brought about by the trunk lines themselves?

A. It was a struggle on the part of the trunk lines to hold the entire oil business, and they avowed it to me not once, but many times, that it was their firm intention never to recognise the Tidewater to the seaboard.

Q. And during that struggle they actually carried it at fifteen cents a barrel?

A. I should have said twenty or twenty-five cents. I knew it was a ridiculously low rate.

NUMBER 39 A (See page 24)

AGREEMENT BETWEEN STANDARD AND TIDEWATER REFINERIES

[From manuscript presented to the Industrial Commission by Lewis Emery, Jr.]

This agreement, made and entered into the ninth day of October, A.D. 1883, by and between the Standard Oil Company, a corporation of Ohio, the Standard Oil Company of New York, a corporation of New York, and the Standard Oil Company of New Jersey, a corporation of New Jersey, who collectively constitute the party of the first part, and the Ocean Oil Company, a corporation of New Jersey, the Chester Oil Company, a corporation of Pennsylvania, and Ayres, Lombard and Company, a corporation of New York, who collectively constitute the party of the second part.

Witnesseth: That in consideration of the mutual covenants and agreements hereby made and entered into, the said parties do hereby covenant and agree to and with each other as follows:

First.—That for the purpose of this contract the business of refining petroleum is defined to mean the distillation of crude petroleum within the United States, without regard to where the crude is obtained; the quantity of crude petroleum received at each refinery, except for export in its crude state, shall be regarded as thê quantity refined by it.

Second.—That in said business the refineries named in schedule "A" and schedule "B" (which schedules are hereto attached and made a part of this agreement) shall respectively be entitled to have and do the following percentage or proportionate part of the aggregate business of all refineries named in both schedules, viz.: The refineries named in Schedule "A," eighty-eight and one-half (88½) per cent. thereof, and the refineries named in Schedule "B," eleven and one-half (11½) per cent. thereof.

Third.—The refineries named in Schedule "A" and the refineries named in Schedule "B" shall respectively do as nearly as practicable their said proportion or percentage of said business; and is agreed that,

A.—If in any calendar month the refineries named in Schedule "A" shall receive more than their said percentage of the said aggregate of crude petroleum received except for export in its crude state, the party of the first part hereto will pay to the party of the second part hereto, twenty (20) cents per barrel on the quantity so received in excess of their said percentage.

B.—If in any calendar month the refineries named in Schedule "B" shall receive more than their said percentage of the said aggregate of crude petroleum received except for export in its crude state, the party of the second part hereto will pay to the party of the first part hereto twenty (20) cents per barrel on the quantity so received in excess of this said percentage.

C.—If in any year the refineries named in Schedule "A" shall neglect or refuse to do eighty (80) per cent. of their said percentage of said business, then the party of the first part shall return and repay the party of the second part the sums received under the provisions of this paragraph in excess of the sums paid under the same provisions during the same year.

D.—If in any year the refineries named in Schedule "B" shall neglect or refuse to do eighty (80) per cent. of their said percentage of said business, then the party of the second part shall return and repay to the party of the first part the sums received under the provisions of this paragraph in excess of the sums paid under the same provisions during the same year.

Fourth.—Each party hereto shall make to the other daily reports showing all crude petroleum received at the refineries named in said schedule, and when, where and from whom received, and all crude petroleum exported therefrom, and when, where and to whom delivered. The reports of the party of the first part shall show the crude received at and exported from refineries named in Schedule "A," and the reports of the party of the second part shall show the crude received at and exported from refineries named in Schedule "B." The correctness of such reports shall, if required of either party, be verified by the party making them.

Fifth.—A settlement shall be made, on or before the fifteenth day of each month, of all business done under this agreement during the preceding month, and payments shall then be made of all such sums as under the terms hereof shall be found payable by either party to the other.

Sixth.—All refineries now owned or controlled by those owning or controlling a majority of the refineries embraced in Schedule "A" are or shall be included in Schedule "A," and all refineries which may hereafter be acquired or controlled in the same interest shall, as acquired or controlled, be added to said Schedule "A," and by such addition be included in the terms of this agreement. All refineries now owned or controlled by those owning or controlling a majority of the refineries embraced in Schedule "B," and all refineries which may hereafter be acquired or controlled in the same interest shall, as acquired or controlled, be added to said Schedule "B," and by such addition be included in the terms of the agreement.

Seventh.—It is understood that forty-two gallons constitute a barrel.

Eighth.—A year, whenever used in this contract, is understood to mean a calendar year.

Ninth.—This agreement shall take effect on the first day of October, 1883, and remain in force for fifteen (15) years from said date.

Provided, however, and it is agreed that it shall not remain in force longer than a certain other agreement of even date herewith between the National Transit Company and the United Pipe Lines of the first part, and the Tidewater Pipe Company, Limited, of the second part, shall remain in force, and that a termination of said other agreements shall at the same time terminate this one.

In Witness Whereof, the said parties have caused their common and corporate seals to be hereto attached and to be attested by the signature of their proper officers the day and year first aforesaid.

Standard Oil Company, by

O. H. PAYNE, *Vice-President.*

[S. O. C., Cleveland] Attest: W. P. THOMPSON, *Secretary.*

Standard Oil Company of New York, by

WILLIAM ROCKEFELLER, *President.*

[S. O. C., New York] Attest: GEORGE H. VILAS, *Secretary.*

Standard Oil Company of New Jersey, by

J. A. McGEE, *President.*

[S. O. C., New Jersey] Attest: GEO. H. VILAS, *Secretary.*

NUMBER 39 B (See page 24)

AGREEMENT BETWEEN STANDARD AND TIDEWATER PIPE LINES

[From manuscript presented to the Industrial Commission by Lewis Emery, Jr.]

This agreement, entered into the ninth day of October, A.D. 1883, by and between the National Transit Company and the United Pipe Lines, each being a corporation of the state of Pennsylvania, parties of the first part, and the Tidewater Pipe Company, Limited, a limited partnership association formed under the laws of the state of Pennsylvania, party of the second part.

Witnesseth: That in consideration of the mutual covenants and agreements hereby made and entered into, the said parties do hereby covenant and agree to and with each other as follows:

First.—That for the purposes of this contract the business hereinafter referred to is divided into departments, one known as the "Gathering Department," one known as the "Transporting Department," one known as the "Interior Export Department," and one known as the "Seaboard Export Department."

All crude petroleum received directly or indirectly from wells located in the state of New York or state of Pennsylvania, and into the system of pipes and tanks now owned or controlled, or which may hereafter be owned or controlled by any party hereto, either directly or indirectly, shall constitute gathering, and the business of so receiving crude petroleum is the business of said gathering department. All deliveries from local lines of pipe of crude petroleum gathered as aforesaid, to or for any of the refineries then embraced in Schedule "A" or Schedule "B" (which schedules are hereto attached and made part of this agreement), and also all deliveries of crude petroleum from any of the trunk lines of pipe now owned or controlled, or which may hereafter be owned or controlled, by any party hereto, either directly or indirectly, and the getting of such crude petroleum to the point of delivery shall constitute transporting, and the business of so getting and delivering crude petroleum is the business of said transporting department, except, and it is agreed, that whatever petroleum gathered as aforesaid shall be delivered to or for any party hereto, or to or for any refinery or refining company then embraced in either of said schedules, for export in its crude state, whether the same shall be delivered from a local line of pipe or a trunk line of pipe, shall not be included in transporting, nor in the business of said transporting department.

All petroleum gathered as aforesaid and delivered from local lines of pipe for export in its crude state (other than deliveries to trunk lines of pipe of such petroleum for export in its crude state) by or for any party hereto or by or for any refinery or refining company then embraced in either of said schedules, shall constitute interior exporting and the business of receiving and exporting such petroleum in its crude state shall be the business of said interior export department.

All petroleum gathered as aforesaid and delivered from trunk lines of pipe for export in its crude state by or for any party hereto or by or for any refinery or refining company then embraced in either of said schedules shall constitute seaboard exporting, and the business of receiving and exporting such petroleum in its crude state shall be the business of said seaboard export department.

All pipes used for gathering and delivering at points in the oil-producing regions are herein called local lines.

All lines of pipe used for transporting beyond the oil-producing regions are herein called trunk lines.

Second.—That in each said department of the business the respective parties hereto shall be entitled to do the following percentage or proportionate part of the aggregate business done by all parties hereto then in said department, viz.: The said parties of the first part eighty-eight and one-half (88½) per centum thereof, and the said party of the second part eleven and one-half (11½) per centum thereof.

Third.—Each party hereto shall do as nearly as practicable its said proportion or percentage of said business. And it is agreed that:

A.—If in any calendar month either party shall gather more than its said percentage of said aggregate of crude petroleum gathered, as gathering is herein defined, it shall pay to the other party on the quantity gathered in excess of its said percentage an amount per barrel equal to three-fourths of the then current full rate per barrel charged for collecting and delivering crude petroleum in the oil-producing regions—commonly called local pipage;

Provided, however, and it is hereby agreed that this clause shall not be applicable to crude petroleum gathered as aforesaid prior to September 1, 1884.

And provided, further, That the excess over its said percentage gathered prior to September 1, 1884, by either party shall on demand of the other be delivered to the other party at some point or points in the oil-producing regions convenient to both the party receiving and the party delivering (the means and places to be mutually agreed upon) when and as often as the said excess amounts to ten thousand (10,000) barrels, upon legal orders or certificates with storage and assessments thereon paid to date of delivery being presented therefor, or upon the payment of the then market price of United Pipe Line certificates for a like quantity. The party receiving shall pay the party delivering the same a gathering charge of ten (10) cents per barrel upon all petroleum so delivered.

B.—If in any calendar month either the parties of the first part or the party of the second part shall transport and deliver more than their or its said percentage of the said aggregate of crude petroleum transported, as transporting is herein defined, they or it shall pay to the other party twenty-five (25) cents per barrel upon the quantity transported and delivered in excess of their or its said percentage.

Provided, That the amount payable under this clause shall not exceed the amount it would cost to bring said excess from the mouth of a local pipe in the oil-producing regions to either the port of New York or the port of Philadelphia at the then current rate of transportation by any route or method not owned or controlled directly or indirectly by any party hereto.

C.—If in any calendar month either party shall do more than its said percentage of business in either the exterior export department or the seaboard export department, it shall pay to the other party twenty-five (25) cents per barrel upon the quantity so exported in excess of its said percentage.

Provided, however, That the amount per barrel payable under this clause shall not exceed the amount per barrel which would be payable under Clause B and its proviso at the same time for excess in the transporting department.

D.—If in any year either party shall neglect or refuse to do eighty (80) per centum of its said proportion or percentage in any department of said business, then the party so doing less than eighty (80) per centum of its said proportion shall return or repay to the other party the sums received in that department under the provisions of this paragraph in excess of the sums paid in the same department under the same provisions during the same year.

Fourth.—Each party shall make to the other daily reports showing:

1st. All crude petroleum gathered, as gathering is herein defined.

2nd. All crude petroleum delivered from local lines other than deliveries to trunk lines, stating when, where and to whom delivered.

3rd. All crude petroleum delivered from local lines to trunk lines, stating when, where and to which line delivered.

4th. All crude petroleum delivered from trunk lines, stating when, where and to whom delivered.

5th. All crude petroleum exported in the crude state, stating when, where and from whom received, so as to distinguish between receipts from local lines and receipts from trunk lines, and when, where and to whom delivered for export. The correctness of such reports shall, if required by either party, be verified by the party making them.

Fifth.—On all deliveries of crude petroleum from local lines made by said parties of the first part or either of them, other than such deliveries as constitute transporting, as transporting is hereinbefore defined, the parties of the first part will account for and pay to the party of the second part eleven and one-half (11½) per centum of the then current full rate of local pipage, first deducting from such full rate ten (10) cents per barrel for the work of gathering and delivering such petroleum.

On all deliveries of crude petroleum from local lines made by said party of the second part other than such deliveries as constitute transporting as hereinbefore defined, the party of the second part will account for and pay to the parties of the first part eighty-eight and one-half (88½) per centum of the then current full rate of local pipage, first deducting from such full rate ten (10) cents per barrel for the work of gathering and delivering such petroleum.

Sixth.—It is agreed that in case of excess of deliveries over the quantity gathered, as gathering is herein before defined, by all the parties hereto, the stocks in custody of the respective parties shall to the extent of such excess be diminished in the ratio of eighty-eight and one-half (88½) per centum thereof from the stocks in custody of said parties of the first part, and eleven and one-half (11½) per centum thereof from the stocks in custody of said party of the second part; and to this end it is agreed that whenever and as often as under the working of this agreement the depletion of the stocks in the custody of either of the respective parties shall amount to ten thousand (10,000) barrels in excess of such party's percentage of depletion, then the other party shall and will on demand deliver, and the party whose stocks are so depleted will when tendered receive, said ten thousand (10,000) barrels at some point or points in the oil-producing regions convenient to both the party receiving and the party delivering (the means and place to be mutually agreed upon), upon legal orders or certificates with storage and assessments thereon paid to date of delivery being presented therefor, or upon the payment of the then market price of United Pipe Line certificates for a like quantity. The party receiving shall pay to the party delivering a gathering charge of ten (10) cents per barrel upon all petroleum gathered.

Seventh.—A settlement shall be made on or before the fifteenth day of each month of all business done under this agreement during the preceding month, and payment shall then be made of all such sums as under the terms hereof shall be found payable by either party to the other.

Eighth.—If in any year the profits of the party of the second part added to the profits of the several refineries then embraced in Schedule "B" shall in the aggregate amount to less than five hundred thousand (500,000) dollars (excluding from the calculations all profits realised and losses sustained from speculation and the value of property destroyed by fire), then the said party of the second part shall have the right within three months from the time the profits of such year shall have been ascertained to cancel this agreement.

Provided, however, That the said right shall not exist or shall not be exercised under the following circumstances, to wit:

1st. If the average of such profits during the said year and all previous years from the beginning of this agreement shall equal five hundred thousand (500,000) dollars per year.

2nd. If the said parties of the first part or either of them shall contribute to the said party of the second part such sums of money as together with the said profits for the said year will make the average profit five hundred thousand (500,000) dollars per year.

And provided, further, That in exercising the right of cancellation the said party of the second part must give to one or both of said parties of the first part three (3) months' written notice of said cancellation, which notice must be accompanied by a statement of the said profits of the party of the second part, and of said refineries then embraced in Schedule "B," and any contributions made as aforesaid must be made within the said three (3) months.

The party receiving said notice shall have the right to verify the statement by an examination of the books of said party of the second part, and books of said refineries.

Ninth.—All refineries now owned or controlled by those owning or controlling a majority of the refineries embraced in Schedule "A" are or shall be included in Schedule "A"; and all refineries which may hereafter be acquired or controlled in the same interest shall, as acquired or controlled, be added to said Schedule "A," and by such addition be included in the terms of this agreement.

All refineries now owned or controlled by those owning or controlling a majority of the refineries embraced in Schedule "B" are or shall be included in Schedule "B"; and all refineries which may hereafter be acquired or controlled in the same interest shall, as acquired or controlled, be added to said Schedule "B," and by such addition be included in the terms of this agreement.

Tenth.—It is agreed that any business done in either the interior export department or the seaboard export department by any of the refineries or refining companies then embraced in Schedule "A" shall be treated for the purpose of this agreement as if done by the parties of the first part; and that any business done in either of said export departments by any of the refineries or refining companies then embraced in Schedule "B" shall be treated for the purposes of this agreement as if done by the party of the second part.

Eleventh.—It is understood that forty-two (42) gallons constitute a barrel.

Twelfth.—A year, whenever used in this contract, is understood to mean a calendar year.

Thirteenth.—This agreement shall take effect as of the first day of October, 1883, and unless sooner cancelled, as provided in the eighth paragraph, shall remain in force for fifteen (15) years from said first day of October, 1883.

In Witness Whereof, the said parties of the first part have caused their common and corporate seals to be hereto attached and to be attested by the signatures of their proper officers; and the said party of the second part has caused the same to be signed in its name and on its behalf by two of its managers, the day and year first aforesaid.

NATIONAL TRANSIT COMPANY,
[Nat. Tran. Co. Seal.] (Signed by) BENJAMIN BREWSTER, *Vice-President.*
Attest: JOHN BUSHNELL, *Secretary.*

UNITED PIPE LINES,
[U. P. L. Seal.] (Signed by) J. J. VANDERGRIFT, *President.*
Attest: H. D. HANCOCK, *Secretary.*

SCHEDULE OF REFINERIES REFERRED TO IN THE ATTACHED AGREEMENT

SCHEDULE "A"

Atlas Refining Co	Works	at Buffalo, N. Y.
Acme Oil Co. of Pennsylvania	"	" Titusville, Pa.
Acme Oil Co. of New York	"	" Olean, N. Y.
Atlantic Refining Co	"	" Philadelphia, Pa.
American Lubricating Oil Co	"	" Cleveland, Ohio.
Baltimore United Oil Co	"	" Canton, Md.
Bush Denslow Mfg. Co	"	" South Brooklyn, N. Y.
Camden Consolidated Oil Co	"	" Parkersburg, W. Va.
" " " "	"	" Canton, Md.
Central Refining Co., Limited	"	on Newtown Creek, L. I.
Empire Refining Co., Limited	"	" " " "
Eclipse Lubricating Co., Limited	"	at Franklin, Pa.
" " " "	"	" Olean, N. Y.
Eagle Oil Co	"	" Communipaw, N. J.
Galena Oil Works, Limited	"	" Franklin, Pa.
Imperial Refining Co	"	" Oil City, Pa.
Pratt Mfg. Co	"	" Bushwick Creek, L. I.
Jenny & Son, S	"	" Wallabout Land.
Donald & Co., James	"	" Newtown Creek, L. I.
Portland Kerosene Co	"	" Portland, Me.
Paine, Ablett & Co., Limited	"	" Smith's Ferry.
" " " " "	"	" Freedom, Pa.
Sone Fleming Mfg. Co., Limited	"	" Newtown Creek, L. I.
Standard Oil Co. of New York	"	" " " "
" " " "	"	" Hunter's Point, L. I.
" " " New Jersey	"	" Bayonne, N. J.
" " " Pennsylvania	"	" Pittsburg, Pa.
" " " Ohio	"	" Cleveland, Ohio.
Union Refining Co., Limited	"	" Oil City, Pa.
Vacuum Oil Co	"	" Rochester, N. Y.

SCHEDULE "B"

Chester Oil Co	Works	at Chester, Pa.
Ocean Oil Co	"	" Bayonne, N. J.
Seaboard Oil Co	"	" " "
Solar Oil Co	"	" Buffalo, N. Y.

NUMBER 40 (See page 28)

TWO AGREEMENTS OF EVEN DATE, AUGUST 22, 1884, BETWEEN THE PENNSYLVANIA RAILROAD COMPANY AND THE NATIONAL TRANSIT COMPANY

[Report of the Industrial Commission, 1900. Volume I, pages 663–666.]

Memorandum of a traffic agreement, made this twenty-second day of August, 1884, between the Pennsylvania Railroad Company, hereinafter designated the railroad company, and the National Transit Company, hereinafter designated the transit company, *Witnesseth:*

That for consideration mutually interchanged, the parties hereto agree, each with the other, as follows:

First.—The transit company owns an extended system of local pipes in the Oil Regions of Pennsylvania and New York, which are grouped into a separate division, known as the United Pipe Lines Division of the National Transit Company. This division will be hereinafter designated as the Transit Company's Local Division.

The business of this division is to collect oil from producer, store it in tanks, and deliver it, as may be desired, to any through carrier of petroleum, which will transport the same to where it is to be refined or otherwise disposed of.

The transit company also own certain through or trunk line pipes, extending from several points of connection with the aforesaid local pipe division to various refining and terminal points.

With these latter pipes, which will be hereinafter entitled the Transit Company's Trunk Line Division, it competes in the through carriage of petroleum with all other through carriers, whether pipe or rail.

The business of its local division is therefore entirely distinct from the business of its through trunk line division.

It undertakes and agrees that its local division will deliver into cars furnished by the railroad company at any of its regular delivery points and under its regular delivery rules whatever petroleum the owners thereof may desire to have so delivered, and as the railroad may furnish cars to transport, and will make no discrimination in its local charges for carriage, storage, and other services, or in the use of any of its local facilities, against such oil, but will at all times treat it in the said respects as favourably as it at the same time treats any other petroleum which may be delivered to its own trunk line division or to any other through carriers.

Second.—The transit company agrees that all petroleum brought to the Atlantic seaboard by all existing carriers, whether rail or pipe, now engaged in transporting such property, or which may hereafter engage in such transportation in conjunction with the transit company's pipe-lines, shall be ascertained monthly, and so much of it as shall have been shipped in the refined state shall be reduced to its equivalent in crude oil by considering that one and three-tenths ($1\frac{3}{10}$) gallons of crude are required to make one (1) gallon of refined oil. It further undertakes and agrees that if of the total so transported the railroad company shall not have moved in its cars twenty-six (26) per centum thereof, the transit company shall cause to be delivered to cars furnished by the railroad company at Milton, Pa., such quantity of crude petroleum as shall, when added to the amount which has been actually transported by the railroad company to the seaboard in said month, make the total transported by the railroad company in said month equal to said twenty-six (26) per centum.

The railroad company agrees to furnish the needful cars and facilities, and promptly transport the oil which the transit company agrees in this contract to deliver to it at Milton:

Provided, That if during any month the railroad company is not able to assign from its oil equipments a sufficient number of cars to the traffic of the transit company to move the proportion of oil herein provided to be delivered at Milton, then during that month the transit company shall only be required to so deliver to the railroad company such quantity of oil as the railroad company shall be able to transport, and shall not be required to make up any deficiency that may occur during said month.

Efforts shall be made by the transit company to deliver so much during each month as will probably be necessary to make the total carried by the railroad company equal to said percentage.

Shortages, if not due to short supply of cars, and such excesses as may be found to have occurred in any month, shall be adjusted in the following month, or as soon afterwards as shall be possible.

Third.—It is agreed that the proportion of petroleum which the transit company is to deliver under the second section of this agreement shall be considered as petroleum transported from Coalgrove, Pa., *via* Milton, Pa., to the Atlantic seaboard, and that the railroad company shall be entitled to one-half of the current through rates thereon.

It is agreed that whenever the through rates shall be so low that the railroad company shall suspend the movement of oil by its cars, at other points than Milton, the transit company shall during such suspension not be bound to deliver to the railroad company any oil at Milton.

Fourth.—All joint rates for the joint transportation of oil from any delivery point of the local pipe division aforesaid to any refining or terminal point shall be fixed by the railroad company, subject to the advice and concurrence of the transit company.

It is agreed that said joint through rates shall be uniform to all parties. The railroad company stipulates that it will make no discrimination whatever, either in rates or facilities, against the transit company or against the oil which the said transit company herein covenants to deliver to it.

It is agreed that the joint through rates to Philadelphia shall always be five cents less per barrel on crude oil, or its refined equivalent, than shall be currently charged to New York harbour.

It is agreed that the joint through rates, which shall be so fixed from time to time, shall be as low as shall be currently made between same and similar points by rival carriers of petroleum, and shall not be higher than an approximate mileage proportion of rates current on petroleum produced south of Oil City, nor than rates from Olean and similar points.

It is also agreed that rates on refined oil and other products of crude oil shall be fixed by the railroad company upon the following basis, viz.:

From railroad stations in the Oil Regions to which oil is delivered by local pipes the rate to any point east thereof on a barrel of refined oil or other products shall be one and three-tenths ($1\frac{3}{10}$) times the current rate on a barrel of crude oil to the same point.

From Pittsburg the rate to any point east thereof on a barrel of refined oil or other products shall be one and three-tenths ($1\frac{3}{10}$) the rate currently charged on crude oil to any such eastern point from rail points south of Oil City:

Provided, That one and three-tenths times the charges for moving a barrel of crude oil by rail or through pipe from the local pipe to Pittsburg shall first be deducted therefrom.

From Cleveland and Buffalo the net rate on a barrel of refined oil or other products to any point east thereof shall be not less than is currently charged to the same point from Pittsburg.

Fifth.—Whenever the term barrel is used herein, unless otherwise specified, it means forty-five gallons of crude petroleum; and whenever the term oil is used herein, unless otherwise specified, it means crude petroleum.

Sixth.—The transit company hereby agrees that it will not make any more favourable terms with any other rail line connecting with any of its pipes than the terms which under this agreement are given to the railroad company; or if for any reason it should desire to do so, it hereby agrees to modify this contract so as to give the said "more favourable terms" to the railroad company.

Seventh.—All existing contracts between the parties hereto shall be deemed to have been accomplished, and shall become void and of no effect upon the day this contract goes into operation.

Eighth.—This contract shall take effect as of the first day of August, 1884, and shall continue until terminated under the provisions hereof. It may be terminated after August 1, 1889, by either party hereto giving ninety days' written notice to the other of a desire that it shall end, at the expiration of which notice it shall cease and determine.

In Witness Whereof, the parties hereto have executed this agreement under their corporate seals the day and date above written.

THE PENNSYLVANIA RAILROAD COMPANY,
[L.S.] By FRANK THOMSON, *Second Vice-President.*
Attest: JOHN C. SIMS, JR., *Secretary.*

THE NATIONAL TRANSIT COMPANY,
[L.S.] By C. A. GRISCOM, *President.*
Attest: JOHN BUSHNELL, *Secretary.*

Memorandum of agreement, made this twenty-second day of August, 1884, between the Pennsylvania Railroad Company, hereinafter designated the railroad company, and the National Transit Company, hereinafter designated the transit company.

Witnesseth: That for considerations mutually interchanged the parties hereto hereby agree with each other as follows:

Whereas, The parties hereto have made an agreement of even date herewith, in which, among other things, it is stipulated that under certain circumstances the transit company shall deliver certain crude petroleum into cars furnished by the railroad company at Milton, Pa.; and

Whereas, It has been proposed that the railroad company shall contract with the transit company to the effect that the transit company shall transport through its pipe-lines the aforesaid crude oil, which, under the other contract aforesaid, it has undertaken to deliver into the cars of the railroad company at Milton.

Now, therefore, this agreement witnesseth:

First.—The railroad company agrees that instead of delivering said crude oil to said cars at Milton, the transit company shall transport the same through its pipes to destination, and the transit company undertakes and agrees to do such transportation. It is mutually agreed that the compensation to the transit company for doing said work shall be as follows:

Whenever the through rate for transporting a barrel of crude petroleum from Olean to Philadelphia shall be forty cents, the transit company shall receive eight cents per barrel as such compensation for so much of said oil as under the provisions hereof shall be considered as Philadelphia oil.

For each five cents of increase or diminution in said rates from Olean to Philadelphia the said compensation on Philadelphia oil shall be increased or diminished one cent per barrel.

Provided, however, That the transit company shall not be obliged to accept less than six cents per barrel, and shall not receive more than ten cents per barrel on such Philadelphia oil.

It is agreed that the said compensation on the oil, which under the provisions hereof is to be deemed New York oil, shall be one cent per barrel greater than it currently shall be on Philadelphia oil.

Whenever, and from time to time, as the said joint through rates shall be so low that the said minimum compensation to the transit company of six cents per barrel shall be as much or more than the railroad company's share of said joint through rates, this contract may, at the option of either party hereto, be suspended during all or any part of the time such low rates shall prevail. During such suspension the aforesaid other contract shall alone remain in force; but whenever, and from time to time, as said joint through rates shall again be high enough to make the said minimum compensation, under said sliding scale, less than the said share of said joint through rates, this contract shall again resume its force and effect.

Second.—The transit company agrees to account for, and pay to the railroad company, on or before the twentieth of each month, the latter's share of the joint rates on joint business *via* Milton (as provided in said other contract) during the next preceding month, first retaining, however, the proportion of such share which it is hereinbefore agreed the transit company is to have for its services in pumping said oil to the seaboard.

It is agreed that all such joint business shall be considered as having transported from Coalgrove *via* Milton, Pa., to the Atlantic seaboard, and that it shall be considered as having gone either to Baltimore, Philadelphia, or New York, or partly to each. The proportion thereof which has constructively gone to New York shall be determined upon the following basis:

The total amount of oil transported in any month by the railroad company to New York shall be compared with fifty (50) per centum of the total oil which the railroad company is entitled to carry in said month under the aforesaid other agreement. If the amount which has been in such month carried by cars to New York shall be less than fifty (50) per centum, then the difference shall be considered as having been moved by the pipe to New York, at New York rates, and shall be accounted for accordingly. The remainder of the oil *via* Milton shall be accounted for at Philadelphia rates.

This contract shall commence and terminate simultaneously with said other contract.

Witness the corporate seals of said parties duly attested the day and date above written.

THE PENNSYLVANIA RAILROAD COMPANY,
[L.S.] By FRANK THOMSON, *President.*
Attest: JOHN C. SIMS, *Secretary.*

THE NATIONAL TRANSIT COMPANY,
[L.S.] By C. A. GRISCOM, *President.*
Attest: JOHN BUSHNELL, *Secretary.*

NUMBER 41 (See page 60)

TABLE SHOWING PRICES OF OIL AT COMPETITIVE AND NON-COMPETITIVE POINTS IN 1892

[Trust Investigation of Ohio Senate, 1898. Appendix, pages 43–44.]

Territories and States.	Prime White Oil.						Water-White Oil.						Per Gallon			
	Non-competitive prices per gallon.			Competitive prices per gallon.			Non-competitive prices per gallon.			Competitive prices per gallon.						
	Barrels.	Case.	Bulk.	Barrels.	Case.	Bulk.	Barrels.	Case.	Bulk.	Barrels.	Case.	Bulk.	Highest.	Lowest.	Difference.	Difference per tank car 6,000 gallons.
Arizona	....	....	....	....	....	....	....	31	....	....	....	....	31	....	....	.....
Arkansas	14	....	13	8	....	7½	16	....	17	....	....	....	17	7½	9½	$570
Alabama	13	....	8½	8¼	....	6½	17	....	12	10¾	....	....	17	6½	10½	630
California	....	....	16	13	....	12½	....	26½	....	13	17½	11½	26½	11½	15	900
Colorado	....	26	21	10	15	7	....	31	25	....	....	....	31	7	24	1,440
Florida	13½	16	12	....	....	....	17	18½	....	....	....	....	18½	12	6½	390
Georgia	14	....	9½	9½	....	6½	17	....	14	....	....	....	17	6½	10½	630
Idaho	22½	29	....	....	....	....	22½	30	17	....	....	....	30	17	13	780
Illinois	10	....	8	7½	....	5½	15	....	....	7¾	....	5½	15	5½	9½	570
Indiana	....	....	....	6¼	....	5	12½	....	....	6½	....	....	12½	5	7½	450
Iowa	9½	....	8	....	....	7	12	....	...	10½	....	8	12	7	5	300
Kansas	10½	....	9½	8½	....	....	16½	....	....	9½	....	....	16½	8½	8	480
Kentucky	9½	....	8¾	7	....	6½	12	....	....	8½	....	....	12	6½	5½	330
Louisiana	12	....	10	7¼	....	7	16	....	14	7¾	...	7½	16	7	9	540
Michigan	8½	....	6¾	6¾	....	3½	8½	....	7	7½	....	3²/5	8½	3½	5	300
Minnesota	....	....	9	7½	...	5	13	....	11	8	....	5½	13	5	8	480
Mississippi	13½	....	....	7¼	....	....	15½	....	....	9½	....	....	15½	7¼	7¼	435
Missouri	12	....	....	6	....	5½	17	....	....	7¾	....	5½	17	5½	11½	690
Montana	....	....	20	...	....	13	21	33	25	....	....	....	33	13	20	1,200
Nebraska	18	....	....	7½	....	....	27	....	....	8½	....	....	27	7½	19½	1,170
Nevada	....	37½	....	....	....	....	....	....	....	....	....	....	37½	....	..	......
New Mexico	....	31	26	....	....	....	....	32	28	....	....	....	32	26	6	360
North Dakota	15½	....	....	12½	...	....	18	....	14	12¼	....	11¼	18	11¼	6¾	405
Oregon	....	21	14	....	19	13	....	24	....	....	23	....	24	13	....	660
Oklahoma	....	....	15	9½	....	....	....	....	17	....	....	....	17	9½	7½	450
South Carolina	12½	....	....	8	....	....	13½	....	....	9	....	....	13½	8	5½	330
South Dakota	11½	....	....	....	....	8	....	....	12	....	....	8	12	8	4	240
Tennessee	11½	....	8½	7¾	....	6	17	....	....	8½	....	....	17	6	11	660
Texas	25	27½	19	8	14	9	30	33½	24	12	16½	8	33½	8	25½	1,530
Utah	23	28	25	13	....	....	....	....	....	....	....	....	28	13	15	900
Washington	16	20½	15	....	....	....	....	25½	....	....	....	....	25½	15	10½	630
Wisconsin	9	....	....	7½	....	6	15¼	....	....	7½	...	6	15¼	6	9¼	555
Wyoming	20	25	15	.. .	....	....	21	35	29	8	16	15	35	8	27	1,620

PRIME WHITE OIL

The table shows that this grade of oil ranges in price as follows:

In barrels........................ 6 to 25 cents per gallon
In cases........................14 to 37½ " "
In bulk........................ 3½ to 25 " "

WATER-WHITE OIL

This table also shows that this grade of oil ranges in price as follows:

In barrels	6½ to 30	cents per gallon
In cases	16 to 35	" "
In bulk	3½ to 29	" "

A comparison of these two grades of oil shows:

A difference of 24 cents per gallon on barrelled oil
" " " 21 " " " case oil
" " " 25½ " " " bulk oil

NUMBER 42 (See page 69)

STANDARD OIL COMPANY'S PETITION FOR RELIEF AND INJUNCTION

[In the case of the Standard Oil Company *vs.* William C. Scofield *et al.*, in the Court of Common Pleas, Cuyahoga County, Ohio, 1880.]

The said plaintiff, the Standard Oil Company, now comes and says that on the twentieth day of July, A.D. 1876, it was and still is a corporation organised and existing under and by virtue of the laws of the state of Ohio, and that at the same time the said defendants, William C. Scofield, Charles W. Scofield, Daniel Shurmer and John Teagle, were and still are partners doing business in the firm name of Scofield, Shurmer and Teagle, and the said plaintiff complains of the said defendants, and says: That on the said twentieth day of July, A.D. 1876, the said plaintiff and the said defendants as such partners were each separately engaged in the business of refining and dealing in crude petroleum and its products, said plaintiff having a number of refining establishments at Cleveland, Ohio, and the said defendants owning and operating one refinery only, also located at Cleveland, Ohio, on the line of the Atlantic and Great Western Railroad, and while so engaged and on the said twentieth day of July, A.D. 1876, the said plaintiff and the said defendants as such partners entered into a joint arrangement in writing in and by which it was, amongst other things, agreed between the said plaintiff and the said defendants individually and as such partners that the said defendants would continue their then business in the firm name of Scofield, Shurmer and Teagle of buying, refining and selling crude petroleum and its products as theretofore carried on by them, for a period of ten years from July 20, A.D. 1876, and furnish for the conducting of said business their refinery aforesaid with all tanks, fixtures, buildings, erections, tools, and all mechanical appliances then or theretofore used by them in their said business, together with the land on which the same are situated, and also within five days from the date of said agreement furnish for the use of said joint business adventure the sum of ten thousand dollars in cash to be used continuously in said business until July 20, A.D. 1886. That the said William C. Scofield, Charles W. Scofield, Daniel Shurmer and John Teagle, in and by said agreement for conducting said joint adventure, further covenanted and agreed with the plaintiff to devote all their time and personal attention necessary to conduct the said business for the period aforesaid, and that during the existence of said adventure they would not nor would either of them as a firm or as individuals directly or indirectly engage or be concerned in any business connected with petroleum or any of its products in Cuyahoga County or elsewhere, except in connection with the parties of the first part under this agreement, nor would they or either of them enter into any new business which would interfere with the time necessary to be devoted to the full and faithful conduct of the business of said adventure.

That the said William C. Scofield, Charles W. Scofield, Daniel Shurmer and John Teagle, in and by said agreement for conducting said joint adventure, further covenanted and agreed with said plaintiff that the amount of crude petroleum to be distilled by them in the business of said adventure should not exceed annually eighty-five thousand barrels of forty-two gallons each in any year, but the same should be distributed as nearly as practicable in equal quantities of 42,500 barrels of forty-two gallons each, each and every six months from the twentieth day of July, A.D. 1876, but the said 42,500 barrels might be run in a less period than six months.

That in and by said agreement for conducting the business of said joint adventure it was stipulated and agreed by both parties, amongst other things, that from the net profits of the business of said joint adventure the said defendants should first be entitled to retain and be paid the sum of $35,000 per annum while the said agreement was in force and operation, and in the case the net profits should not amount to $35,000 for any year that said agreement for conducting said joint adventure was in force and operation, then at the expiration of any such year the plaintiff should on demand pay to the said defendants a sum of money sufficient to make that amount, viz., $35,000 for any year that said agreement should be in force and operation. That all net profits over the amount of $35,000 so stipulated to belong to said defendants annually should belong and be paid to said plaintiff until the plaintiff should receive therefrom as much as said defendants had received from the net profits under the provisions of said agreement, and all net profits in excess of $70,000 annually should be divided equally between the parties thereto.

That in consideration thereof and in and by said agreement for conducting said joint adventure, the said plaintiff stipulated and agreed with the said defendants, amongst other things, that on or before the twenty-fifth day of July, A.D. 1876, it would furnish to the said defendants for them to use in the business of said joint adventure the sum of $10,000 in cash, which sum was so paid in as agreed and still remains in the business.

That the said plaintiff would receive, dock, and sell in the city of New York all oil and the products of petroleum consigned to it for sale at New York by said firm of Scofield, Shurmer and Teagle at actual cost of brokerage and handling without commissions.

That the said plaintiff would and did in said agreement guarantee to the said defendants that their share of the net profits arising from the business of said joint adventure should for ten years from July 20, A.D. 1876, to July 20, A.D. 1886, amount to the sum of $35,000 annually, during the operation of this contract, as hereinbefore stated. The plaintiff further says that between July 20, 1876, and the present time, the said defendants have repeatedly violated their said agreement in this, to wit: that every year since the making of said agreement the said defendants have distilled over 85,000 barrels of crude petroleum; that during the year from July 20, 1876, to July 20, 1877, they distilled 89,983.34-42 barrels; that during the year from July 20, 1877, to July 20, 1878, they distilled 87,754.4-42 barrels; that during the year from July 20, 1878, to July 20, 1879, they distilled 100,246.25-42 barrels, and from July 20, 1879, to July 20, 1880, they distilled 90,082.34-42 barrels.

That up to the present time the defendants have distilled more than by the terms of their said agreement they have a right to distil up to January 20, 1881, and have purchased large quantities of crude petroleum and are distilling portions thereof, and threaten to distil the balance without regarding their said contract. That the crude petroleum so as aforesaid distilled by the defendants has not by them been distributed as nearly as practicable in equal quantities of 42,500 barrels of forty-two gallons each, each and every six months as they agreed to do, but in violation of their said agreement they distilled from July 20, 1876, to January, 1, 1877, 43,509.36-42 barrels; from January 1, 1877, to July 20, 1877, 46,473.40-42 barrels; from July 20, 1877, to January 1, 1878, 50,416.12-42 barrels; from January 1, 1878, to July 20, 1878, 37,337.34-42 barrels; from July 20, 1878, to January 1, 1879, 56,974.15-42 barrels; from January 1, 1879, to July 20, 1879, 43,272.10-42 barrels; from July 20, 1879, to January 1, 1880, 57,499.35-42 barrels; that on or about the twentieth day of July, 1879, the plaintiff having discovered that the said defendants had in violation of said agreement distilled about 22,984 barrels of oil more than they were entitled to by the terms of said agreement, the plaintiff objected and complained to the defendants in regard thereto, and thereupon the defendants admitted the violation of the contract in that respect, and it was agreed between the parties that the defendants would and should during the then coming year diminish their manufacture sufficiently to bring the entire amount of manufacture under said contract within the terms of said agreement.

That during the then coming year from July 20, 1879, to July 20, 1880, the said defendants did not diminish their distillation below the 85,000 barrels as they had agreed to do, but from July 20, 1879, to January 1, 1880, they distilled 57,499.35-42 barrels, and from January 1, 1880, to July 20, 1880, they distilled 32,582.41-42 barrels, making a total of 90,082.34-42 barrels for the year, thus increasing their distillation over the 85,000 barrels 5,082 barrels, instead of diminishing it as they had agreed to do.

That the defendants threaten to and have informed the plaintiff that they will hereafter wholly disregard said contract and continue to distil crude petroleum without regard to quantity.

The plaintiff further says that since the making of said agreement and within the past year the said Daniel Shurmer and John Teagle have in violation of their said contract engaged and been connected in constructing a refinery at Buffalo, New York, for the purpose of distilling crude petroleum with others than the plaintiff under said agreement and are now so engaged.

That within the past year the said Daniel Shurmer and John Teagle and each of them have invested money to the amount of $10,000, and are now engaged and connected in constructing refineries for the purpose of distilling crude petroleum and its products with others in no way connected with the plaintiff or under said agreement, but intending thereby to establish and prosecute with others the same business as that contemplated and conducted under said agreement, and thereby establishing and conducting a rival business to the business of said adventure and tending to involve the plaintiff in loss by reason of its guarantee that the profits of said adventure should amount to the sum of $35,000 annually to defendants, and have during the past year been at said Buffalo and other places giving the said business their time and personal attention, and have done so at times when their time and personal attention was needed and was requisite to properly conduct the business of said adventure under said agreement at Cleveland.

The plaintiff further says that because of the said failures and refusals of the defendants to carry out their said agreement it has already sustained great damage and will sustain further damage if the said defendants are permitted to continue their said violation of said agreement. That the said plaintiff has no adequate remedy therefor at law for the reason that the damages arising therefrom are so remote and difficult of ascertainment, and constantly recurring would necessitate a multiplicity of suits and would involve the plaintiff in the increased hazards of losses arising from such increased manufacture and deprive it of all the benefits of said contract.

The plaintiff therefore prays that the said William C. Scofield, Charles W. Scofield, Daniel Shurmer and John Teagle may by proper process be made defendants herein and compelled to answer this petition; that a preliminary injunction and restraining order be granted restraining the said William C. Scofield, Charles W. Scofield, Daniel Shurmer and John Teagle, and each of them individually and as partners in the name of Scofield, Shurmer and Teagle, until the further order of the court, from distilling at their said works at Cleveland, Ohio, more than 85,000 barrels of crude petroleum of forty-two gallons each in every year, and also from distilling more than 42,500 barrels of crude petroleum of forty-two gallons each, each and every six months, and also from distilling any more crude petroleum until the expiration of six months from and after July 20, 1880, and also from directly or indirectly engaging in or being concerned in any business connected with petroleum or any of its products, except in connection with the plaintiff under their said agreement, and that on the final hearing of this case the said defendants may in like manner be restrained and enjoined from doing any of said acts until the expiration of said agreement, and for such other and further relief in the premises as equity can give.

M. R. Keith,
R. P. Ranney,
Attorneys for Plaintiff.

NUMBER 43 (See page 70)

ANSWER OF WILLIAM C. SCOFIELD *ET AL.*

[In the case of the Standard Oil Company *vs*. William C. Scofield *et al*., in the Court of Common Pleas, Cuyahoga County, Ohio, 1880.]

That the so-called agreement is and at all times has been utterly void and of no effect, as being by its terms in restraint of trade and against public policy.

These defendants further say that they deny that through any action of theirs said plaintiff has sustained or will sustain any damage whatever, but these defendants say that their business of distilling oil has been carried on at a large profit, and that the same is now attended with large profits, and the price of refined oil is now so high, and there is such a large margin between the price of crude oil and refined, that the manufacture and sale of refined oil is attended with large profit; that it is impossible to supply the demand of the public for oil if the business and refineries of both plaintiff and defendant are carried on and run to their full capacity, and if the business of defendants were stopped as prayed for by plaintiff it would result in a still higher price for refined oil and the establishment of more perfect monopoly in the manufacture and sale of the same by plaintiff.

These defendants further say that said plaintiff has constantly and persistently violated the terms of said so-called written agreement in that it has intentionally failed to give and has withheld from the defendants the benefits of the advantages therein agreed to be given, and that it has not given to defendants the benefits of its contracts relating to freight on crude and refined oil, but these defendants have been constantly required to pay more and larger freights than said plaintiff, and that said plaintiff has not allowed to defendants the same rebate that it has received with different carriers; and, further, that said plaintiff has recently constructed a pipe-line to the Oil Regions of Pennsylvania through which its oil has been pumped to Cleveland at an expense of about twelve cents a barrel, but has charged defendants for pumping their oil through the same pipe twenty cents per barrel.

The defendants further say that at the time when said writing was signed said plaintiff was endeavouring by contracts with divers persons to establish a monopoly in the manufacture of refined oil in the state of Ohio and in the United States, and that, for the purpose of monopolising the trade in refined oil and enhancing the price thereof, and maintaining an unnaturally high price, said plaintiff entered into said so-called agreement under the form of a joint arrangement or adventure, and for no other purpose, and contributed to the capital of said so-called adventure the sum of $10,000, whereas those defendants contributed thereto the sum of $73,000 and their time and attention, and their refinery had the capacity for refining 180,000 barrels of crude oil per year, as plaintiff well knew, and said plaintiff thereby, and by said other contracts made with the same design, succeeded in creating a substantial monopoly and averting competition and maintaining an unnaturally high price for refined oil, and that said so-called agreement is therefore in restraint of trade and against public policy, and void.

These defendants further say that defendants have from time to time paid to plaintiff their full share of the profits of said so-called adventure, and at no time has plaintiff been required to pay any sum whatever to defendants, but has realised large profits from said business, and on the fourth day of March, 1880, with full knowledge of how much oil in excess of 85,000 barrels per year had been manufactured by defendants, demanded of said defendants that they should pay to plaintiff the entire profits upon said excess, and claimed that its monopoly was so perfect that it would have sold said excess if defendants had not, and defendants did pay to plaintiff the one-half of the profits on said excess.

NUMBER 44 (See page 71)

AFFIDAVIT OF JOHN D. ROCKEFELLER

[In the case of the Standard Oil Company *vs*. William C. Scofield *et al*., in the Court of Common Pleas, Cuyahoga County, Ohio, 1880.]

John D. Rockefeller being duly sworn, says that for about eighteen years past he has been engaged in the business of refining crude petroleum; that from about the year 1863 to 1870 he was engaged as a member of firms in such refining, and from January, 1870, he has been and still is engaged in such refining business as president of said plaintiff, the Standard Oil Company; that during said time he has given the business personal attention and has thereby become familiar with the general business of refining crude petroleum, with the amount of crude petroleum produced, with the amount of crude petroleum refined, so far as the same can be ascertained, and especially with the business of the Standard Oil Company.

Affiant says the said Standard Oil Company owns and operates its refineries at Cleveland, Ohio, and its refinery at Bayonne, New Jersey; that it has no other refineries nor any interest in any other refineries, nor does the Standard Oil Company operate or control in the United States any other refineries of crude petroleum; that there are in Ohio, West Virginia, Pennsylvania, New York, and New Jersey a large number of refineries of crude petroleum that are not owned or controlled by said Standard Oil Company, and in which the said Standard Oil Company has no interest whatever, directly or indirectly, which are now and for years past have been refining crude petroleum and selling it in the open market; that the amount of crude petroleum refined by the said Standard Oil Company does not exceed thirty-three per cent. of the total amount refined in the United States.

Affiant further says that the capacity of all the refineries in the United States is more than sufficient to supply the markets of the world, and in the judgment of affiant if all the refineries were run to their full capacity they would refine at least twice as much oil as the markets of the world require; that this difference between the capacity of refineries and the demands of the market has existed for at least seven years past, and during that period the refineries of the Standard Oil Company have not been run to their full capacity, and in the judgment of affiant not to exceed one-half of their capacity.

Affiant further says that during all the period of time that he has been engaged in the business of refining oil he has been familiar with the price of crude oil and with the price of refined oil and with the profits to be derived therefrom, and from such experience he states that the average price of refined oil and the average profits to the manufacturer per gallon on same since 1876 have been much less than the average profit for several years previous to 1876; that said Standard Oil Company has no means now and never has had any of influencing the price of refined oil, save by the sale of its product in the open market.

Affiant further says that the Standard Oil Company has not nor did it ever have any interest in any oil property or any control over the production of crude petroleum; that it does not own any oil wells or land producing oil, and never did; nor has it any control over the price of crude petroleum, but relies upon obtaining its supplies, as all others do, by purchase in the open market and at the prices paid by others at the same time; that the said Standard Oil Company is not now nor has it ever been a stockholder in any railroad, pipe-line, or other common carrier for the transportation of oil, but within the year past it has for its own convenience constructed, and owns and is now operating, a pipe-line from Cleveland to the western line of the state of Pennsylvania for the purpose of bringing oil to its refineries at Cleveland; that said pipe-line is now insufficient to supply the demands of the Standard Oil Company for crude oil for its own refineries, and for that reason it has been and is now compelled to bring crude oil to Cleveland in cars to supply its wants.

That from the deponent's experience in business he knows it to be true that a large manufacturer always has an advantage in cheapness of manufacture over a small manufacturer; that all the advantages derived by the Standard Oil Company are legitimate business advantages, due to the very large volume of supplies which it purchases, its long continuance in the business, the experience it has thereby acquired, the knowledge of all the avenues of trade, the skill of experienced employees, the possession and use of all the latest and most valuable mechanical improvements, appliances and processes for the distillation of crude oil, and in the manufacture of its own barrels, glue, etc., etc., by reason of which it is enabled to put the oil on the market at a cost of manufacture much less than by others not having equal advantages. These advantages, by reason of which the Standard Oil Company is enabled to refine oil cheaper than smaller manufacturers, are not exclusive to the Standard Oil Company, but are open to every person doing business under similar circumstances. That this state of facts has been detrimental to smaller refineries and has prevented them from making as much profit as they desired, and in some cases compelled them to suspend refining, and this constitutes the only foundation for the oft-repeated expressions "crushed out," "squeezed out," and "bulldozing."

Affiant says he has examined the answer of the defendants, Shurmer and Teagle, and his attention has been called to various statements contained in it. In regard to the statement made therein that "if the business of the defendants were stopped as prayed for by plaintiff, it would result in a still higher price for refined oil and the establishment of a more perfect monopoly in the manufacture and sale of the same by plaintiff." The same is untrue, as there is not, never has been, and never can be a monopoly in the manufacture of refined oil, nor has the limitation in said agreement as to quantity to be manufactured affected, nor will the stoppage by the defendants of their manufacture, as prayed for in plaintiff's petition, in the least affect the price of refined oil, for the reason that leaving out the entire capacity of the refinery of defendants there would still remain a large excess of capacity for supplying all the demands of the public, and hence there would be no opportunity for advancing the price, nor would it tend to create a monopoly of the business by the plaintiff.

Affiant further says that it is not true that the said plaintiff has at any time or in any manner violated the terms of said agreement as alleged in said answer or in any other manner. That it is not true that plaintiff has intentionally or otherwise withheld from the defendants the benefit of the advantages agreed upon in said contract to be given them, nor is it true that the plaintiff has not given to defendants the benefit of its contracts relating to freight on crude and refined oil, but the plaintiff has given to the defendants privileges not required by the agreement. That it is not true that the defendants have ever been required to pay larger rates of freight than were paid by the plaintiff when the defendants made any shipments of oil in accordance with the terms of the contract; nor is it true that the plaintiff has not allowed to defendants the same rebates that it has received from different carriers upon any shipments of oil made in accordance with the terms of the contract.

That it is true that the plaintiff has recently constructed a pipe-line from Cleveland to the western line of the state of Pennsylvania, through which its oil has been pumped to Cleveland since the spring of 1880, but it is not true that it is the owner of the said pipe-line from the western line of the state of Pennsylvania to the Oil Regions. That it is true that to promote the interest of the defendants, the plaintiff has furnished to defendants crude oil through said pipe-line and charged them twenty cents per barrel for the transportation of same; but it is not true that said pipe-line was constructed for the purpose of transporting oil for others than the plaintiff, nor is it true that under the terms of said agreement the defendants are entitled to the transportation of oil through said pipe-line, nor is it true that the charge of twenty cents per barrel is an unreasonable price for transporting oil through said pipe-line from the Oil Regions to Cleveland; but affiant avers it to be true that during the time it so furnished the oil through the pipe-line at twenty cents per barrel, of forty-two gallons each, the railroads were charging freight at the rate of from thirty-five to fifty cents per barrel, of forty-five gallons each.

Plaintiff continued to deliver defendants through the pipe-line, and at twenty cents per barrel, until they had received all they were entitled to manufacture under the contract dated July 20, 1876.

Affiant says that it is not true that "at the time when said agreement was signed, said plaintiff was endeavouring by contracts with divers persons to establish a monopoly in the manufacture of refined oil in the state of Ohio and in the United States." Affiant avers that it has made but one other contract with other persons like the one made with defendants, and that was a contract made at the same date, viz., July 20, 1876, with the Pioneer Oil Company of the City of Cleveland, of which the defendants had full knowledge. Affiant further says that he was present and participated in the negotiations which resulted in the formation of the contract with these defendants, and that it is not true that said contract was entered into for the purpose of monopolising the trade in refined oil or for the purpose of enhancing the price thereof and maintaining an unnaturally high price for the same; and affiant says that it is not true that plaintiff by said contract, and by the said other contract made with the same design, succeeded in creating a substantial monopoly and averting competition, and maintaining an unnaturally high price for refined oil; but said contract was made, as is therein stated, for the purpose of equalising the business of manufacturing oil and giving to each of said contracting parties their due proportion thereof, and that the amount of 85,000 barrels per annum to which the distillation of defendants is by said contract limited is, as agreed, a relative proportion to their full capacity, as is the amount distilled by plaintiff per annum since said contract was entered into to its total capacity for refining oil; and it is not true that said agreement is in restraint of trade and against public policy, as alleged in the said answer of defendants, Shurmer and Teagle. Affiant says that on or about the first day of October, 1879, it came to his knowledge that the defendants had, in violation of said agreement, distilled about 22,984 barrels of oil more than they were entitled to by the terms of said agreement, and thereupon he had an interview with defendants, W. C. Scofield and John Teagle, who admitted the defendants had distilled in excess of the quantity stipulated in the contract, and agreed to reduce the quantity distilled during the year following, July 20, 1879, by the amount they had already distilled in excess up to that date, but requested they might be allowed to distribute said reduction equally over each six months of the year instead of wholly in either the first or last six months of the year following July 20, 1879, to which request affiant assented.

Affiant says that it is not true that "the plaintiff, on the fourth day of March, 1880, with full knowledge of how much oil in excess of 85,000 barrels per year had been manufactured by defendants and plaintiff, demanded of said defendants that they should pay to plaintiff the entire profits upon said excess," other than as is hereinafter stated; and it is not true that plaintiff, at the time it demanded said profits, claimed that it had any monopoly, or that its monopoly was so perfect that it would have sold said excess if defendants had not, or that it was entitled to said profits in consequence of any monopoly; but affiant says that it did claim the profits upon the oil sold in excess of said 85,000 barrels, because defendants had broken their agreement with said plaintiff, and the profits on such excess the plaintiff at that time was willing to accept as compensation for such breach of said contract.

Affiant says that he does not know what contracts for the sale of oil defendants may have made, or what contracts for the manufacture or for the construction of barrels they may have entered into, or what obligations they may be under to their customers; but he says that for a long time past the defendants have had notice that plaintiff would insist upon the performance by them of their obligations under their said contract, and that if they have entered into contracts for the sale of oil as alleged by them and entered into other obligations, they have done so with the full knowledge that they were thereby violating and continuing the violation of said agreement of July 20, 1876.

I have read the affidavit of H. L. Taylor, filed in this case October 18, 1880, in which he says "that he has been for some six or eight years last past acquainted with Mr. Rockefeller, Mr. Flagler, Mr. Payne, and others; that he has had conversations with some of these parties with regard to the control by the Standard Oil Company of the distilling and refining business in the state of Ohio and in the United States, and that he has heard them say in substance that the Standard Oil Company intended to wipe out all the refineries in the country except theirs, and to control the entire refining business in the United States." Affiant says that he has been acquainted with H. L. Taylor for several years past, that all the foregoing statements so far as they relate to him are false, and that he never made to said Taylor or to any person in his hearing any such statement, nor statements in substance to that effect. Affiant further says that he never in company with said Taylor visited any of the cities or places mentioned in his affidavit for the purpose of inspecting or examining refineries, though he may have met said Taylor incidentally at various places, but that he never showed him refineries that were formerly under the control of others and running independently and stated that the same had passed under the control of the Standard Oil Company, nor did anybody else make such statements to Taylor in his hearing.

Affiant says that it has not come to pass, as sworn to by said Taylor, that said Standard Oil Company has "wiped out" the refining business of the United States or that it to-day controls it, but affiant believes that at the time said Taylor made his affidavit he knew there were very many refineries running independently of and in no way connected with the Standard Oil Company, and that said Taylor was himself then interested in the profits of a large refining business represented by a number of refiners who were large competitors of the Standard Oil Company.

With respect to the assertion of said Taylor that "in many instances to his knowledge the Standard Oil Company has bought refineries and taken them down," affiant says that several years ago when the business was very much scattered, in several instances and for greater economy in manufacturing, the Standard Oil Company dismantled refineries unfavourably located and utilised the construction, machinery, and appliances of the same to increase its manufactory at Cleveland.

It is true that in many cases persons who had been unsuccessfully engaged in refining, but had experience, were to some extent employed by the Standard Oil Company in its business of refining, but that with respect to the averment in said Taylor's affidavit that "in other cases said company employed men who had refineries, at large salaries and at the same time gave them no absolute employment," the same is untrue. But it is true that it has restricted its employees from entering the business of refining and distilling oil except under said company's direction.

But none of these things were done by the plaintiff for the purpose of creating and maintaining a monopoly of the business of refining, but were done for the purpose of conducting its business more efficiently.

And affiant says that it is not true, as sworn to by said Taylor, that the Standard Oil Company during a large portion of the time that he refers to, to wit, six or eight years past, or for any length of time, has substantially controlled the transportation of oil; that it is not true that said Standard Oil Company ever had, or that it now has, any contract with any lines of transportation in which it was stipulated that it should have a lower rate of freight than other shippers undertaking the same obligations and furnishing equal terminal facilities; that in all the contracts ever had with the railroads, the railroad companies have reserved the right to charge others the same rate of freight as that paid by the Standard Oil Company; and affiant further says that even those contracts with the railroad companies which gave the Standard Oil Company a commission for facilities furnished have long been abrogated and abandoned.

Affiant says that with respect to the statement in said Taylor's affidavit that "other language has been used to him—said Taylor—by the officers of said Standard Oil Company to the effect that the said company intended to have all the refineries and aimed at having entire control of the oil market," the same, so far as it related to him, is wholly untrue.

Affiant says that it is not true that the plaintiff got control of the refineries of the firm of Logan Brothers of Philadelphia, Octave Oil Company, Easterly and Davis, and Bennett, Warner and Company of Titusville, Pennsylvania; R. S. Waring and Citizens' Oil Works of Pittsburg, or of either of them. The statement of H. L. Taylor that "the principal way by which these independent refineries came under the control of the Standard Oil Company was from the fact that said company had such rates of transportation that the small companies could not compete with it, and when said company had such in its power it would make such arrangements with parties engaged in these refineries as would prevent them from thereafter competing with the Standard Oil Company," is false in its facts and its inferences. Affiant has already correctly stated the facts as to the purchase of refineries by the Standard Oil Company of Cleveland, what led to such purchases, and that persons engaged in such refineries were in some cases employed by said company; and any statement or inference to the effect that by illegal means or unfair influences the plaintiff "squeezed out" or "crushed out" small refiners and prevented them from again entering into the business of refining, is untrue.

Affiant further responding to the affidavit of said Taylor, says that with reference to the statement therein contained that "the effect of the control of the refining business by the Standard Oil Company upon the oil market is to largely increase the price to consumers beyond what they ought to pay," the same is untrue, and he avers again that since the date of the contract with defendants the average price to consumers of refined oil has been lower than for years previous.

As to the allegation of said Taylor that "if the business was distributed among the independent refineries it would furnish employment to a much larger number of persons than at present, and the interests of the country would be decidedly promoted by having the refining business in the hands of competent parties," in so far as the same implies that there are not independent competing refineries outside of the works of said plaintiff, the same is untrue, and that it is a fact that a larger number of persons are now employed in connection with the business of refining oil than ever before.

Affiant says that with reference to the language used by the said Heisel in his affidavit that he, Heisel, was not afraid, to which Mr. Rockefeller replied, "You may not be afraid to have your head cut off, but your body will suffer," "and that this was said by affiant prior to the time that he sold his interest in the refining business to Bishop and was said for the purpose of inducing affiant to sell out to the Standard Oil Company," that affiant has no recollection of ever using any such language to said Heisel, and so far as said statement implies threats or inducements held out to said Heisel to procure the control of the works of Bishop and Heisel by the Standard Oil Company, the same is wholly false in spirit and effect.

Affiant says respecting the statement in said Heisel's affidavit, that "the effect resulting from the control by this one company—the Standard Oil Company—of the entire refining business in Cleveland has been to largely increase the price of refined oil to consumers, to lessen its production, to reduce the number of hands employed in the refining business, and to reduce the price paid labourers for their work, and thereby to largely injure the public," the same, so far as it alleges that there is a control by the Standard Oil Company of the entire refining business, is false; and that so far as it undertakes to state consequences of said alleged control by the Standard Oil Company, it is also false.

I have read the affidavit of Mrs. B. filed in this case on October 18, 1880. Said affidavit is incorrect, erroneous and in many respects false.

The first interview that I ever had with Mrs. B. was at her house, when she sent for Mr. Flagler and myself to consult with her in reference to selling out her establishment to one of her employees. This occurred during the year 1876. She stated to us the terms of an offer that she had received from the said employee, and expressed an earnest desire to dispose of the business and to be free from its perplexities and annoyances, and evinced a disposition to accept the offer, and we advised her to accept providing the payments were made secure. I did not see her again until the fall of 1878, more than two years later. Then at her urgent request I met her at her house, at which time she made reference to the conversation she had had with Mr. Jennings, and desired me to pursue negotiations with her with reference to the sale of her property, which I positively declined, stating to her that I knew nothing about her business or the mechanical appliances used in the same, and that I could not pursue any negotiations with her with reference to the same, but that if, after reflection, she yet desired to do so, some of our people familiar with the lubricating oil business would take up the question with her. She was very desirous to begin negotiations, but I declined to negotiate and advised her not to take any hasty action, as from her own statements there was no such change in the condition of the business as to discourage the expectation that she could do as well in the future as she had in the past. When she responded expressing her fears about the future of the business, stating that she could not get cars to transport sufficient oil, and other similar remarks, I stated to her that though we were using our cars and required them in our own business, yet we would loan her any number she required or do anything else in reason to assist her, and I saw no reason why she could not prosecute her business just as successfully in the future as in the past. This is the last interview I had with her.

Affiant thinks it is true that Mrs. B. stated in the course of the conversation in substance that "the B. Oil Company was entirely in the power of the Standard Oil Company, and that all she could do would be to appeal to affiant's honour as a gentleman and to his sympathy to do with her the best that he could do." To the statement that she was in the power of the Standard Oil Company, affiant made a positive denial, and stated to her there was no foundation for the fears she expressed, and in this connection made the offer to her to furnish her with cars. He cannot remember what was said by Mrs. B. at this interview in relation to an agreement upon the part of the Standard Oil Company not to touch the lubricating branch of the trade. It is true that the Standard Oil Company had a contract with the B. Oil Company, made early in 1873, terminable on sixty days' notice by either party, in reference to carbon oil only—which contract had been voluntarily assumed by the B. Oil Company—and it was entirely optional with the said B. Oil Company to discontinue said contract upon a notice of sixty days and thereby relieve itself from its obligations if it so desired; but said contract was continued in full force and effect up to the time of the sale by Mrs. B. of her interest in said B. Oil Company; but the Standard Oil Company had no contract with B. Oil Company by which it "agreed not to touch the lubricating branch of the trade," nor did it have any contract with the said B. Oil Company having reference in any particular to the lubricating oil business, nor did affiant have any such contract. While affiant declined to enter into a negotiation with the said Mrs. B., it may be true that during the interview alluded to he said to her that in case a sale were made she could retain whatever stock in the B. Oil Company she desired. As a result of the negotiations, in which affiant took no part, the construction and good-will of the B. Oil Company was purchased for sixty thousand dollars, which was at least twenty thousand dollars in excess of its value, and largely in excess of the value placed upon it by Mrs. B. in the interview above referred to between Mr. Flagler and affiant with her in 1876. In addition to the construction and good-will which was purchased for the sum of sixty thousand dollars, there was purchased of the B. Oil Company its entire stock of oils on hand at the full market value, and the sum paid for same amounted to $19,144.49, making an aggregate of $79,144.49, and did not include any other assets of the company, such as cash, accounts receivable and accrued dividends.

With respect to the allegation in said affidavit that "Mrs. B., seeing that the property had to go, asked that she might, according to the understanding with the president of the company, retain fifteen thousand dollars of her stock," so far as said statement implies that she was parting with her property under any duress, restraint, or undue influence, or was forced thereto by any acts of the Standard Oil Company, the same is absolutely false; and it is also false that she ever had any understanding with the president of the Standard Oil Company that she should retain fifteen thousand dollars of the stock of the B. Oil Company, nor was there any reference to that subject save as is hereinbefore stated; and if the said Mrs. B. refers to this affiant in that connection wherein she says that "to this request the reply was, 'No outsider can have any interest in this concern' and 'that said Standard Oil Company had dallied as long as it would over this matter, that it must be settled up that day or go, and insisted upon her signing the bond above referred to,'" the same is also false; nor has he any knowledge that during said negotiation any such language was ever used, or that the negotiations were ever carried on or closed in any such spirit.

Affiant says that it is not true that he made any promises that he did not keep in the letter and spirit; and it is not true that he was instrumental to any degree in her being obliged to sell the property much below its true value; and he avers that she was not obliged to sell out, and that such sale was a voluntary one upon her part and for a sum far in excess of its value, and that the construction which was purchased of her could be replaced for a sum not exceeding twenty thousand dollars.

On Saturday, the ninth day of November, 1878, the negotiations were closed and payments made to Mrs. B. Affiant had no knowledge of dissatisfaction upon her part until the receipt of a letter dated Monday, November 11, which reached him on the 12th, and on November 13 the reply thereto was made, copy of which is as follows:

November 13, 1878.

Dear Madam: I have held your note of 11th inst., received yesterday, until to-day, as I wished to thoroughly review every point connected with the negotiation for the purchase of the stock of the B. Oil Company, to satisfy myself as to whether I had unwittingly done anything whereby you would have any right to feel injured. It is true that in the interview I had with you I suggested that if you desired to do so you could retain an interest in the business of the B. Oil Company by keeping some number of its shares, and I then understood you to say that if you sold out you wished to go entirely out of the business. That being my understanding, our arrangements were made in case you concluded to make the sale, that precluded any other interests being represented, and therefore when you did make the inquiry as to your taking some of the stock our answer was given in accordance with the facts noted above, but not at all in the spirit in which you refer to the refusal in your note. In regard to the reference that you make as to my permitting the business of the B. Oil Company to *be taken* from you, I say that in this, as in all else that you have written in your letter of 11th inst., you do me most grievous wrong. It was of but little moment to the interests represented by me whether the business of the B. Oil Company was purchased or not. I believe that it was for your interest to make the sale, and am entirely candid in this statement, and beg to call your attention to the time, some two years ago, when you consulted Mr. Flagler and myself as to selling out your interests to Mr. Rose, at which time you were desirous of selling at *considerably less price*, and upon time, than you have now received in cash, and which sale you would have been glad to have closed if you could have obtained satisfactory security for the deferred payments. As to the price paid for the property, it is certainly three times greater than the cost at which we could now construct equal or better facilities; but wishing to take a liberal view of it, I urged the proposal of paying the sixty thousand dollars, which was thought much too high by some of our parties. I believe that if you would reconsider what you have written in your letter, to which this is a reply, you must admit having done me great injustice, and I am satisfied to await upon your innate sense of right for such admission. However, in view of what seems your present feelings, I now offer to restore to you the purchase made by us, you simply returning the amount of money which we have invested and leaving us as though no purchase had been made. Should you not desire to accept this proposal, I offer to you one hundred, two hundred, or three hundred shares of the stock at the same price that we paid for the same with, this addition that if we keep the property we are under engagement to pay into the treasury of the B. Oil Company an amount which, added to the amount already paid, would make a total of $100,000, and thereby make the shares one hundred dollars each.

That you may not be compelled to hastily come to conclusion, I will leave open for three days these propositions for your acceptance or declination, and in the meantime, believe me,

Yours very truly,

JOHN D. ROCKEFELLER.

To which letter no reply was ever received, and since which time affiant has had no communication with Mrs. B. upon any subject.

Affiant says that he has had his attention called to the affidavit of Daniel Shurmer, filed in this case October 18, 1880, and to the language as follows: "That the Standard Oil Company had already squeezed out one refining concern with which he was connected, whereby he had lost over twenty thousand dollars." Affiant says that the same is false, as nothing of the kind ever occurred.

Affiant says that he conducted most of the negotiations which led to the making of the contract with defendants, and that at no time previous or during the same were any threats made by him or any officer of the Standard Oil Company or agent to his knowledge to the effect that the firm of Scofield, Shurmer and Teagle would be ruined if they did not make such a contract, and no promises were made by him nor anybody else in behalf of said Standard Oil Company to said Shurmer or any of the defendants, that if said contract was signed the Standard Oil Company and defendants would control and monopolise the whole refining business in Cleveland; nor is it true, as alleged by said Shurmer, that he was reluctant to enter into said agreement, but, so far as affiant knows, the said Shurmer was anxious to make the arrangement, believing it to be a profitable one for the defendants. That some time in the year 1878, when the refining business of the City of Cleveland was in the hands of a number of small refineries and was unproductive of profit, it was deemed advisable by many of the persons engaged therein, for the sake of economy, to concentrate the business and associate their joint capital therein. The state of the business was such at that time that it could not be retained profitably at the City of Cleveland by reason of the fact that points nearer the Oil Regions were enjoying privileges not shared by refiners at Cleveland, and could produce refined oil at a much less rate than could be made at this point. That it was a well-understood fact at that time among refiners that some arrangement would have to be made to economise and concentrate the business or ruinous losses would not only occur to the refiners themselves, but ultimately Cleveland as a point of refining oil would have to be abandoned. At that time those most prominently engaged in the business here consulted together, and as a result thereof several of the refiners conveyed to the plaintiff their refineries and had the option in pay therefor to take stock in the Standard Oil Company at par or to take cash. That at this time the Standard Oil Company, by reason of its facilities and large cash capital, was agreed upon as the one best adapted to concentrate the business, and for no other reason whatsoever. That said Standard Oil Company had no agency in creating this state of things which made that change in the refining business necessary at that time, but the same was the natural result of the trade; nor did it in the negotiations which followed use any undue or unfair means, but in all cases, to the general satisfaction of those whose refineries were acquired, the full value thereof either in stock or cash was paid, as the parties preferred.

Since that time the Standard Oil Company, by diligent and faithful attention to its business, by the exercise of the most rigid economy, by promptly taking advantage of all legitimate business opportunities, has acquired large and valuable property at Cleveland with a capacity to refine oil largely in excess of any local refinery, but he denies that from 1872 to the present time, by any conclusion, conspiracy, or undue means from first to last, the present standing and capacity of the Standard Oil Company has been acquired, or that it seeks to maintain its hold upon business through any purpose to create or maintain a monopoly.

John D. Rockefeller.

NUMBER 45 (See page 72)

FINDINGS OF FACT

[Transcript of record, Supreme Court of the United States, October term, 1886. Number 1,290. The Lake Shore and Michigan Southern Railway Company, plaintiff in error, *vs.* Scofield, Shurmer and Teagle, in error to the Supreme Court of the state of Ohio, pages 14–21.]

This cause came on to be heard upon the pleadings, exhibits, and testimony, and was argued by counsel; in consideration whereof the plaintiffs, having moved for a reservation to the Supreme Court, the judges are unanimously of opinion that important and difficult questions exist in the case, making it proper that the same should be reserved to the Supreme Court for decision, which questions embrace the following propositions:

1st. Is this a case upon the face of the petition and under the laws of the state in which the court ought to interfere by injunction?

2nd. Whether such remedy by injunction will apply as well to the case of shipments over the defendants' road alone, as to cases of through shipments over such road and connecting roads?

3rd. What are the duties and obligations of common carriers at common law as distinguished from the statutory provisions of this and other states and countries?

4th. Are the defendants at common law obliged to carry freight at the same price for all parties or members of the public, without regard to quantity or circumstances connected with the transportation?

5th. May the defendant, as a common carrier and a corporation organised for that purpose, contract with a party controlling $\frac{90}{100}$ or more of all the freight of a particular class, at a given city or point, to carry the same for less than general tariff rates, in consideration that it shall receive all the freight thus controlled by such party?

6th. May the defendant, as a common carrier, in consideration of receiving all the freight of such party, that the quantity shall not be diminished, and that terminal facilities as to loading, unloading, and delivering the freight shall be furnished different from regular or usual freight and with less expense and risk to the carrier, contract to carry such freight, with such convenience and benefits, for less than general tariff rates to the public?

7th. May the defendant, as common carrier, transport over its road large quantities of oil, amounting to many full car-loads per day, for a less price per car-load than it charges the public generally per barrel or for single car-loads or less, provided all persons are charged like prices for like quantities?

8th. May defendant, as common carrier, make any distinction in prices for carrying like freight on the ground of quantity and covenants to continue the same if thereby it can make a greater profit than to charge the same prices for quantities small and great? Is defendant, under all circumstances, obliged to charge the same prices per ton or other quantity, for the same distance, to all persons tendering freight of the same class, or may it, in good faith and without intention to injure other producers or patrons, contract to carry for one party at a less price than general rates if thereby it can secure a large and profitable business which would otherwise be diverted from it, in whole or part?

8½. Should decree be rendered for plaintiffs; and, if so, to what extent should it be enforced—only within the bounds of the state or to all parts of the country within or without the state, to all points reached by defendant and connecting lines?

9th. Was section 3373 of the Revised Statutes intended to apply to cases like the present, and under it is there any authority for the injunction relief prayed for in this action?

10th. Whether upon such shipments so made by the defendant's cars by the barrel, either in car-load lots or in less amounts, the plaintiffs are, either by common law or by the Ohio statutes on the subjects, entitled to have their said products carried at the same rate of charge between like points of shipment as are allowed to said Standard Oil Company or other shippers, either to points on its line or branches of said road beyond?

11th. Whether the defendant, as a common carrier, may exact from the plaintiffs upon such shipments in barrels any amount greater than the amount charged to said Standard Oil Company upon shipment of like amounts by such tank-cars so long as the plaintiffs offer to ship by their own tank-cars on substantially like terms?

12th. Whether, if such defendant can be required to give to said plaintiffs equal rates of freight upon its shipments with those allowed said Standard Oil Company to points upon its line and branches, it can be required to give as low a rate to terminal points as the rate it receives for its proportion of the service to such points, on shipments to points beyond, and on its connecting lines on a through rate fixed by it, and such connecting line or lines for the through shipment?

13th. Whether the fact of the existence of such arrangement, and the fact of the said Standard Oil Company being a shipper in amounts larger than the plaintiffs, is any justification for the making of such charges to the plaintiffs in excess of such charges made to said Standard Oil Company? And in order that the same may be legally presented to said Supreme Court, the District Court do find the facts as follows:

1st. The court find the plaintiffs are, and since 1875 have been, partners, carrying on, in a large way, at Cleveland, Ohio, where this refinery is situated, the business of refining petroleum and selling the refined product mainly throughout the territory west and northwest of Cleveland, and extending throughout the Western and Northwestern states, this business being one in which they have invested a large amount of capital, and in which they have established a large and profitable trade throughout such territory, which constitutes the natural market for the sale of such products manufactured at Cleveland, the cost of plaintiffs' refining being about $70,000, with a refining capacity of about 150,000 barrels per year.

2nd. That the defendant is a consolidated railroad company, owning and operating a railroad extending from Buffalo, in the state of New York, to Chicago, in the state of Illinois, and passing through parts of the states of New York, Pennsylvania, Ohio, Indiana, Michigan, and Illinois, and also owning and operating branches from Toledo, in the state of Ohio, to Detroit, in the state of Michigan, and also from White Pigeon, in the state of Michigan, to Grand Rapids, in the state of Michigan.

3rd. That said railroad, so far as the same is constructed and operated in the state of Ohio, extends from the Easterly line of Ashtabula County to the Westerly line of Williams County; that it is a corporation engaged as common carrier in the business of transporting persons and property for hire and reward over its said line of road and branches.

4th. That it crosses and connects with other lines of railroads at Toledo, Coldwater, and Chicago, over which it can and does forward passengers and freight to their destination and consignment points as requested and directed; that it holds itself out as ready to make and does make the rates to points reached by connecting roads; that defendant, as such common carrier, has been accustomed to receive for transportation property over its line and branches to points beyond the termini of the same by delivering the same at such termini to connecting roads for carriage to the points of consignment.

5th. That the rates for such through freights are fixed by agreement between the different companies owning the lines over which such freights are carried, and not by the defendant alone, and are charged by like agreement, from time to time.

6th. That what are termed local rates, being for property received and delivered at points on the line of defendant's road, are fixed exclusively by the defendant.

7th. That some of the towns and cities on the main line and branches of the defendant's road can only be reached by shippers from Cleveland over its said road and branches; and all of them, as well as the towns on most of its connecting branches, can be most directly reached by means of its line from Cleveland.

8th. That the defendant is sufficiently supplied with cars and engines and appliances for transportation necessary to enable it, in the ordinary course of its business, to receive and carry for the plaintiffs such products from Cleveland to such markets.

9th. That for a period of time extending back beyond the time when plaintiffs commenced the manufacture of oil in the City of Cleveland, the defendant has published for the benefit of the public, tariff rates for local and through freights, which have been frequently changed, and including rates for the carriage of oil in barrels.

10th. The plaintiffs commenced and established their present business in Cleveland in the spring or summer of 1875, and subsequently, in July, 1876, became engaged in the same by arrangement with the Standard Oil Company to the partial extent of their own manufacturing establishment.

10½. That during the time in the petition named the Standard Oil Company, the plaintiffs' principal competitor in business, has also been and still is engaged in a like business with them, it having at Cleveland a large refinery from which it sells like products in the same markets; that the refineries of both are situate on the line of railroads other than that of the defendant, but having like connection with it; that each has switch tracks extending to their refineries from the main lines of its roads on which they are situate, by means of which shipments from them are made, the course of business in making shipments by defendant's road by the car-load (which is the manner in which nearly all the business is done) being for the defendant, on request of either, to furnish its cars, which are switched from its connecting track by the road on which the refineries are situate to the refineries, then loaded by the shippers, and by said road drawn out and placed on the defendant's tracks for shipment by its road. By some traffic arrangement between the roads a switching charge per car for such service is charged by the local road against the defendant, which is by it at its discretion charged against the shippers with its general freight charge. Upon shipments in less than car-load lots delivery is made to the defendant's freight depot.

11th. That the Standard Oil Company was then, and ever since has been, engaged in the same business at Cleveland and elsewhere, and did then and ever since has manufactured and shipped more than ninety one-hundredths of all the illuminating oil and products of petroleum manufactured and shipped at and from the City of Cleveland.

12th. The court further find that prior to 1875 it was a question whether the Standard Oil Company would remain in Cleveland or remove its works to the oil-producing country, and such question depended mainly upon rates of transportation from Cleveland to market; that prior thereto said Standard Company did ship large quantities of its products by water to Chicago and other lake points, and from thence distributed the same by rail to inland markets; that it then represented to defendant the probability of such removal; that water transportation was very low during the season of navigation; that unless some arrangement was made for rates at which it could ship the year round as an inducement, it would ship by water and store for winter distribution; that it owned its tank-cars and had tank-stations and switches or would have at Chicago, Toledo, Detroit, and Grand Rapids, on and into which the cars and oil in bulk could be delivered and unloaded without expense and annoyance to defendant; that it had switches at Cleveland leading to its works at which to load cars, and would load and unload all cars; that the quantity of oil to be shipped by the company was very large, and amounted to 90 per cent. or more of all the oil manufactured or shipped from Cleveland, and that if satisfactory rates could be agreed upon it would ship over defendant's road all its oil products for territory and markets west and northwest of Cleveland, and agree that the quantity for each year should be equal to the amount shipped the preceding year; that upon the faith of these representations the defendant did enter into the contract and arrangement substantially as set forth in defendant's answer; that the rates were not fixed rates, but depended upon the general card tariff rates as charged from time to time, but substantially to be carried from time to time for about ten cents per barrel less than tariff rates, and, in consideration of such reduced rates as to bulk oil, the Standard Company agreed to furnish its own cars and tanks, load them on switches at distributing points, and unload them into distributing tanks, and was also to load and unload oil shipped in barrels, and without expense to defendant, and with, by reason thereof, less risk to defendant, which entered into the consideration, and was also to ship all its freight to points west and northwest of Cleveland, except small quantities, to lake ports not reached by rail, and to so manage the shipments, as to cars and times, as would be most favourable to defendant; that defendant then agreed to said terms; that said agreement so made in 1875 has remained in force ever since.

13th. That at a cost exceeding $100,000 said Standard Company had and constructed the terminal facilities promised and herein found; that, in fact, the risk of danger from fire to defendant, the expense of handling, in loading and unloading, and in the use of the standard tank-cars is less (but how much the testimony does not show) than upon oil shipped without the use of such or similar terminal facilities; that said Standard Company commenced by shipping about 450,000 barrels a year over defendant's road, which increased from year to year until, in 1882, the year before the filing the petition in this action, the quantity so shipped on defendant's road amounted to 742,000 barrels, equal to 2,000 barrels or one full train-load per day.

14th. That said arrangement was not exclusive, but was at all times open to others shipping a like quantity and furnishing like service and facilities; that it was not made or continued with any intention on the part of the defendant to injure the plaintiffs in any manner; that plaintiffs knew of an arrangement between defendant and Standard Oil Company years before January 1, 1880, and on or about July 20, 1876, contracted with the Standard Company to give it the control of the shipments of plaintiffs' oil and the plaintiffs the benefit, if any, of any arrangements then existing or that might thereafter exist with the Standard Oil Company upon shipment of oil, and which plaintiffs received until about January 1, 1880, when they ceased operating with the Standard Oil Company, and thereafter were charged and paid the regular tariff rates published by defendant and by it charged and collected from all the public except the Standard Oil Company under the arrangement aforesaid.

15th. That the testimony on behalf of the plaintiffs fails to show the quantity manufactured or shipped by them, and how much they could or would ship by defendant's road if the Standard Company were charged tariff rates, does not appear in the testimony, although the testimony does show that plaintiffs shipped many car-loads, but the court find that the Standard Company have shipped and do ship over defendant's road more than $\frac{90}{100}$ of all the oil manufactured at and shipped from Cleveland.

16th. The court further find that at the time of filing the petition, and at all times after November 29, 1882, the prices charged the Standard Company from Cleveland to Chicago was fifty cents per barrel on oil in barrels, and forty dollars for each tank-car; that at the time of filing the petition, and from and after May 19, 1883, the tariff rate between the points aforesaid was sixty cents per barrel, while from November 20, 1882, to May 19, 1883, the tariff was seventy cents per barrel; that prior to the dates aforesaid the tariff rates and rates to the Standard frequently changed, and the difference was frequently greater than after said dates; that sixty-one barrels constitute a car-load and eighty barrels are estimated to the tank, but that some tanks hold one hundred and some one hundred and twenty barrels, and that at no time were tariff rates made or published for tank-cars carried by defendant with refined oil except when furnished by said Standard Company.

17th. That after said May 19th, 1883, about the same difference of ten cents per barrel existed between tariff rates and the prices charged to the Standard Oil Company to the different points along the line and consignment points beyond the termini of defendant's road; that five barrels of oil make a ton, and that the prices charged the Standard after November, 1882, from Cleveland to Chicago, amounted to $\frac{70}{100}$ of one cent per ton, per mile, and tariff rates to $\frac{83}{100}$ of one cent per ton per mile; that the contract of arrangement made with defendant has been largely profitable to defendant; that during the season of water navigation the Standard Company could have shipped to said distributing points on vessels by the lakes and river barreled oil for a less sum than the rates charged to it by defendant—to plaintiffs and the public were reasonable rates in themselves.

18th. That the defendant from time to time published and still does publish and hold forth to the public a certain printed tariff of rates of charge for the shipment and delivery of all classes of freight, including the products of the plaintiffs' refinery, between Cleveland aforesaid and the various towns and cities upon its said line, branches, and connecting lines, and has refused and still does refuse to ship such products for the plaintiffs to any of such points named in its tariff or schedule except for the prices therein named; and that such schedule fixes the prices for oil shipment at so much per barrel to the public, irrespective of their being shipped in barrels by ordinary freight cars or in bulk by means of tank-cars.

19th. That the plaintiffs have since December, 1879, frequently applied to the defendant both for reduced rates upon such tariff rates and for like rates with those made to such Standard Oil Company, both upon their general shipments by the ordinary freight cars of the defendant and also upon shipments to be by them made in bulk by means of tank-cars owned by them, they proposing to load and unload the same at terminal points, and to assume all risks by fire or leakage; but that the defendant has and still does refuse to allow them by either course of shipment rates less than such tariff rates, the tariff charged and demanded upon such shipments in bulk being on the basis of eighty barrels allowed to be shipped by each tank-car.

20th. The defendant has received ever since the first day of December, 1879, and still does receive from said Standard Oil Company at Cleveland and ship for *him*, like products to those of the plaintiffs at rates much less than such schedule rates, and receives and ships for said Standard Oil Company oil for shipment in bulk to such points by means of tank-cars of said Standard Company at rates much less than said schedule rates and much less than the rates allowed to said company for the shipment of oil by barrels in ordinary freight cars, and that such reduced rates to said Standard Oil Company by means of such tank-cars are allowed both by the making to it a lower rate upon its shipments by the defendant's cars in barrels, and also by means of its being allowed to ship by means of its said tank-cars to their full capacity, running from 80 to 120 barrels each, and averaging over 100 barrels each, and the reduced rate being charged on a basis of 80 barrels per car. The defendant charged the plaintiffs the switching charge, and omitted to charge the same to the Standard Oil Company; that it was a further part of such understanding, that should the defendant give to other shippers like rates, said Standard Oil Company would as far as possible withdraw from it its shipments; and that for the purpose of effectually securing at least the greater part of said trade, the defendant, on the completion of the New York, Cleveland and St. Louis Railway, a competing line from Cleveland to the West, in the year 1883 entered into a traffic arrangement with it, giving to it a portion of the shipments of said Standard Oil Company west, on a condition of its uniting with it in the carrying out of such understanding as to reduced rates to said Standard Company, which arrangements still exist.

21st. That upon the shipment made by the defendant for said Standard Oil Company of such products the rates paid for shipment to points of delivery upon the defendant's connecting lines and beyond its line have been and are less for the ratable amount of carriage charged for the distance transported over its own line, than said schedule rates or than the lower rates charged to said Standard Oil Company for shipments to the terminal points at which said shipments went from said road to its connecting line; how much less the defendant has refused to state.

22nd. That the reduced rates charged to said Standard Oil Company upon its shipments are arrived at by charging upon such shipments full tariff rates, and afterward, in accordance with some prearranged method agreed on with said Standard Oil Company, refunding to it a portion of the freight so charged and collected, the amount refunded being known as a "drawback" or "rebate."

23rd. That the evidence does not establish the fact whether or not all the various advantages claimed as secured to defendant by its contract with the Standard Oil Company are the equivalent for the discrimination made to it in freights.

NUMBER 46 (See page 80)

LETTER OF EDWARD S. RAPALLO TO GENERAL PHINEAS PEASE, RECEIVER CLEVELAND AND MARIETTA RAILROAD COMPANY

[Proceedings in Relation to Trusts, House of Representatives, 1888. Report Number 3,112, pages 576–577.]

32 NASSAU STREET, NEW YORK, March 2, 1885.

GENERAL PHINEAS PEASE,

Receiver Cleveland and Marietta Railroad Company.

Dear Sir: My opinion is asked as to the legality of your making such an arrangement with the Standard Oil Company as set forth below.

The facts, as I understand them, are as follows:

The Standard Oil Company proposes to ship or control the shipping of a large amount of oil over your road, say a quantity sufficient to yield to you $3,000 freight per month. That company also owns the pipes through which oil is conveyed from the wells owned by individuals to your railroad, except those pipes leading from the wells of George Rice, which pipes are his own. The company has, or can acquire, facilities for storing all its oil until such time as it can lay pipes to Marietta, and thus deprive your company of the carriage of all its oil.

The amount of oil shipped by Mr. Rice is comparatively small, say a quantity sufficient to yield $300 per month for freight.

The Standard Oil Company threatens to store, and afterward pipe all oil under its control unless you make the following arrangements, viz.: You shall make a uniform rate of thirty-five cents per barrel for all persons excepting the Standard Oil Company; you shall charge them ten cents per barrel for oil and also pay them twenty-five cents per barrel out of the thirty-five cents collected from other shippers.

It may render the subject less difficult of consideration to determine, first, those acts which you cannot with propriety do as receiver.

You are by the decree vested with all the powers of receiver, according to the rules and practice of the court; are directed to continue the operations of the railroad and can safely make disbursements from such moneys as come into your hands for such purposes only as the decree directs, viz.: wages, interest, taxes, rents, freights, mileage on rolling stock, traffic balances and certain debts for supplies.

In my opinion this would not protect you in collecting freight from one shipper and paying it over to another.

All moneys received, therefore, from any person for freight over your road, must pass into your hands and there remain to be disbursed by proper authority. After an examination of your statute, however, I find no prohibition against your allowing a discount, or charging a rate less than a schedule rate to a shipper on account of the large amount shipped by him.

As you are acting, therefore, in the interest of the company, and endeavouring to increase its legitimate earnings as much as possible, I find nothing in the statutes to prevent your making a discrimination, especially where the circumstances are such that a large shipper declines to give your road his freight unless you allow him to ship at less than the schedule rates. Therefore, there is no legal objection to the making of an arrangement which in practical effect may be the same as that proposed, provided the objections pointed out above are obviated.

You may with propriety allow the Standard Oil Company to charge twenty-five cents per barrel for all oil transported through their pipes to your road, and I understand from Mr. Terry that it is practicable to so arrange the details that the company can, in effect, collect this direct, without its passing through your hands. You may agree to carry all such oil of the Standard Oil Company or of others delivered to your road through their pipes, at ten cents per barrel. You may also charge all other shippers thirty-five cents per barrel freight, even though they delivered oil to your road through their own pipes, and this I gather from your letter and from Mr. Terry would include Mr. Rice.

You are at liberty, also, to arrange for the payment of a freight by the Standard Oil Company calculated upon the following basis, viz.:

Such company to be charged an amount equal to ten cents per barrel, less an amount equivalent to twenty-five cents per barrel upon all oil shipped by Rice, the agreement between you and the company thus being that the charge to be paid by them is a certain sum ascertained by such a calculation. If it is impracticable so to arrange the business that the Standard Oil Company shall, in effect, collect the twenty-five cents per barrel from those persons using the company's pipes from the wells to the railroad without its passing into your hands, you may properly also deduct from the price to be paid by this company an amount equal to twenty-five cents per barrel upon the oil shipped by such persons provided your accounts, bills, vouchers, etc., are consistent with the real arrangement actually made, you will incur no personal responsibility by carrying out such an arrangement as I suggest. It is possible that by a proper application to the court, some person may prevent you in the future from permitting any discrimination. Even if Mr. Rice should compel you, subsequently, to refund to him the excess charged over the Standard Oil Company, the result would not be a loss to your road, taking into consideration the receipts from the Standard Oil Company, if I understand correctly the figures. There is no theory, however, in my opinion under the decisions of the courts, relating to this subject, upon which, for the purpose, an action could be successfully maintained in this instance.

Yours truly,

EDWARD S. RAPALLO.

NUMBER 47 (See page 84)

TESTIMONY OF F. G. CARREL, FREIGHT AGENT OF THE CLEVELAND AND MARIETTA RAILROAD COMPANY

[In the case of Parker Handy and John Paton, Trustees, *vs.* The Cleveland and Marietta Railroad Company *et al.*, Circuit Court of the United States, Southern District of Ohio, Eastern Division.]

Q. The auditor reports it (the $340) remitted on October 29, 1885. Please state by whom it was held from the first of May to that time.

A. We might as well go back of that, and I will make a clean sweep, so far as I am concerned. This overcharge of twenty-five cents was held by the Macksburg Pipe Line Company. Whether this was my fault or the fault of the general agent I am not able to say. I know no difference between Mr. Rice's oil and the Pipe Line Company's.

Q. The books of the company show from the 26th of March, 1885, until April 28, 1885, Mr. Rice shipped from Macksburg to Marietta 1,360 barrels; that upon these shipments $340, or twenty-five cents per barrel, were reported to the auditor of the Cleveland and Marietta Railway upon the 29th of October. Who sent the money—$340—to the railroad company, and who reported the amount of money to the auditor?

A. If I understand correctly, if it is the amount I think it is, that is the amount for overcharge. It came through my office.

Q. In whose hands had the $340 been from the time paid by Mr. Rice until it was sent by you to the bank at Cambridge?

A. I received check from Pipe Line.

Q. How soon did you send money to Cambridge after receiving check?

A. I think the next day.

Q. How did you come to get that check?

A. I don't understand.

Q. Did you go after it?

A. No, sir; it was sent to me by mail.

Q. Where was it mailed?

A. Oil City, I think.

Q. By whom was the check signed?

A. By the treasurer, J. R. Campbell, I think.

. . . .

Q. If I understand the arrangement during the month of April, 1885, you collected thirty-five cents per barrel for all oil shipped by George Rice, and paid ten cents to the receiver of the railroad company and twenty-five cents to the Macksburg Pipe Line?

A. Yes, sir; as long as Mr. Rice shipped.

Q. Afterwards the Macksburg Pipe Line Company sent the money thus paid to it to you, and you paid the money into the depository of the railroad company on the 29th of October, 1885?

A. Yes, sir.

NUMBER 48 (See page 84)

REPORT OF THE SPECIAL MASTER COMMISSIONER GEORGE K. NASH TO THE CIRCUIT COURT

[In the case of Parker Handy and John Paton, Trustees, *vs.* The Cleveland and Marietta Railroad Company *et al.*, Circuit Court of the United States, Southern District of Ohio, Eastern Division.]

To the Honoured the Circuit Court of the United States,
Southern District of Ohio, Eastern Division.

By an order of your court made on the 18th day of December, 1885, in the case of Parker Handy and John Paton, Trustees, vs. The Cleveland and Marietta Railroad Company *et al.*, I was appointed a special master commissioner to investigate and report to the court for its action what discriminations have been made in freights by Receiver Pease, or during his administration by those under him, and to this end I was authorised to summon and examine witnesses and to cause their testimony to be reduced to writing so far as in my discretion it might be necessary. I was also required to inquire fully and particularly into the facts and report to the court what discriminations had been made, under what arrangements and to what extent, and to report fully all the facts and show to what extent and under what circumstances discriminations have been made against shippers as well as in favour of shippers, and by whom such discriminations were authorised and by whom made. In compliance with this order I proceeded to examine the matters therein referred to, and in the course of such examination called the following-named persons as witnesses:

T. D. Dale, C. C. Pickering (auditor of the Cleveland and Marietta Railroad Company under Receiver Pease), F. G. Carrel, J. E. Terry, Daniel O'Day, George Rice, H. L. Wilgus, W. H. Slack, W. J. Cramm, George Best, Jr., and J. C. McCarty, whose evidence I caused to be reduced to writing by A. C. Armstrong, a stenographer, and is herewith submitted.

I find from the evidence that soon after General Pease was appointed receiver of the Cleveland and Marietta Railroad, an arrangement was entered into with Daniel O'Day and W. T. Scheide, by which it was agreed that the rate to be charged by Receiver Pease and his subordinates upon all crude oil shipped from Macksburg and vicinity upon the line of the Cleveland and Marietta Railroad Company to Marietta should be thirty-five cents per barrel; that the agent of the receiver at Marietta should also pay the agent of the parties represented by O'Day and Scheide; that his compensation was to be $85 per month, $60 of which was to be paid by Receiver Pease and $25 by the parties represented by O'Day and Scheide; that it was the duty of this joint agent (one F. G. Carrel) to collect from all shippers the sum of thirty-five cents per barrel, and to account to Receiver Pease for ten cents of this sum, and to the parties represented by O'Day and Scheide for the balance. This arrangement went into force on the 20th day of March, 1885, and continued in force until September, 1885, at which time one George Rice made complaint to your court that discriminations were being made by the receiver against oil shippers.

Negotiations for this arrangement were opened in the City of Toledo on the 8th day of February, 1885, at a meeting which was attended by Daniel O'Day, W. T. Scheide, A. G. Blair (acting general freight and passenger agent of the receiver of the Wheeling and Lake Erie Railroad Company), and J. E. Terry (general freight and passenger agent of Pease, the receiver of the Cleveland and Marietta Railroad Company). The agreement above referred to was substantially reached at this meeting. Mr. Terry reported the same to General Pease, receiver of the Cleveland and Marietta Railroad Company, who thereupon wrote a letter to his general counsel in New York, asking advice in regard thereto, which letter was transmitted to said counsel by J. E. Terry in person. E. S. Rapallo, an attorney in New York City, replied to the letter of General Pease, and a copy of his letter is now on file in your court and is a part of a report filed by General Pease in November, 1885. This arrangement seems to have been entered into with full knowledge of General Pease, the receiver, and after consultation with his counsel, and with the full knowledge of his general freight and passenger agent, J. E. Terry.

George Rice was the owner of certain oil wells in the Macksburg Oil Region and he also purchased some oil from the owners of certain other wells in the same district. The oil which he produced and also the oil which he purchased he was in the habit of transporting to his refinery at Marietta, Ohio, by means of the Cleveland and Marietta Railroad. Before the arrangements to which I have referred went into effect he had been charged upon the shipment made by him the sum of seventeen and one-half cents per barrel. After the 20th of March, 1885, he was charged thirty-five cents per barrel upon all oil shipped by him. Between the 20th of March and the 30th of April following, Mr. Rice shipped from Macksburg to Marietta over the Cleveland and Marietta Railroad, 1,360 barrels of oil. Upon this oil he was charged thirty-five cents per barrel, or the sum of $476. This money was collected by F. G. Carrel, the agent of the receiver and also the agent of the parties represented at Toledo by O'Day and Scheide. This money was divided according to the agreement, and $136 was sent by Carrel to the bank of the receiver at Cambridge, Ohio, and the remaining $340, or twenty-five cents for each barrel of oil shipped by Rice, was sent by Carrel to the oil parties who had their headquarters at Oil City, Pennsylvania. On or about the 29th of October, 1885, this $340 was returned to Mr. Carrel at Marietta, by a check from Oil City, which check was signed by one J. R. Campbell, treasurer. This money was sent by Carrel to the bank in Cambridge in which the receiver made his deposits. It will be observed that this money was returned from Oil City some ten or twelve days after Judge Baxter made his order directing the receiver to make a report showing what discriminations, if any, had been made by him in the shipments of oil, which order had been obtained upon the complaint of George Rice. It was also returned after a consultation had by J. E. Terry with Daniel O'Day in the City of Cleveland. Mr. Terry states that the receiver was made acquainted with the steps taken by him in connection with this transaction. The receiver did not submit himself to an examination in regard to this matter, but filed an affidavit with me which I attach to this report, in which he states in substance that he did not know at the time he filed his reports with your court that that part of the agreement between himself and the oil parties which required that twenty-five cents per barrel of the moneys collected by him should be paid to the oil parties had been carried out, or that the money thus paid by Rice, and by Carrel paid over to the oil parties, had been returned. The reason given by Receiver Pease and by Mr. Terry for entering into this agreement was that the parties represented by O'Day and Scheide were threatening to put down a pipe-line from Macksburg to Parkersburg, through which to transport the oil produced by them in this region to the latter city, and that if this threat was carried out, the Railroad Company would be prevented from carrying oil produced by them to Marietta. They further stated that in consideration of the arrangement to which I have referred, the parties represented by O'Day and Scheide agreed not to put down a pipe-line, but to ship their oil over the Cleveland and Marietta Railroad.

As soon as George Rice found that the rates on oil had been raised from seventeen and one-half to thirty-five cents per barrel, and that he could not get any better terms for his shipment from the railroad, he commenced to lay a pipe-line from his wells in the Macksburg field to Lowell, on the Muskingum River. This line was completed about the first of May, 1885, and from that time he transported all his oil through this pipe to Lowell, and thence shipped it to Marietta by boat on the Muskingum River. As soon as the parties represented by O'Day and Scheide ascertained that Rice was putting down a pipe-line, they proceeded also to lay a pipe-line from the Macksburg oil field to Parkersburg, in West Virginia. Since the completion of their pipe-line all the oil sent to Parkersburg and Marietta has been sent through this pipe-line. For several months they continued to ship some of their oil North over the Cleveland and Marietta Railroad to Cleveland, but during the last two months these shipments have ceased, and all the oils now produced by the parties represented by O'Day and Scheide are sent by them through their pipe-line to Parkersburg.

Mr. Rice, since the completion of his pipe-line, has shipped through it to Marietta more than forty-five thousand barrels of oil. The shipments by Mr. Rice might have been retained for the benefit of the railroad had the rate of seventeen and one-half cents per barrel been continued. It is probable that had not the arrangement which we have been considering been entered into, a line would have been put down by the parties represented by O'Day and Scheide, but without the arrangement the patronage of Mr. Rice could have been retained. The result of the arrangement seems to be that the railroad has lost the patronage not only of the parties represented by O'Day and Scheide, but also of Mr. Rice, and it is not to-day carrying a barrel of oil.

The Argand Oil Works and the Argand Refining Company, two corporations located at Marietta, Ohio, have made complaint that from the eighteenth day of February until the fourteenth day of October, 1885, they were shippers of oil from the Macksburg Oil Region, over the Cleveland and Marietta Railroad, and that they were discriminated against by the receiver and his agents. I conceived that the order of your court referring this subject to me was broad enough to cover the complaint made by these corporations and I accordingly called W. H. Slack, W. J. Cramm, C. C. Pickering, and F. G. Carrel as witnesses in regard to this complaint, and their testimony is herewith submitted, together with the account presented by these two corporations and the receipted bills taken by them in payment of freight. From the evidence of these witnesses it appears that these corporations, during the time covered by the complaint, were engaged in refining oil at Marietta, Ohio. They purchased their crude oil of the parties represented by O'Day and Scheide at Macksburg. Their purchases were made by ordering their oil when needed by telegraph from a man by the name of Seep, located at Oil City, Pennsylvania, and they were charged therefor the market price of oil at Oil City on the day when the telegraphic order was given. The oil was then shipped to them over the Cleveland and Marietta Railroad and a bill for freight presented to them in the form following: "The Argand Oil Works, Marietta, Ohio, To the Cleveland and Marietta Railroad Company, Dr."

In these bills they were charged for all oil shipped at the rate of thirty-five cents per barrel. This amount was paid by them to Carrel, the agent of the receiver, at Marietta, Ohio. Of this amount Carrel paid to the receiver ten cents, and to the parties represented by O'Day and Scheide, twenty-five cents. I am of the opinion that these parties were in the same position as George Rice, with the exception that Mr. Rice produced his oil from the ground and shipped it over the Cleveland and Marietta Railroad, and these parties bought their oil instead of producing it from the ground. I cannot see as this difference modifies in any way the discrimination made against them. They claim that from February 18, 1885, until October 14, 1885, they shipped $3{,}679\frac{6}{10}$ barrels of oil, for which they were charged $1,232.06 as freight, and that the discriminations against them amounted to $888.70. From their bill certain reduction should be made. All shipments made prior to March 20, 1885, should be excluded for the reason that the discriminating arrangement entered into between the receiver and the parties represented by O'Day and Scheide did not go into effect until the 20th of March, 1885. Two shipments, one made on the 7th of August, and the other made on the 21st of September, from Dexter City, should also be excluded for the reason that all oils shipped from Dexter City were charged for at the same rates as these complainants were taxed. After making these deductions, I find that under the contract complained of, the Argand Oil Works and the Argand Refining Company shipped from the 20th of March until the 14th of October, 2,695 barrels of oil; that they were required to pay upon these shipments the sum of $894.59, and that of this sum Carrel, the agent of the receiver at Marietta, paid to the receiver the sum of $245.44, and to the parties in Pennsylvania represented by O'Day and Scheide the sum of $649.15.

A complaint of a similar character is made by the Marietta Oil Works, a partnership engaged in the business of refining oils at Marietta, Ohio. Upon their complaint, I examined George C. Best, Jr., J. C. McCarty, W. H. Slack, C. C. Pickering, and F. G. Carrel as witnesses, and their evidence is submitted herewith in full, together with the account presented by this partnership and the receipted bills presented by the Cleveland and Marietta Railroad and paid by them. Their case in all respects seems to be precisely like that of the Argand Oil Works and the Argand Refining Company. They claim that from the 1st day of April until the 31st day of August, 1885, inclusive, they shipped 2,717 barrels of oil, for which they were charged as freight $950.95, and that they were discriminated against to the extent of $679.25. From their bill I think that there should be excluded two shipments from Dexter City, one made on the 12th day of June, and the other on the 18th day of June, for the reason that no discriminations were made in freights, by the receiver, of oils shipped from Dexter City. After taking into account these two shipments, I find that the Marietta Oil Works shipped from Macksburg and Elba on their account 2,547 barrels of oil; that the freights paid by them upon these shipments amounted to the sum of $891.45, and that out of this sum Carrel, the agent at Marietta, paid to the receiver the sum of $251.70, and to the parties represented by O'Day and Scheide the sum of $639.75.

I find that during the receivership of General Pease, no oils were shipped from Macksburg North over the Cleveland and Marietta Railroad except such as were shipped by the parties represented by Messrs. O'Day and Scheide.

I have purposely referred to the parties who entered into this arrangement with Receiver Pease and his freight agent, J. E. Terry, as "the parties represented by O'Day and Scheide," for the reason that I have not been able to ascertain who or what the parties are. It appears from the evidence that during the time that M. D. Woodford had control as manager of the Cleveland and Marietta Railroad, one W. J. Brundred and T. D. Dale conceived the idea of running pipes to all the wells in the Macksburg Oil Regions, and then by concentrating them together convey all the oils thus gathered through the main line to the Cleveland and Marietta Railroad and deposit it in tanks, and with this end in view entered into a contract in writing with said

Woodford, a copy of which contract is attached to the report of Receiver Pease, filed in your court in November, 1885. After this contract was entered into, they organised a corporation known as the Ohio Transit Company, with T. D. Dale as president and W. J. Brundred as vice-president, to which corporation this contract was assigned. This company continued in the business until January, 1885. Mr. Dale, the president, states that "We said we could not compete with the Standard Oil Company, and for that reason we sold out at a fair price." When asked to whom his company sold their property, Mr. Dale answered, "I don't know what company, but my recollection is that it might have been the National Transit Company." "It was done in their office. I don't know whether the bill of sale was made to Mr. O'Day or to Mr. Scheide." Mr. Dale further states that "Mr. O'Day was vice-president of the National Transit Company, and that Mr. Scheide was its general manager; it, however, is conjecture on my part." In another place Mr. Dale states that the gentleman managing the National Transit Company bought the property of the Ohio Transit Company, and gives as their names Daniel O'Day, W. T. Scheide, and J. R. Campbell. The corporation or partnership, or whatever it is which now manages the pipe-line system in Macksburg oil fields, and extending from there to Parkersburg, is known as the Macksburg Pipe Line. One Daniel O'Day, now having his headquarters at Macksburg, is the manager of this pipe-line. When O'Day was asked, "To whom does the Macksburg Pipe Line belong?" he answered, "I do not believe I can answer that; I do not know." When asked, "Who has general control of it?" he answered, "Mr. Scheide, Mr. O'Day, and J. R. Campbell." He stated that "Mr. Scheide lives in Titusville, Mr. Campbell at Oil City, and Mr. O'Day at Buffalo." He also stated that these gentlemen were officers of the National Transit Company and the United Pipe Line, a division of the National Transit Company; that Mr. O'Day is general manager of the National Transit Company, and when asked whether the Macksburg Pipe Line is also a branch of the same system, he answered, "Really, I am not well enough posted to know, but I presume it is." Daniel O'Day also stated that the National Transit Company is a corporation organised under the laws of New York, and that its principal office is located in New York City. He also stated that "its property is located throughout the state of New York and the state of Pennsylvania, and some in Ohio." The line located in Ohio he described as running from Parker's Landing, in Pennsylvania, to Cleveland. He also stated that the United Pipe Line is a division of the National Transit Company which runs from wells to railroad points or pumping stations, and that the wells to which he referred are located in Alleghany County, New York, and throughout a large portion of Pennsylvania. He also stated that the Macksburg Pipe Line controls, by lease and deed, sixty or seventy acres of land in this state of the line of the Cleveland and Marietta Railroad Company, and that the lease and deeds for this land are in the name of one Benjamin Brewster, of New York City, and that said Brewster is the vice-president of the National Transit Company.

When Mr. O'Day was asked, "What relation does the National Transit Company and the United Pipe Line Company sustain to the Standard Oil Company?" he answered, "I believe that people having stock in the National Transit Company or the United Pipe Line can hold stock, and do hold stock, in the Standard Oil Company, but I do not know what further relations they have."

. . . .

I have attempted to summarise in a very brief manner the evidence which has been taken by me under the order of your court, but in order to obtain a full understanding of the situation, it will perhaps be necessary to read all the evidence which is herewith submitted in full, in connection with the reports and exhibits filed by General Pease, in November, 1885.

Respectfully submitted,

(Signed) George K. Nash,

Special Master Commissioner.

NUMBER 49 (See page 120)

A STATEMENT FROM AN OIL-PRODUCER'S STAND-POINT FOR 1886

[Circular used in the campaign against the Billingsley Bill.]

Total production for the year, 25,145,088 barrels.

Average price per barrel, .71½.

The gross income from the entire Oil Regions, based on these figures, $17,978,237.

The cost of producing the above amount of oil was as follows:

Wells drilled, 3,525—at an average cost of $3,000 each	$10,575,000	
Cost of pumping and raising the oil to the surface and keeping rigs and wells in repair, estimated at .25 per barrel of production....	6,286,272	
Add estimated cost of royalty, one-eighth....	2,247,342	
Total expenditures....	$19,108,614	
Deduct total income of the entire Oil Regions	17,978,737	
Net loss to oil producers during the year		$1,129,877

If the estimated value of the one-eighth royalty be not added, then the value of five acres of land should be added to the cost of each well and the result would be practically the same.

The daily production January 1, 1886, was 59,603 barrels, valued at $750 per barrel....	$44,702,250
The daily production January 1, 1887, was 66,383 barrels, valued at $500 per barrel....	33,191,500
Showing a shrinkage in value of the producing territory for the year 1886 to be	$11,510,750

NOTE.—To make it more clear to the uninitiated, the foregoing means that producing territory was bought and sold in 1885 on the basis of $750 to each barrel of production, and in 1886 on the basis of $500. It is on this basis that the value of oil-producing territory is estimated. A well producing one barrel a day at the present time is valued at $500; one year ago it was worth $750.

The valuation of the stock of the Standard Oil Company at the present time is $150,000,000, or nearly five times as great as the entire Oil Region country valuation. The profits of the Standard Oil Company for the year 1886 were over $26,000,000.

Strangers may ask, Why is there no competition in pipage and storage of oil if the profits are so great? We answer, that with rebates, drawbacks, discrimination, and conspiracies the Standard Oil Company has been able to freeze out and suppress nearly every attempt at competition.

Does not the foregoing array of figures, showing as it does the terrible shrinkage which the property of the oil producers has sustained, amounting to nearly twenty-five per cent. in one year, demand such relief in pipage, storage, and shrinkage, as is contemplated by the Billingsley Bill, now before the Senate of Pennsylvania?

NUMBER 50 (See page 121)

THE BILLINGSLEY BILL

[Legislature of Pennsylvania. File of the House of Representatives. Number 104, session of 1887.]

An act to punish corporations, companies, firms, associations and persons and each of them engaged in business of transporting by pipe-lines or lines or storing petroleum in tank or tanks, under certain restrictions and penalties from charging in excess of certain fixed rates for receiving, transporting, storing, and delivering petroleum, and to regulate deductions for losses caused to petroleum in pipe-lines and storage tanks by lightning, fire, storm, or other unavoidable causes.

SEC. 1. Be it enacted by the Senate and House of Representatives of the Commonwealth of Pennsylvania in general assembly met, and it is hereby enacted by authority of the same: That no corporation, company, firm, association, person or persons who are now, or shall hereafter engage in the business of transporting or storing crude or refined petroleum by means of pipe-line or pipe-lines, or storage by tank or tanks, shall demand or receive any rate of charge in excess of ten cents per barrel, reckoning forty-two gallons for each barrel, for all services performed within this commonwealth in receiving petroleum from tank or tanks or other receptacle on the lease or farm at the place of its production and transporting and delivering the same, or petroleum of like kind and quantity in every essential particular in the division of such pipe-line within which the same shall have been received at any shipping point in said division which may be designated by the holder, owner, or purchaser of said petroleum, whether said petroleum is held by certificate, voucher, receipt, credit balance, accepted order or otherwise. And such corporation, company, firm, association, person or persons, and each of them are hereby required immediately upon this act becoming a law to erect and establish, if not already established, and maintain thereafter at least one shipping point within each pipe-line division within this commonwealth of sufficient dimensions, capacity and equipment to accommodate the entire trade within each such pipe-line division.

SEC. 2. No such corporation, company, firm, association, person or persons shall demand or receive from any person or persons, firms, association, company or corporation owning or holding a credit balance for petroleum in line or tank within this commonwealth, any rate of charge whatever for the tankage or storage of petroleum owned or so held by credit balance for the first thirty days from the date of said credit balance. And no corporation, company, firm, association, person or persons who are now engaged or shall hereafter engage in the business of transporting or storing crude or refined petroleum by means of pipe-line or pipe-lines, or storage tank or tanks, shall demand or receive, from any source whatever, for the tankage of crude or refined petroleum within this commonwealth any rate of charge in excess of one-sixtieth of one cent per barrel of forty-two gallons a day or fractional part thereof so long as said petroleum shall thereafter be held and stored in tank.

SEC. 3. Such corporation, company, firm, association, person or persons are hereby obliged and required, and it is hereby made the duty of such corporation, company, firm, association, person or persons, and each of them, to hold and store in tank any and all petroleum offered for storage or transportation, or any and all petroleum received and transported by them or either of them for the owner thereof; or for the person or persons holding certificate, voucher, receipt, credit balance or accepted order thereof, for a period of one year or for any shorter period than one year from the time when said petroleum was first received by such corporation, company, firm, association, person or persons for storage, if requested so to do by the owner thereof, or by the person or persons holding certificate, voucher, receipt, credit balance or accepted order therefor, at and for the rate of charge of one-sixtieth of one cent per barrel of forty-two gallons for each day, or fractional part thereof thereafter. Except that when said petroleum is held by credit balance, no rate of charge whatever shall be made or charged on said credit balance for the first thirty days from the date of said credit balance.

SEC. 4. Such corporation, company, firm, association, person or persons shall be allowed to make a deduction from the crude petroleum received, transported or stored, not to exceed one-half of one per cent. of said petroleum so received, transported or stored, on account of water, sediment, evaporation, waste, and the like. The deduction mentioned in this section shall be made when the petroleum is first run or transported by such corporation, company, firm, association, person or persons, from the tank or receptacle on the lease or farm where produced, and it is hereby declared to be unlawful for such corporation, company, firm, association, person or persons to make the reduction in this section provided for at any other time or place than as above provided.

SEC. 5. Any corporation, company, firm, association, officer or officers, agent or agents, person or persons, engaged in the business of transporting or storing crude or refined petroleum within this commonwealth by means of pipe-line or pipe-lines or storage tank or tanks shall, upon application of the owner of any well or wells, lay pipe or pipes to any well or wells on any lease or leases in any locality where there is any oil on any farm or farms in this commonwealth, and receive the oil therefrom and transport the same through their pipe-line or pipe-lines and store the same in their storage tank or tanks, in any division or in any place in any division designated by the owner or purchaser of said petroleum, and hold the same subject to the owner or purchaser at the rate or charge prescribed in the preceding sections.

SEC. 6. Such corporation, company, firm, association, person or persons shall be liable for all loss caused by lightning, fire, storm, or other unavoidable cause to the petroleum received, transported or stored by them, and in the event of any such loss the same shall be charged by said corporation, company, firm, association, person or persons, *pro rata*, upon and deducted from all petroleum in the custody of such corporation, company, firm, association, person or persons, at the date of such loss.

SEC. 7. Any corporation, company, firm, association, officer or officers, agent or agents thereof, person or persons engaged in the business of transporting or storing crude or refined petroleum within this commonwealth by means of pipe-line or pipe-lines or storage tank or tanks, who shall demand or receive any rate of charge in excess of ten cents per barrel, reckoning forty-two gallons for each barrel, for all services performed within this commonwealth for receiving petroleum from tank or tanks or other receptacle on the lease or farm at the place of its production and transporting and delivering the same or petroleum of like kind and quality in every essential particular in the division of the pipe-line within which the same shall have been received at the shipping points designated by the holder, owner or purchaser of said petroleum, or who shall fail or neglect to erect and establish immediately upon this act becoming a law—if not already established—and maintain thereafter at least one shipping point within each pipe-line division within this commonwealth of sufficient dimensions and capacity and properly equip the same to accommodate the entire trade within each such district, or who shall demand or receive for the storage of petroleum within this commonwealth any rate of charge in excess of one-sixtieth of one cent a barrel of forty-two gallons a day or a fractional part thereof so long as said petroleum shall thereafter be held and stored in tank, or who shall demand or receive from any person or persons, firm, association, company, or corporation owning or holding a credit balance for petroleum in line or tank within this commonwealth, any rate of charge whatsoever for the tankage or storage of petroleum so owned or held by credit balance for the first thirty days commencing from the date of said credit balance, or who shall refuse to hold and store in tank any and all petroleum received and transported by them or either of them for the owner thereof, or for the person or persons holding certificate, voucher, receipt, credit balance or accepted order therefor for the period of one year, or for any shorter period than one year from the time when said petroleum was first received, by such corporation, company, firm, association, person or persons for storage if requested so to do by the owner thereof, or by the person or persons holding certificate, voucher, receipt, credit balance or accepted order therefor, at and for the rate of charge of one-sixtieth of one cent per barrel of forty-two gallons for each day or fractional part thereof thereafter—but no rate of charge whatever shall be had or made for the first thirty days from date of credit balance when oil is held by credit balance—or who shall make any deduction on account of water, sediment, evaporation, waste, or the like, in excess of one-half of one per cent. of the petroleum received, transported, and stored, or who shall violate any or either of the provisions or requirements of any or either of the first sections of this act, shall be deemed guilty of a misdemeanour, and on conviction thereof shall be sentenced to pay a fine of not less than one thousand dollars nor more than two thousand dollars for the first offense, and for the second and any subsequent offenses to pay a fine of not less than two thousand dollars nor more than five thousand dollars, and to undergo an imprisonment of not less than sixty days and not exceeding one year, one-half of any such fine or fines to be paid to the prosecutor and the other one-half to be for the use of the county in which such offence or offences shall have been committed, and in addition to the penalties hereinbefore provided shall be liable in any action of debt to any person or persons, firm, company, association, or corporation thereby aggrieved for double the amount of the damage sustained by reason of the violation of any of the provisions of this act.

SEC. 8. No contract heretofore made or now existing for receiving, transporting, or storing petroleum within this commonwealth shall be in any manner impaired or affected by the provisions of this act.

SEC. 9. All acts and parts of acts inconsistent herewith are hereby repealed.

SEC. 10. This act shall take effect immediately upon its becoming a law.

NUMBER 51 (See page 130)

EXTRACTS FROM TESTIMONY OF H. H. ROGERS

[Report of Special Committee on Railroads, New York Assembly, 1879. Volume III, pages 2613–2618.]

Q. Was your firm's business sold out to the Standard Oil Company?

A. I would like to have the question explained.

Q. Was there a sale or transfer made of your business to the Standard Oil Company, by which practically the Standard Oil Company really controlled your business?

A. I will answer this much of the question, by saying that the Standard Oil Company does not practically control our business.

Q. Do they control the rates at which your business gets the transportation of oil?

A. That I don't know anything about; I don't know anything about the rates of transportation.

By the Chairman.

Q. Was not your firm taken in with the Standard Oil Company upon some agreed basis or arrangement, whether you regard it as a purchase or transfer or not?

A. We worked in harmony with the Standard Oil Company for a number of years.

Q. Upon an agreed basis of general business?

A. Our interest was in common, to a certain extent.

. . . .

Q. Has your firm any contract with the Standard Oil Company?

A. That I cannot answer.

Q. What member of your firm would be able to answer that?

A. I think Mr. Pratt would, if he were here.

Q. When was it that your firm began to work in harmony with the Standard Oil Company?

A. I cannot say exactly how long ago; seven or eight years ago we got up a refining association here; that was the first, and then we got up another, and we got up another, and we have always been trying to get into some relations with all the refiners, so that we might make some money out of the business.

Q. Had you difficulty before you entered into relations with the Standard Oil Company to make money out of the business?

A. The competition was always very sharp, and there was always some one that was willing to sell goods for less than they cost, and that made the market price for everything; we got up an association, and took in all the refiners until some of them went back on us, and that would break up the association; we tried that two or three times.

Q. Then finally you entered the Standard Oil arrangement?

A. Then we made an alliance or association with some of the refiners about here, and it was more successful.

Q. What are the refiners about here with whom that alliance was made, and are they or are they not all of them covered by the Standard Oil arrangement?

A. They would come in and then they would go out; there is no refiner that I know of, with one exception, about New York but what has been in the association.

Q. What are the refiners that are now in association of the Standard Oil?

A. The people that are working in harmony with us comprise about, I should think, 90 or 95 per cent. of the refiners.

Q. Now tell us their names, the leading ones.

A. Some of the leading ones? The Standard Oil Company; Charles Pratt and Company; the Sone and Fleming Manufacturing Company; Warden, Frew and Company of Philadelphia; the Standard Oil Company of Pittsburg; the Acme Oil Refining Company of Titusville; the Imperial Refining Company of Oil City; the Baltimore United Oil Company of Baltimore.

. . . .

Q. You said that substantially 95 per cent. of the refiners were in the Standard arrangement?

A. I said 90 to 95 per cent. I thought were in harmony.

Q. When you speak of their being in harmony with the Standard, what do you mean by that?

A. I mean just what harmony implies.

Q. Do you mean that they have an arrangement with the Standard?

A. If I am in harmony with my wife, I presume I am at peace with her, and am working with her.

Q. You are married to her, and you have a contract with her?

A. Yes, sir.

Q. Is that what you mean?

A. Well, some people live in harmony without being married.

Q. Without having a contract?

A. Yes; I have heard so.

Q. Now, which do you mean? Do you mean the people who are in the Standard arrangement, and are in harmony with it, are married to the Standard or in a state of freedom—celibacy?

A. Not necessarily, so long as they are happy.

Q. Is it the harmony that arises from a marriage contract?

A. Not necessarily, so long as they are happy.

Q. When you speak of their harmony, is it a relation of contract?

A. I mean by harmony that if you and I agree to go on Wall Street and buy a hundred shares of Erie at 33, and we agree to sell it out together at 40, that is harmony. I mean just the same that way—if I go into the Standard Oil office and conclude to buy some oil of them and agree on a fair price to sell it out at, that is harmony.

Q. Is that the harmony that you mean—that you gentlemen have agreed between each other the rate at which you will buy and the rate at which you will sell?

A. Well, not going too far into detail, I would say that the relations are very pleasant.

Q. But we want the detail; we want precisely what that harmony is, what it consists of, and what produces it.

A. Well, is it a railroad abuse, or is it an abuse to be in harmony with people?

Q. No; it is not abuse to be in harmony; there are some kinds of harmony that the law considers conspiracy.

A. Well, I have heard so.

By the Chairman.

Q. What we want to know is this: This Standard Oil Company in itself is, as we understand it, a large organisation, not very extensive, but is made so by contracts with various other organisations, that are not a part of it, by their written contract or verbal contract or understanding, or whatever you term it; we want to know whether that is not the fact, and if that is not what you refer to when you speak about working in harmony.

A. Mr. Chairman, I want to give you all the information that is necessary in this matter for your purposes, but it is a question in my mind whether it is a proper thing for me, even if there is no harm done by it, to divulge my business secrets.

Q. We do not ask you for your secrets; we simply ask you the general nature of this organisation.

A. I have explained it, I think, to you quite as fully as I can.

NUMBER 52 (See page 136)

THE TRUST AGREEMENT OF 1882

[Proceedings in Relation to Trusts, House of Representatives, 1888. Report Number 3,112, pages 307–313.]

This agreement, made and entered upon this second day of January, A.D. 1882, by and between all the persons who shall now or may hereafter execute the same as parties thereto:

Witnesseth: I. It is intended that the parties to this agreement shall embrace three classes, to wit:

1st. All the stockholders and members of the following corporations and limited partnerships, to wit:

Acme Oil Company, New York; Acme Oil Company, Pennsylvania; Atlantic Refining Company of Philadelphia; Bush and Company (limited); Camden Consolidated Oil Company; Elizabethport Acid Works; Imperial Refining Company (limited); Charles Pratt and Company; Paine, Abbett and Company; Standard Oil Company, Ohio; Standard Oil Company, Pittsburg; Smith's Ferry Oil Transportation Company; Solar Oil Company (limited); Sone and Fleming Manufacturing Company (limited).

Also, all the stockholders and members of such other corporations and limited partnerships as may hereafter join in this agreement, at the request of the trustees herein provided for.

2d. The following individuals, to wit:

W. C. Andrews, John D. Archbold, Lide K. Arter, J. A. Bostwick, Benjamin Brewster, D. Bushnell, Thomas C. Bushnell, J. N. Camden, Henry L. Davis, H. M. Flagler, Mrs. H. M. Flagler, John Huntington, H. A. Hutchins, Charles F. G. Heye, A. B. Jennings, Charles Lockhart, A. M. McGregor, William H. Macy, William H. Macy, Jr., estate of Josiah Macy, William H. Macy, Jr., executor, O. H. Payne, A. J. Pouch, John D. Rockefeller, William Rockefeller, Henry H. Rogers, W. P. Thompson, J. J. Vandergrift, William T. Wardwell, W. G. Warden, Joseph L. Warden, Warden, Frew and Company, Louise C. Wheaton, H. M. Hanna and George W. Chapin, D. M. Harkness, D. M. Harkness, trustee, S. V. Harkness, O. H. Payne, trustee; Charles

Pratt, Horace A. Pratt, C. M. Pratt, Julia H. York, George H. Vilas, M. R. Keith, trustees, George F. Chester.

Also, all such individuals as may hereafter join in the agreement at the request of the trustees herein provided for.

3d. A portion of the stockholders and members of the following corporations and limited partnerships, to wit:

American Lubricating Oil Company; Baltimore United Oil Company; Beacon Oil Company; Bush and Denslow Manufacturing Company; Central Refining Company of Pittsburg; Cheesborough Manufacturing Company; Chess, Carley Company; Consolidated Tank Line Company; Inland Oil Company; Keystone Refining Company; Maverick Oil Company; National Transit Company; Portland Kerosene Oil Company; Producers' Consolidated Land and Petroleum Company; Signal Oil Works (limited); Thompson and Bedford Company (limited); Devoe Manufacturing Company; Eclipse Lubricating Oil Company (limited); Empire Refining Company (limited); Franklin Pipe Company (limited); Galena Oil Works (limited); Galena Farm Oil Company (limited); Germania Mining Company; Vacuum Oil Company; H. C. Van Tine and Company (limited); Waters-Pierce Oil Company.

Also, stockholders and members (not being all thereof) of other corporations and limited partnerships who may hereafter join in this agreement at the request of the trustees herein provided for.

II. The parties hereto do covenant and agree to and with each other, each in consideration of the mutual covenants and agreements of the others, as follows:

1st. As soon as practicable a corporation shall be formed in each of the following states, under the laws thereof, to wit, Ohio, New York, Pennsylvania, New Jersey; provided, however, that instead of organising a new corporation any existing charter and organisation may be used for the purpose when it can advantageously be done.

2d. The purposes and powers of said corporations shall be to mine for, produce, manufacture, refine, and deal in petroleum and all its products, and all the materials used in such businesses, and transact other business collateral thereto. But other purposes and powers shall be embraced in the several charters such as shall seem expedient to the parties procuring the charter, or, if necessary to comply with the law, the powers aforesaid may be restricted and reduced.

3d. At any time hereafter, when it may seem advisable to the trustees herein provided for, similar corporations may be formed in other states and territories.

4th. Each of said corporations shall be known as the Standard Oil Company of (and here shall follow the name of the state or territory by virtue of the laws of which said corporation is organised).

5th. The capital stock of each of said corporations shall be fixed at such an amount as may seem necessary and advisable to the parties organising the same, in view of the purpose to be accomplished.

6th. The shares of stock of each of said corporations shall be issued only for money, property, or assets equal at a fair valuation to the par value of the stock delivered therefor.

7th. All of the property, real and personal, assets and business of each and all of the corporations and limited partnerships mentioned or embraced in class first, shall be transferred to and vested in the said several Standard Oil companies. All of the property, assets, and business in or of each particular state shall be transferred to and vested in the Standard Oil Company of that particular state, and in order to accomplish such purpose the directors and managers of each and all of the several corporations and limited partnerships mentioned in class first are hereby authorised and directed by the stockholders and members thereof (all of them being parties to this agreement) to sell, assign, transfer, convey, and make over, for the consideration hereinafter mentioned, to the Standard Oil Company or companies of the proper state or states, as soon as said corporations are organised and ready to receive the same, all the property, real and personal, assets and business of said corporations and limited partnerships. Correct schedules of such property, assets, and business shall accompany each transfer.

8th. The individuals embraced in class second of this agreement do, each for himself, agree for the consideration hereinafter mentioned to sell, assign, transfer, convey, and set over all the property, real and personal, assets and business mentioned and embraced in schedules accompanying such sale, and transfer to the Standard Oil Company or companies of the proper state or states, as soon as the said corporations are organised and ready to receive the same.

9th. The parties embraced in class third of this agreement do covenant and agree to assign and transfer all of the stock held by them in the corporations or limited partnerships herein named, to the trustees herein provided for, for the consideration and upon the terms hereinafter set forth. It is understood and agreed that the said trustees and their successors may hereafter take the assignment of stocks in the same or similar companies upon the terms herein provided, and that whenever and as often as all the stocks of any corporations or limited partnerships are vested in said trustees, the proper steps may then be taken to have all the moneys, property, real and personal, of such corporation or partnership assigned or conveyed to the Standard Oil Company, of the proper state, on the terms and in the mode herein set forth, in which event the trustees shall receive stocks of the Standard Oil companies, equal to the value of the money, property, and business assigned, to be held in place of the stocks of the company or companies assigning such property.

10th. The consideration for the transfer and conveyance of the money, property, and business aforesaid to each or any of the Standard Oil companies shall be stock of the respective Standard Oil Company to which said transfer or conveyance is made, equal at par value to the appraised value of the money, property, and business so transferred. Said stock shall be delivered to the trustees hereinafter provided for, and their successors, and no stock of any of said companies shall ever be issued except for money, property, or business, equal, at least, to the par value of the stock so issued, nor shall any stock be issued by any of said companies for any purpose, except to the trustees herein provided for, to be held subject to the trusts hereinafter specified. It is understood, however, that this provision is not intended to restrict the purchase, sale, and exchange of property by said Standard Oil companies as fully as they may be authorised to do by their respective charters; provided only that no stock be issued therefor except to said trustees.

11th. The consideration for any stocks delivered to said trustees, as above provided for, as well as for stocks delivered to said trustees by persons mentioned or included in class third of this agreement, shall be the delivery by said trustees, to the persons entitled thereto, of trust certificates hereinafter provided for, equal at par value to the par value of the stocks of the said several Standard Oil companies so received by said trustees and equal to the appraised value of the stocks of other companies or partnerships delivered to said trustees.

The said appraised value shall be determined in a manner agreed upon by the parties in interest and said trustees.

It is understood and agreed, however, that the said trustees may, with any trust funds in their hands, in addition to the mode above provided, purchase the bonds and stocks of other companies engaged in business similar or collateral to the business of said Standard Oil companies on such terms and in such mode as they may deem advisable, and shall hold the same for the benefit of the owners of said trust certificates, and may sell, assign, transfer, and pledge such bonds and stocks whenever they may deem it advantageous to said trust so to do.

III. The trusts upon which said stock shall be held, and the number, powers, and duties of said trustees shall be as follows:

1st. The number of trustees shall be nine.

2d. J. D. Rockefeller, O. H. Payne and William Rockefeller are hereby appointed trustees, to hold their office until the first Wednesday of April, A.D. 1885.

3d. J. A. Bostwick, H. M. Flagler and W. G. Warden are hereby appointed trustees, to hold their office until the first Wednesday of April, A.D. 1884.

4th. Charles Pratt, Benjamin Brewster and John Archbold are hereby appointed trustees, to hold their office until the first Wednesday of April, A.D. 1883.

5th. Elections for trustees to succeed those herein appointed shall be held annually, at which election a sufficient number of trustees shall be elected to fill all vacancies occurring either from expiration of the term of the office of trustee or from any other cause. All trustees shall be elected to hold their office for three years, except those elected to fill a vacancy arising from any cause except expiration of term, who shall be elected for the balance of the term of the trustee whose place they are elected to fill. Every trustee shall hold his office until his successor is elected.

6th. Trustees shall be elected by ballot by the owners of trust certificates or their proxies. At all meetings the owners of trust certificates, who may be registered as such on the books of the trustees, may vote in person or by proxy, and shall have one vote for each and every share of trust certificates standing in their names, but no such owner shall be entitled to vote upon any share which has not stood in his name thirty days prior to the day appointed for the election. The transfer books may be closed for thirty days immediately preceding the annual election. A majority of the shares represented at such election shall elect.

7th. The annual meeting of the owners of said trust certificates for the election of trustees, and for other business, shall be held at the office of the trustees in the City of New York, on the first Wednesday of April of each year, unless the place of meeting be changed by the trustees, and said meeting may be adjourned from day to day until its business is completed. Special meetings of the owners of said trust certificates may be called by a majority of the trustees, at such times and places as they may appoint. It shall also be the duty of the trustees to call a special meeting of holders of trust certificates whenever requested to do so by a petition signed by the holders of ten per cent. in value of such certificates. The business of such special meetings shall be confined to the object specified in the notice given therefor. Notice of the time and place of all meetings of the owners of trust certificates shall be given by personal notice so far as possible, and by public notice in one of the principal newspapers of each state in which a Standard Oil Company exists, at least ten days before such meeting. At any meeting, a majority in value of the holders of trust certificates represented consenting thereto, by-laws may be made, amended, and repealed relative to the mode of the election of trustees, and other business of the holders of trust certificates; provided, however, that said by-laws shall be in conformity with this agreement. By-laws may also be made, amended, and repealed at any meeting, by and with the consent of a majority in value of the holders of trust certificates, which alter this agreement relative to the number, powers, and duties of the trustees, and to other matters tending to the more efficient accomplishment of the objects for which the trust is created; provided only, that the essential intents and purposes of this agreement be not thereby changed.

8th. Whenever a vacancy occurs in the board of trustees, more than sixty days prior to the annual meeting for the election of trustees, it shall be the duty of the remaining trustees to call a meeting of the owners of Standard Oil Trust certificates for the purpose of electing a trustee or trustees to fill the vacancy or vacancies. If any vacancy occurs in the board of trustees, from any cause, within sixty days of the date of the annual meeting for the election of trustees, the vacancy may be filled by a majority of the remaining trustees, or, at their option, may remain vacant until the annual election.

9th. If for any reason at any time a trustee or trustees shall be appointed by any court to fill any vacancy or vacancies in said board of trustees, the trustee or trustees so appointed shall hold his or their respective office or offices only until a successor or successors shall be elected in the manner above provided for.

10th. Whenever any change shall occur in the board of trustees, the legal title to the stock and other property held in trust shall pass to and vest in the successors of said trustees without any formal transfer thereof. But if at any such time formal transfer shall be deemed necessary or advisable, it shall be the duty of the board of trustees to obtain the same, and it shall be the duty of any retiring trustee, or the administrator or executor of any deceased trustee, to make said transfer.

11th. The trustees shall prepare certificates which shall show the interest of each beneficiary in said trust and deliver them to the persons properly entitled thereto. They shall be divided into shares of the par value of $100 each, and shall be known as the Standard Oil Trust certificates, and shall be issued subject to all the terms and conditions of this agreement. The trustees shall have power to agree upon and direct the form and contents of said certificates and the mode in which they shall be signed, attested, and transferred. The certificates shall contain an express stipulation that the holders thereof shall be bound by the terms of this agreement and by the by-laws herein provided for.

12th. No certificates shall be issued except for stocks and bonds held in trust as herein provided for, and the par value of certificates issued by said trustees shall be equal to the par value of the stocks of said Standard Oil Company and the appraised value of other bonds and stocks held in trust. The various bonds, stocks, and moneys held under said trust shall be held for all parties in interest jointly, and the trust certificates so issued shall be the evidence of the interest held by the several parties in this trust. No duplicate certificates shall be issued by the trustees, except upon surrender of the original certificate or certificates for cancellation, or upon satisfactory proof of the loss thereof, and in the latter case they shall require a sufficient bond of indemnity.

13th. The stocks of the various Standard Oil companies, held in trust by said trustees, shall not be sold, assigned, or transferred by said trustees, or by the beneficiaries, or by both combined, so long as this trust endures. The stocks and bonds of other corporations held by said trustees may be by them exchanged or sold and the proceeds thereof distributed *pro rata* to the holders of trust certificates, or said proceeds may be held and reinvested by said trustees for the purposes and uses of the trust; provided, however, that said trustees may, from time to time, assign such shares of stock of said Standard Oil Company as may be necessary to qualify any person or persons chosen or to be chosen as directors and officers of any of said Standard Oil companies.

14th. It shall be the duty of said trustees to receive and safely to keep all interest and dividends declared and paid upon any of the said bonds, stocks, and moneys held by them in trust, and to distribute all moneys received from such sources or from sales of trust property or otherwise by declaring and paying dividends upon the Standard Trust certificates as funds accumulate which in their judgment are not needed for the use and expenses of said trust. The trustees shall, however, keep separate accounts of receipts from interest and dividends, and of receipts from sales or transfers of trust property, and in making any distribution of trust funds, in which moneys derived from sales or transfers shall be included, shall render the holders of trust certificates a statement showing what amount of the fund distributed has been derived from such sales or transfers. The said trustees may be also authorised and empowered by a vote of a majority in value of holders of trust certificates, whenever stocks or bonds have accumulated in their hands from moneys purchases thereof, or the stocks or bonds held by them have increased in value, or stock dividends shall have been declared by any of the companies whose stocks are held by said trustees, or whenever, from any such cause, it is deemed advisable so to do, to increase the amount of trust certificates to the extent of such increase or accumulation of values and to divide the same among the persons then owning trust certificates *pro rata*.

15th. It shall be the duty of said trustees to exercise general supervision over the affairs of said several Standard Oil companies, and, as far as practicable, over the other companies or partnerships, any portion of whose stock is held in said trust. It shall be their duty, as stockholders of said companies, to elect as directors and officers thereof faithful and competent men. They may elect themselves to such positions when they see fit so to do, and shall endeavour to have the affairs of all of said companies managed and directed in the manner they may deem most conducive to the best interests of the holders of said trust certificates.

16th. All the powers of the trustees may be exercised by a majority of their number. They may appoint from their own number an executive and other committees. A majority of each committee shall exercise all the powers which the trustees may confer upon such committee.

17th. The trustees may employ and pay all such agents and attorneys as they deem necessary in the management of said trust.

18th. Each trustee shall be entitled to a salary for his services not exceeding $25,000 per annum, except the president of the board, who may be voted a salary not exceeding $30,000 per annum, which salaries shall be fixed by said board of trustees. All salaries and expenses connected with or growing out of the trust shall be paid by the trustees from the trust fund.

19th. The board of trustees shall have its principal office in the City of New York, unless changed by a vote of the trustees, at which office, or in some place of safe deposit in said city, the bonds and stocks shall be kept. The trustees shall have power to adopt rules and regulations pertaining to the meetings of the board, the election of officers, and the management of the trust.

20th. The trustees shall render at each annual meeting a statement of the affairs of the trust. If a termination of the trust be agreed upon, as hereinafter provided, or within a reasonable time prior to its termination by a lapse of time, the trustees shall furnish to the holders of trust certificates a true and perfect inventory and appraisement of all stocks and other property held in trust, and a statement of the financial affairs of the various companies whose stocks are held in trust.

21st. This trust shall continue during the lives of the survivors and survivor of the trustees in this agreement named, and for twenty-one years thereafter: provided, however, that if, at any time after the expiration of ten years, two-thirds of all the holders in value, or if, after the expiration of one year, ninety per cent. of all the holders in value of trust certificates, shall, at a meeting of holders of trust certificates called for that purpose, vote to terminate this trust at some time to be by them then and there fixed, the said trust shall terminate at the date so fixed. If the holders of trust certificates shall vote to terminate the trust as aforesaid, they may, at the same meeting, or at a subsequent meeting called for that purpose, decide by a vote of two-thirds in value of their number the mode in which the affairs of the trust shall be wound up, and whether the trust property shall be distributed, or whether it shall be sold and the values thereof distributed; or whether part, and, if so, what part, shall be divided and what part shall be sold, and whether such sales shall be public or private.

The trustees, who shall continue to hold their offices for that purpose, shall make the distribution in the mode directed; or, if no mode be agreed upon by two-thirds in value, as aforesaid, the trustees shall make distribution of the trust property according to law. But said distribution, however made, and whether it be of property or values, or of both, shall be just and equitable, and such as to insure to each owner of a trust certificate his due proportion of the trust property, or the value thereof.

22d. If the trust shall be terminated by expiration of the time for which it is created, the distribution of the trust property shall be directed and made in the mode above provided.

23d. This agreement, together with the registry of certificates, books of accounts, and other books and papers connected with the business of said trust, shall be safely kept at the principal office of said trustees.

BENJ. BREWSTER; JNO. D. ARCHBOLD; J. A. BOSTWICK; CHAS. PRATT; HENRY H. ROGERS; H. A. PRATT; C. M. PRATT; D. M. HARKNESS, *Trustee*, by H. M. FLAGLER, *Attorney;* THOMAS C. BUSHNELL; W. C. ANDREWS, CHAS. F. G. HEYE; WILLIAM T. WARDWELL; WM. H. MACY; Estate of JOSIAH MACY, JR., WM. H. MACY, JR., *Executor;* WM. H. MACY, JR.; A. M. MCGREGOR; J. N. CAMDEN, by H. M. FLAGLER, *Attorney;* O. H. PAYNE, by H. M. FLAGLER,

Attorney; GEO. F. CHESTER, *Trustee;* GEO. H. VILAS, *Trustee;* W. G. WARDEN; H. M. FLAGLER; JOHN D. ROCKEFELLER; WM. ROCKEFELLER; J. J. VANDERGRIFT; Mrs. H. M. FLAGLER, by H. M. FLAGLER; A. J. POUCH; O. B. JENNINGS; D. M. HARKNESS, by H. M. FLAGLER, *Attorney;* W. P. THOMPSON, by H. M. FLAGLER, *Attorney;* S. V. HARKNESS, by H. M. FLAGLER, *Attorney;* JOHN HUNTINGTON, by H. M. FLAGLER, *Attorney;* LIDE K. ARTER, by H. M. FLAGLER, *Attorney;* H. M. HANNA and GEO. W. CHAPIN, by H. M. FLAGLER, *Attorney;* LOUISE C. WHEATON, by H. M. FLAGLER, *Attorney;* O. H. PAYNE, *Trustee,* by H. M. FLAGLER, *Attorney;* CHAS. LOCKHART; JOS. L. WARDEN, by HENRY L. DAVIS, *Attorney;* JULIA H. YORK, by H. M. FLAGLER, *Attorney;* H. A. HUTCHINS, by H. M. FLAGLER, *Attorney;* M. R. KEITH, *Trustee;* D. BUSHNELL; WARDEN, FREW and COMPANY; HENRY L. DAVIS.

Whereas, in and by an agreement dated January 2, 1882, and known as the Standard Trust agreement, the parties thereto did mutually covenant and agree *inter alia* as follows, to wit: That corporations to be known as Standard Oil companies of various states should be formed, and that all of the property, real and personal, assets, and business of each and all of the corporations and limited partnerships mentioned or embraced in class first of said agreement should be transferred to and vested in the said several Standard Oil companies; that all of the property, assets, and business in or of each particular state should be transferred to and vested in the Standard Oil company of that particular state, and the directors and managers of each and all of the several corporations and associations mentioned in class first were authorised and directed to sell, assign, transfer, and convey, and make over to the Standard Oil Company or companies of the proper state or states, as soon as said corporations were organised and ready to receive the same, all the property, real and personal, assets, and business of said corporations or associations; and

Whereas, it is not deemed expedient that all of the companies and associations mentioned should transfer their property to the said Standard Oil companies at the present time, and in case of some companies and associations it may never be deemed expedient that the said transfers should be made and said companies and associations go out of existence; and

Whereas, it is deemed advisable that a discretionary power should be vested in the trustees as to when such transfer or transfers should take place, if at all. Now, it is hereby mutually agreed between the parties to the said trust agreement, and as supplementary thereto, that the trustees named in the said agreement and their successors shall have the power and authority to decide what companies shall convey their said property as in said agreement contemplated, and when the said sales and transfers shall take place, if at all; and until said trustees shall so decide, each of said companies shall remain in existence and retain its property and business, and the trustees shall hold the stocks thereof in trust as in said agreement provided. In the exercise of said discretion, the trustees shall act by a majority of their number as provided in said trust agreement. All portions of said trust agreement relating to this subject shall be considered so changed as to be in harmony with this supplemental agreement.

In Witness Whereof, the said parties have subscribed this agreement, this fourth day of January, 1882.

BENJAMIN BREWSTER; JOHN D. ARCHBOLD; J. A. BOSTWICK; CHARLES PRATT; HENRY H. ROGERS; H. A. PRATT; C. M. PRATT; D. M. HARKNESS, *Trustee;* D. M. HARKNESS; T. C. BUSHNELL; W. C. ANDREWS; CHARLES F. G. HEYE; WILLIAM T. WARDWELL; WILLIAM H. MACY; Estate of JOSIAH MACY, JR., WILLIAM H. MACY, JR., *Executor;* WILLIAM H. MACY, JR.; A. M. MCGREGOR; J. N. CAMDEN; JULIA H. YORK, by B. H. Y.; O. H. PAYNE; GEORGE F. CHESTER, *Trustee;* M. R. KEITH, *Trustee;* H. M. FLAGLER; JOHN D. ROCKEFELLER; WILLIAM ROCKEFELLER; J. J. VANDERGRIFT; MRS. H. M. FLAGLER; by H. M. FLAGLER; A. J. POUCH; O. B. JENNINGS; W. O. THOMPSON; S. V. HARKNESS; JOHN HUNTINGTON; LIDE K. ARTER; H. M. HANNA; GEORGE W. CHAPIN, H. M. HANNA, *Attorney in Fact;* LOUISE C. WHEATON, by H. M. FLAGLER; O. H. PAYNE, *Trustee;* CHARLES LOCKHART; JOSEPH L. WARDEN; HENRY L. DAVIS; W. G. WARDEN; WARDEN, FREW and COMPANY; D. BUSHNELL; H. A. HUTCHINS; GEORGE H. VILAS, *Trustee.*

NUMBER 53 (See page 153)

LIST OF CONSTITUENT COMPANIES OF THE STANDARD OIL TRUST, WITH ASSETS AND CAPITALISATION IN 1892

[From History of Standard Oil Case in the Supreme Court of Ohio, 1897–1898. Part I, page 112.]

	ASSETS	CAPITALISATION
Anglo-American Oil Co., Limited	$6,913,639.49	$5,000,000
Atlantic Refining Co	8,631,376.67	5,000,000
Buckeye Pipe Line Co	7,941,038.15	10,000,000
Eureka Pipe Line Co.	1,547,055.16	5,000,000
Forest Oil Co	3,528,813.11	5,500,000
Indiana Pipe Line Co	2,014,053.91	1,000,000
National Transit Co	25,796,712.97	25,455,200
New York Transit Co	4,999,300.00	5,000,000
Northern Pipe Line Co	707,067.00	1,000,000
Northwestern Ohio Natural Gas Co	1,396,760.00	3,278,500
Ohio Oil Co	8,260,378.04	2,000,000
Solar Refining Co.	711,793.87	500,000
Southern Pipe Line Co	3,279,018.28	5,000,000
South Penn. Oil Co	3,021,654.87	2,500,000
Standard Oil Co., Indiana	1,038,518.61	1,000,000
Standard Oil Co., Kentucky	3,604,800.78	1,000,000
Standard Oil Co., New Jersey	14,983,943.30	10,000,000
Standard Oil Co., New York	16,772,186.29	7,000,000
Standard Oil Co., Ohio	3,426,014.72	3,500,000
Union Tank Line Co	3,057,187.41	3,500,000
	$121,631,312.63	
Capitalisation twenty corporations	102,233,700.00	
Excess of assets over capitalisation	$19,397,612.63	

NUMBER 54 (See page 154)

FORMS OF MR. ROCKEFELLER'S CERTIFICATE OF HOLDINGS IN THE STANDARD OIL TRUST, WITH ASSIGNMENT OF LEGAL TITLE WHICH TOOK ITS PLACE IN 1892

[From History of Standard Oil Case in the Supreme Court of Ohio, 1897–1898. Part II, pages 53–56.]

KNOW ALL MEN BY THESE PRESENTS

That we, John D. Rockefeller, Henry M. Flagler, William Rockefeller, John D. Archbold, Benjamin Brewster, Henry H. Rogers, Wesley H. Tilford, and O. B. Jennings, Trustees, for winding up the Standard Oil Trust, by W. H. Tilford, our Attorney in Fact, and John D. Rockefeller, of ...., do hereby constitute and appoint John Bensinger, of New York City, our true and lawful attorney for the purposes following, to wit:

Whereas, John D. Rockefeller has placed in the hands of said attorney assignment Number A 365 for $\frac{256,854}{972,500}$ of the amount of corporate shares held by said trustees on the first day of July, 1892, in each of the companies whose stocks were so held.

Now the said attorney is hereby authorised to secure from each of said companies transfer upon their corporate books of said stock and stock certificates for whole shares, and scrip for fractional shares thereof, and when the said certificates and scrip are received from all the companies referred to, the said attorney shall deliver the same to John D. Rockefeller, and the said assignment Number A 365 shall at the same time be delivered to the said trustees.

And the said attorney hereby agrees to obtain the said certificates and scrip and to deliver the same and the said assignment as above specified.

(Signed in print) JOHN D. ROCKEFELLER,
HENRY M. FLAGLER,
WILLIAM ROCKEFELLER,
JOHN D. ARCHBOLD,
BENJAMIN BREWSTER,
HENRY H. ROGERS,
O. B. JENNINGS,
WESLEY H. TILFORD.

(Signed in ink) W. H. TILFORD, *Attorney in Fact*,
JOHN D. ROCKEFELLER, *per* GEO. D. ROGERS,
JOHN BENSINGER.

Received from John Bensinger, Attorney aforesaid, stock certificates and scrip as follows, being in full satisfaction of Assignment Certificate No. A 365 aforesaid:

NAMES OF COMPANIES	SHARES	SCRIP
Anglo-American Oil Co., Limited	6867	465-9725
The Atlantic Refining Co	13205	8375-9725
The Buckeye Pipe Line Co	52823	4325-9725
The Eureka Pipe Line Co	13205	8375-9725
Forest Oil Co	14526	4350-9725
Indiana Pipe Line Co	5282	3350-9725
National Transit Co	134463	131316-9725
New York Transit Co	13205	8375-9725
Northern Pipe Line Co	2641	1675-9725
Northwestern Ohio Natural Gas Co	8659	80890-9725
The Ohio Oil Co	21129	3675-9725
The Solar Refining Co	1320	5700-9725
Southern Pipe Line Co	13205	8375-9725
South Penn. Oil Co.	6602	9056-9725
Standard Oil Co., Indiana	2641	1675-9725
Standard Oil Co., Kentucky	2641	1675-9725
Standard Oil Co., New Jersey	26411	7025-9725
Standard Oil Co., New York	18488	2000-9725
Standard Oil Co., Ohio	9244	1000-9725
Union Tank Line Co	9244	1000-9725

(Signed in ink) JOHN D. ROCKEFELLER,
Per GEO. D. ROGERS.

Received of John Bensinger, Attorney, Assignment Certificate, Number.....

(Signed in ink) JOHN D. ROCKEFELLER,
WILLIAM ROCKEFELLER,
BENJAMIN BREWSTER,
WESLEY H. TILFORD,
HENRY M. FLAGLER,
JOHN D. ARCHBOLD,
HENRY H. ROGERS,
O. B. JENNINGS.

By ...., *Attorney in Fact.*

11-3-92.

Number A 365. JOHN D. ROCKEFELLER.

Received from trustees to liquidate the Standard Oil Trust assignment of legal title to $\frac{256,854}{972,500}$ of the amount of corporate stocks held by them in each of the corporations whose stocks were so held on July 1, 1892, and I do hereby authorise and direct the said trustees, or the survivor or survivors of them, to receive from the respective companies and to pay over to me or my assigns the dividends upon the stocks so assigned, and actual transfer thereof is recorded upon the books of the respective corporations.

(Signed) JOHN D. ROCKEFELLER,
Per GEO. D. ROGERS.

There is pasted to this stub the original assignment of legal title for the transfer of Mr. Rockefeller's trust certificates into corporate stock of the respective companies. This has been returned and marked "cancelled" and attached to the original stub, and is as follows:

Number A 365.

STANDARD OIL TRUST COMPANY

Assignment of Legal Title to Stocks Heretofore Represented by 256,854 shares.

Whereas, John D. Rockefeller is the owner of the equitable title to $\frac{256,854}{972,500}$ of the amount of corporate stocks held by the trustees of the Standard Oil Trust in each of the several corporations whose stocks were held by said trust on the first day of July, A.D. 1892, which equitable ownership was represented by 256,854 shares of Standard Oil Trust surrendered for cancellation. Now, we, the trustees in whose names the legal title to said stock stands, do hereby assign and transfer to John D. Rockefeller and his assigns the legal title to the aforesaid amount of the said stocks and authorise the proper officers of the several corporations to transfer upon their books and to issue corporate certificates for the required amount of their respective capital stocks upon presentation and cancellation of this assignment. The several corporations will issue stock certificates for whole shares and scrip for fractions of shares and upon presentation of fractional share scrip sufficient for the purpose, certificates for whole shares will be issued. When transfer of stock upon the corporate books is desired by virtue of this assignment, it must be placed in the hands of an attorney in fact, both for the assignee and the undersigned trustees, and said attorney shall first obtain the proper certificates and scrip from all the several companies, and thereupon shall deliver the certificates to the trustees and the stock certificates and scrip to the party or parties entitled thereto.

(Signed in print) JOHN D. ROCKEFELLER,
WILLIAM ROCKEFELLER,
HENRY M. FLAGLER,
JOHN D. ARCHBOLD,
BENJAMIN BREWSTER,
HENRY H. ROGERS,
WESLEY H. TILFORD,
O. B. JENNINGS, *Trustees.*
(Signed in writing) H. M. FLAGLER, *Secretary.*
W. H. TILFORD, *Attorney in Fact.*

On the left-hand corner of this same certificate this indorsement appears:

Cancelled November 7, 1892. Transfer Number 4833. Certificate issued.

There appears on the back of this assignment of legal title the following:

For value received, I hereby assign the corporate stocks mentioned or referred to in the within assignment, and authorise their transfer upon the respective corporate books to myself or my heirs.

(Signed in writing) JOHN D. ROCKEFELLER.

NUMBER 55 (See page 160)

AGREEMENT OF 1887 BETWEEN THE STANDARD OIL COMPANY AND PRODUCERS

[Proceedings in Relation to Trusts, House of Representatives, 1888. Report Number 3,112, pages 69–70.]

Memorandum of agreement, made this first day of November, 1887, between the Standard Oil Company of New York and the following-named persons, partnerships, and corporations, producers of crude petroleum, Thomas W. Phillips and others, whose names will be found in the schedule hereto attached and made part of this agreement, as follows:

Whereas, there has accumulated in past years an excessive stock of crude petroleum, which is deteriorating in quality, and a portion of which each year becomes sediment, valueless for any purpose, and the carrying of which excessive stock requires the expenditure of vast sums annually; and

Whereas, in consequence of the existence of said stock the price of crude petroleum has for the past year been largely below the cost at which the same was produced; now, in order as far as possible to preserve the said stock from further waste, and to conserve the public interest and our own, this agreement *witnesseth:*

That the Standard Oil Company of New York will set apart at sixty-two cents per barrel, and hold for the use of the above-named producers and those who shall hereafter become parties to this agreement, as hereinafter provided, 5,000,000 barrels of merchantable crude petroleum, of forty-two gallons each, to be sold and disposed of in the manner hereinafter provided. The said 5,000,000 barrels of petroleum to be subject, until sold by the said producers, to the usual assessments, storage charges, and interest upon the same, as also interest on the price of said petroleum, at sixty-two cents per barrel; said assessments, charges, and interest to be added to the price aforesaid.

In consideration of which the above-named producers agree to limit their production of petroleum, that for the year next ensuing from this date, they or any number of them shall, for said year, collectively produce at least 17,500 barrels of crude petroleum less per day than they or any number of them collectively produced per day for the months of July and August, 1887, and that they will use every reasonable endeavour to control their production so that the same shall be in the aggregate 30,000 barrels less per day than it was during the said period of July and August, 1887.

If at the end of three months from the date hereof the said reduction of 17,500 barrels per day shall be attained, to be measured by taking the average production of the above-named producers for the months of December and January next, and comparing the same with their average production for the months of July and August, 1887, a statement of the same being hereto attached and made part of this agreement, then the said 5,000,000 barrels of petroleum shall be delivered as fast as the same shall be sold by, upon the order, and for the account of said producers through their executive committee appointed by agreement between themselves, and hereinafter named, to be paid for with interest and storage as delivered; that the profits aforesaid upon said 5,000,000 barrels of petroleum as sold, in accordance with the provisions of this agreement, shall, by said Standard Oil Company and said producers' executive committee, be deposited with the United States Trust Company in New York City, until the expiration of one year from the date hereof, in trust, in accordance with and subject to the provisions of this agreement; and in case the above-named producers or any number of them shall not have lessened their production 17,500 barrels per day for said year as aforesaid, then all of said profits upon said 5,000,000 barrels of petroleum shall belong and be paid to the Standard Oil Company of New York; and in case the said above-named producers or any number of them collectively shall have lessened their production 17,500 barrels per day for the said year as aforesaid, then the entire profits aforesaid upon the 5,000,000 barrels of petroleum shall be paid to said producers' executive committee, to be by it distributed in accordance with agreements between themselves to such of said producers as have fulfilled the terms of this agreement, and all agreements between themselves relating to such distributions.

The said producers are guaranteed by said Standard Oil Company of New York against loss within said year upon said 5,000,000 barrels of petroleum. The lessening of 17,500 barrels per day above provided shall embrace and include any reduction or lessening of production by producers who shall sign contracts not to use means to increase their production by drilling or otherwise.

Producers may become parties to this agreement within the year the contract is to operate by signing the agreement between producers authorising the executive committee to sign this contract on their behalf, and having their names added hereto as parties by said executive committee.

The following-named persons constitute the executive committee above referred to, to wit:

(Names omitted by consent of the chairman.)

NUMBER 56 (See page 187)

JOHN D. ARCHBOLD'S STATEMENT TO THE INDUSTRIAL COMMISSION CONCERNING THE STANDARD'S OPPOSITION TO THE BUILDING OF THE UNITED STATES PIPE LINE

[Report of the Industrial Commission, 1900. Volume I, page 529.]

Mr. Lee makes a statement regarding the difficulty of his pipe-line, the United States Pipe Line, in crossing railroads and securing right of way to the seaboard, and makes a general statement implying that we have instituted and carried out great obstruction to their progress. I want to make general denial of this statement. We have not at any time had any different relations with reference to any obstruction or effort at obstruction of their line than would attach to any competitor in a line of business engaging against another. With reference to the special features referred to by Mr. Lee, and which he attempts, by implication at any rate, to connect us with, in the crossing of the Delaware and Lackawanna Railroad in New Jersey, I want to say that the contention in that respect was entirely at the hands of the railroad, and not at our hands in any possible respect. They went there surreptitiously and endeavoured to force their way, on a Sunday, over a line where they had no right, either by private purchase or by public franchise. Having accomplished the crossing of the road in that surreptitious way, they stationed there an armed force to prevent the railroad company from asserting its rights and taking out their lines, and kept that force there for a long period. The railroad went about it in a peaceful way, in the courts, and the final result is that the decision is against the line, after the case has been carried up finally to the supreme court of the state, and they must, of course, remove their line. But any statement on Mr. Lee's part, or any other witness, that we had anything to do with that matter, or with reference to any of the difficulties interposed in their progress to the seaboard, is absolutely false.

By Mr. Phillips.

Q. Did your company own in fee simple the tract of ground, and was a roadway reserved by the landholder? Was that purchased by them?

A. It was not my case, and I am not conversant with the details regarding it. The fact that, after having been fought in the newspapers and in the courts for a term of years, seeking the sympathy of the judges as well as the public, the supreme court of the state has ruled against them, is the best evidence, I think, that the right was against them. I want to say with reference to our pipe lines, that we never endeavoured to cross any man's right of way without first seeing him about it.

Q. Still, did they not go through the railroad on their own ground, and was not this the final decision, that they had not the right to lay a pipe line where a man had reserved a right of way under the ground?

A. It was not only decided that they had no right there, but they were ordered to remove.

NUMBER 57 (See page 194)

TABLES OF YEARLY AVERAGE PRICES OF CRUDE AND REFINED

[All quotations up to 1899 are from the Oil City Derrick; all quotations for 1900–1903 are from the New York Commercial.]

TABLE OF YEARLY AVERAGE PRICE OF CRUDE

In the following table is presented the highest and lowest price of oil, the months in which these quotations occurred, and the general average for each year. The "average" as estimated is usually the mean price between the highest and lowest quotation of a given time. It is sufficiently accurate for general purposes of comparison. It would be an almost impossible task to determine a "true average" from the reports of the daily sales that are now on record. Previous to 1875 the quotations are given for points along Oil Creek, and they hardly represent what the producer actually realised for oil at the wells. From 1875 onward the trading in oil was placed on a more satisfactory basis by the general adoption of pipe-line certificates, and the exchange quotations show very closely the value of the oil at the wells. When the certificate was finally purchased by the refiner, it was subject to a uniform charge for pipage of the oil from the wells to the nearest shipping point.

YEAR	Highest Month	Price	Lowest Month	Price	Average	YEAR	Highest Month	Price	Lowest Month	Price	Average
1859	Sept.	$20.00	Dec.	$20.00	$20.00	1882	Nov.	$1.37	July	$0.49¼	$0.78½
1860	Jan.	20.00	Dec.	2.00	9.60	1883	June	1.24¾	Jan.	.83¼	1.05⅞
1861	Jan.	1.75	Dec.	.10	.52	1884	Jan.	1.15⅝	June	.51¼	.83⅝
1862	Dec.	2.50	Jan.	.10	1.05	1885	Oct.	1.12⅝	Jan.	.68	.88⅜
1863	Dec.	4.00	Jan.	2.00	3.15	1886	Jan.	.92¼	Aug.	.59¾	.71⅜
1864	July	14.00	Feb.	3.75	8.15	1887	Dec.	.90	July	.54	.66⅝
1865	Jan.	10.00	Aug.	4.00	6.59	1888	Mar.	1.00	June	.71⅜	.87
1866	Jan.	5.50	Dec.	1.35	3.75	1889	Nov.	1.12½	April	.79½	.94⅛
1867	Oct.	4.00	June	1.50	2.40	1890	Jan.	1.07⅝	Dec.	.60¾	.86⅝
1868	July	5.75	Jan.	1.70	3.62½	1891	Feb.	.81⅜	Aug.	.50	.66⅞
1869	Jan.	7.00	Dec.	4.25	5.60	1892	Jan.	.64⅛	Oct.	.50	.55½
1870	Jan.	4.90	Aug.	2.75	3.90	1893	Dec.	.80	Jan.	.52⅞	.64
1871	June	5.25	Jan.	3.25	4.40	1894	Dec.	.95¾	Jan.	.78½	.83¾
1872	Oct.	4.55	Dec.	2.67½	3.75	1895	April	2.60	Jan.	.95¼	1.35¼
1873	Jan.	2.75	Nov.	.82½	1.80	1896	Jan.	1.50	Dec.	.90	1.19
1874	Feb.	2.25	Nov.	.62½	1.15	1897	Mar.	.96	Oct.	.65	.78⅜
1875	Feb.	1.82½	Jan.	.75	1.24¾	1898	Dec.	1.19	Jan.	.65	.91⅛
1876	Dec.	4.23¾	Jan.	1.47½	2.57⅝	1899	Dec.	1.66	Feb.	1.13	1.29⅜
1877	Jan.	3.69⅜	June	1.53¾	2.39⅜	1900	Mar.	1.68	Nov.	1.07	1.35¼
1878	Feb.	1.87½	Sept.	.78¾	1.17⅛	1901	Nov.	1.30	June	1.05	1.21⅛
1879	Dec.	1.28¾	June	.63⅛	.85⅝	1902	Dec.	1.44½	Mar.	1.15	1.23
1880	June	1.24¾	April	.71¼	.94⅛	1903	Dec.	1.88	Mar.	1.50	1.58¾
1881	Sept.	1.01¼	July	.72½	.85¼						

TABLE OF YEARLY AND MONTHLY AVERAGE PRICE OF REFINED

In the following table is given the average monthly and yearly prices of refined oil per gallon, in barrels, in New York, from January, 1863, to December, 1903. During the years when a tax was levied on this article of domestic production the quotations do not include the tax:

	1863	1864	1865	1866	1867	1868	1869	1870	1871	1872
Jan	.40	.46⅝	.70	.57⅞	.31	.24¾	.34⅛	.31⅜	.24⅝	.22⅝
Feb	.38¼	.47⅛	.67¼	.48⅝	.28¼	.25	.36⅜	.29⅞	.25⅛	.21¾
March	.34¾	.49⅛	.58¾	.41⅞	.27½	.25¾	.32⅛	.27	.24⅛	.22⅝
April	.33¼	.54⅛	.52⅞	.40⅛	.27	.26¼	.32¼	.26½	.23¼	.21¾
May	.39½	.59½	.51⅛	.43	.26¾	.29⅝	.31½	.27½	.24⅝	.23⅜
June	.44½	.72	.51½	.41⅞	.24¾	.31⅜	.31	.27	.25¾	.23
July	.49	.86⅛	.52⅛	.39½	.30⅞	.34¼	.32¼	.26	.25¾	.22⅜
Aug	.53½	.84⅞	.52	.44⅜	.29¼	.33	.32½	.25	.24⅜	.22⅜
Sept	.58	.75	.58¼	.44⅝	.31¾	.31	.32¼	.26⅛	.24⅛	.24⅛
Oct	.52½	.63¾	.61¾	.40⅝	.34½	.30	.32⅞	.24⅝	.23¾	.26
Nov	.41½	.70	.62⅝	.35¾	.27½	.30⅞	.34	.23	.22⅜	.27
Dec	.46½	.72¾	.65¼	.31¼	.24¾	.32¼	.31⅛	.23	.23	.26
Yearly average	.44¾	.64¾	.58¾	.42½	.28⅜	.29⅛	.32¾	.26⅜	.24¼	.23⅝

	1873	1874	1875	1876	1877	1878	1879	1880	1881	1882
Jan	.22⅛	.13½	.12⅜	.14⅛	.24	.12⅛	9	7⅞	9¼	7
Feb	.19⅝	.15	.14	.14¼	.18⅝	.12¼	9⅜	7⅞	9¼	7⅜
March	.19	.14⅞	.15	.14½	.16	.11⅝	9¼	7¾	8½	7⅜
April	.20	.15⅝	.13⅞	.14	.15¾	.11⅜	9⅛	7⅝	7¾	7⅜
May	.19¾	.13⅞	.12¾	.14⅛	.14½	.11¼	8½	7⅝	8	7½
June	.19	.12⅞	.12⅝	.14¾	.13¾	.11¼	7½	9⅝	8⅛	7½
July	.18⅛	.12⅛	.11½	.16⅞	.13⅜	.10¾	6¾	9⅞	7⅞	6¾
Aug	.16½	.11¾	.11¼	.19⅞	.13⅝	.10⅞	6⅝	9	7¾	6⅞
Sept	.16½	.12⅛	.12¾	.26	.14½	.10¼	6⅞	10⅝	8	7½
Oct	.16¼	.11⅞	.14⅛	.26	.14⅝	9⅝	7½	12	7¾	8
Nov	.14⅛	.10¾	.13	.26¼	.13¼	9⅛	8	10½	7½	8¼
Dec	.13½	.11¼	.12¾	.29⅜	.13⅛	8⅝	8⅝	9½	7⅛	7⅝
Yearly average	.18¼	.13	.13	.19⅛	.15¾	.10¾	8⅛	9⅛	8	7⅜

	1883	1884	1885	1886	1887	1888	1889	1890	1891	1892
Jan.............	7¾	9⅜	7¾	7¾	6¾	7¾	7	7½	7.42	6.45
Feb.............	7⅞	9⅛	7¾	7⅝	6⅝	7¾	7⅛	7½	7.48	6.42
March..........	8	8½	8	7⅜	6⅝	7¾	7	7¼	7.31	6.32
April............	8¼	8⅝	7⅞	7⅜	6⅝	7⅜	6⅞	7⅛	7.18	6.10
May............	7⅞	8½	7¾	7¼	6¾	7½	6⅞	7¼	7.20	6.06
June............	8	8⅛	8	7⅛	6⅝	7⅛	6⅞	7⅛	7.13	6.00
July.............	7⅝	7⅞	8¼	7	6½	7¼	7¼	7⅛	7.02	6.00
Aug.............	7⅞	8	8⅜	6¾	6½	7⅝	7¼	7¼	6.70	6.08
Sept.............	8⅛	7⅞	8⅜	6⅝	6¾	7¾	7⅛	7⅜	6.42	6.10
Oct..............	8⅜	7⅞	8½	6¾	6¾	7⅝	7⅛	7½	6.45	6.03
Nov.............	8¾	7⅞	8½	6⅞	7	7¼	7½	7½	6.40	5.80
Dec.............	9⅛	7¾	8	6⅞	7¼	7¼	7½	7¼	6.44	5.45
Yearly average...	8⅛	8¼	8⅛	7⅛	6¾	7½	7⅛	7⅜	6.93	6.07

	1893	1894	1895	1896	1897	1898	1899	1900	1901	1902	1903
Jan.......	5.33	5.15	5.87	7.85	6.13	5.40	7.43	9.90	7.58	7.20	8.27
Feb......	5.30	5.15	6.00	7.35	6.26	5.48	7.40	9.90	7.81	7.20	8.20
March....	5.34	5.15	6.75	7.40	6.36	5.82	7.33	9.90	8.00	7.20	8.21
April	5.52	5.15	9.12	7.00	6.13	5.67	7.05	9.51	7.68	7.30	8.35
May......	5.20	5.15	8.20	6.75	6.23	6.00	7.01	8.98	7.04	7.40	8.47
June......	5.21	5.15	7.83	6.85	6.14	6.16	7.20	7.88	6.90	7.40	8.55
July......	5.15	5.15	7.65	6.55	5.87	6.27	7.61	7.90	7.15	7.40	8.55
Aug......	5.18	5.15	7.10	6.65	5.75	6.44	7.82	8.05	7.50	7.21	8.55
Sept......	5.15	5.15	7.10	6.85	5.74	6.60	8.63	7.98	7.50	7.20	8.55
Oct	5.15	5.15	7.10	6.90	5.55	7.21	9.00	7.48	7.65	7.26	9.01
Nov......	5.15	5.15	7.88	7.15	5.40	7.35	9.40	7.33	7.65	7.71	9.36
Dec......	5.15	5.61	7.77	6.35	5.40	7.40	9.85	7.28	7.43	8.12	9.45
Yearly average..	5.24	5.19	7.36	6.98	5.91	6.32	7.98	8.50	7.49	7.38	8.62

NOTE.—In the above tables the quotations down to 1890, inclusive, are noted in cents and fractional parts of a cent; from 1891 to 1903 the prices are given in cents and decimal parts of a cent, *i.e.*, 7.42 signifies seven and forty-two hundredths cents, and 9⅜ means nine and three eighths cents per gallon. The above are New York quotations in barrels; bulk oil is generally 2.50c. below these prices. Philadelphia and Baltimore quotations are five points below New York; for instance, if New York price was 5.75c., the Philadelphia and Baltimore price would be 5.70 c.

NUMBER 58 (See page 225)

JOHN D. ARCHBOLD'S STATEMENT ON THE PRICES THE STANDARD RECEIVES FOR REFINED OIL

[Report of the Industrial Commission, 1900. Volume I, pages 569–570.]

Q. Now, the general result then is this: By virtue of your greater power you are enabled to secure prices that on the whole could be considered steadily somewhat above competitive rates?

A. Well, I hope so. I think we have better merchandising facilities, better marketing facilities, better distributing facilities, and better talent than a competitor can have.

Q. I am not asking with reference to your power of making profits, but it is with reference to getting the prices from the consumer.

A. Prices are what make the profit. If we had a better average price, we could get a better profit.

Q. You think, generally speaking, that you get prices for oil slightly above competitive prices?

A. Well, I should think so; I could not answer—that is a very general question, and very difficult to answer. I could not answer that specifically. I hope that we do.

Q. Of course, in this investigation, we are seeing if we can get some general principles on which legislation might be based, and these questions are to bring out, if we can, the power that so great an organisation has in fixing prices. Would you say, then, that in the case of an organisation that controls perhaps eighty per cent. of the markets of the country, there is a monopolistic element that enters in which enables them to hold prices above the regular rate? Is there a monopolistic power that comes merely from the power of capital itself?

A. Undoubtedly, there is an ability, and when that ability, as I have said, is unwisely used, it is sure to bring its own defeat.

Q. If that ability goes to get an exorbitant price, of course it will invite competition, but when that ability is kept within modest limits, would you still say that it was in the power of such an organisation to get the benefit of the monopolistic power that comes merely from the power of capital itself?

A. Well, I should say that that would be a very restricted power, a very restricted limit. The competitors in this country are very active.

Q. What?

A. The competitors are very active; they are alert at all points with their small offerings in the hope to find just such a condition as you describe.

Q. Certainly.

A. But as I say, as business is and as it has been for many years, we could not have that ability to any considerable extent as merchants.

Q. If the ability were operative only to a slight extent, would it still be enough, do you think, to make a difference between what we may call a moderate dividend, say 6 or 7 per cent., and a pretty high dividend of between 15 and 20 per cent.?

A. Well, that involves so nice a question that I could hardly undertake to answer it; but generally as to the effect on the community, I should say——

Q. Generally on the prices in the United States?

A. I should say that the lessened cost incident to doing business in a large volume would more than compensate the consumer for any ability in getting higher prices.

Q. Then that leads to this point, whether the large capital does itself give an organisation the power to get a somewhat higher price than it could in the market provided the competitors were substantially equal in power?

A. Oh, it may be so, but that is a difficult question to answer.

NUMBER 59 (See page 254)

W. H. VANDERBILT'S CHARACTERISATION OF STANDARD OIL MEN

[Report of the Special Committee on Railroads, New York Assembly, 1879. Volume II, pages 1668–1669.]

Q. Can you attribute, or do you attribute, in your own mind, the fact of there being one refiner instead of fifty, now, to any other cause except the larger capital of the Standard Oil Company?

A. There are a great many causes; it is not from their capital alone that they have built up this business; there is no question about it but that these men—and if you come in contact with them I guess you will come to the same conclusion I have long ago—I think they are smarter fellows than I am, a good deal; they are very enterprising and smart men; never came in contact with any class of men as smart and able as they are in their business, and I think a great deal is to be attributed to that.

Q. Would that alone monopolise a business of that sort?

A. It would go a great way toward building it up; they never could have got in the position they are in now without a great deal of ability, and one man would hardly have been able to do it; it is a combination of men.

Q. Wasn't it a combination that embraced the smart men in the railways, as well as the smart men in the Standard Company?

A. I think these gentlemen from their shrewdness have been able to take advantage of the competition that existed between the railroads for their business, as it grew, and that they have availed themselves of that there is not a question of doubt.

Q. Don't you think they have also been able to make their affiliations with railroad companies and railroad officers?

A. I have not heard it charged that any railway official has any interest in any of their companies, only what I used to see in the papers some years ago, that I had an interest in it.

Q. Your interest in your railway is so large a one that nobody would conceive, as a matter of personal interest, that you would have an interest antagonistic to your road?

A. When they came to do business with us in any magnitude; that is the reason I disposed of my interest.

Q. And that is the only way you can account for the enormous monopoly that has thus grown up?

A. Yes; they are very shrewd men; I don't believe that by any legislative enactment or anything else through any of the states or all of the states, you can keep such men as them down; you can't do it; they will be on top all the time; you see if they are not.

Q. You think they get on top of the railways?

A. Yes; and on top of everybody that comes in contact with them; too smart for me.

NUMBER 60 (See page 259)

FACSIMILE OF ONE OF MR. KEMPER'S SHARES

[From History of Standard Oil Case in Supreme Court of Ohio, 1897–1898. Part II, page 271.]

No. S. 11

$\frac{509,104}{972,500}$ of one share.	Incorporated under the laws of the State of Pennsylvania.	Whole Shares $50 each.

NATIONAL TRANSIT COMPANY

This certifies that J. L. Kemper is the owner of Five Hundred Nine Thousand One Hundred and Four 972,500ths of one share of stock in the National Transit Company. The holder or assignee of this Scrip will be entitled to a Certificate of Stock, and to have his name entered on the corporate books as a stockholder, on presentation of sufficient fractional Scrip to entitle him to one full share.

Witness the corporate seal of said Company, attested by the signatures of its President and Treasurer at Philadelphia, Pa., this 20th day of February, 1896.

H. H. Rogers,
President.

Geo. W. Colton,
Treasurer.
[Seal]

[On the reverse side.]

For value received............hereby sell, assign, and transfer unto........... 972,500ths of one share of the Capital Stock represented by the within Certificate of Scrip, and do hereby irrevocably constitute and appoint............Attorney to transfer the said Scrip on the books of the within named company, with full power of substitution in the premises.

Dated,......

J. L. Kemper.

In the presence of Harwood R. Pool.

Notice.—The signatures to this assignment must correspond with the name as written upon the face of the certificate in every particular, without alteration or enlargement or any change whatever.

NUMBER 61

GENERAL BALANCE SHEET, STANDARD

[In the case of James Corrigan *vs.* John D. Rockefeller in

ASSETS			
	Plant	Other Assets	Total
Anglo-American Oil Co., Lim.	$6,111,436.75	$10,877,942.53	$16,989,379.28
Atlantic Refining Co	4,879,636.08	6,637,750.39	11,517,386.47
Buckeye Pipe Line Co	4,559,213.27	8,593,413.44	13,152,626.71
Eureka Pipe Line Co	1,489,533.37	5,050,615.30	6,540,148.67
Forest Oil Company	4,236,370.10	800,482.59	5,036,852.69
Indiana Pipe Line Co	992,426.01	2,222,381.90	3,214,807.91
National Transit Co	6,800,056.66	42,529,353.39	49,329,410.05
New York Transit Co	1,860,334.55	5,171,303.80	7,031,638.35
Northern Pipe Line Co	639,001.65	583,766.46	1,222,768.11
N. W. Ohio Nat. Gas. Co.	118,679.71	204,480.33	323,160.04
Ohio Oil Co., The	4,832,307.19	310,705.42	5,143,012.61
Solar Refining Co., The	537,797.54	1,323,374.92	1,861,172.46
Southern Pipe Line Co	1,527,175.80	2,074,374.05	3,601,549.85
South Penn Oil Co	11,300,603.72	1,735,979.54	13,036,583.26
Standard Oil Co., Indiana	3,105,001.95	4,918,025.18	8,023,027.13
" Kentucky	474,352.83	4,236,638.24	4,710,991.07
" New Jersey	5,469,277.44	13,864,446.39	19,333,723.83
" New York	4,957,545.26	56,822,284.95	61,779,830.21
" Ohio	1,166,013.90	2,752,274.01	3,918,287.91
Union Tank Line Co	2,615,594.64	340,563.75	2,956,158.39
Total Plant	$67,672,358.42		
Other Assets	....	$171,050,156.58	
Total Assets	....	....	$238,722,515.00
Less Actual Liabilities	....	....	....
Total Net Value	....	....	....
Capital Stock	....	....	....
Total Undivided Profits	....	....	....
Total Capital and Surplus	....	....	....
Other Assets S. O. Trust	....	....	....

(See page 266)

OIL INTERESTS, DECEMBER 31, 1896

the Court of Common Pleas, Cuyahoga County, Ohio, 1897.]

NOMINAL LIABILITIES				
Liabilities	Net Value	Capital Stock	Surplus or Impairment.	Net Value
$8,997,759.61	$7,991,619.67	$2,530,666.66	$5,460,953.01	
357,691.56	11,159,694.91	5,000,000.00	6,159,694.91	
302,998.58	12,849,628.13	10,000,000.00	2,849,628.13	
352,320.90	6,187,827.77	5,000,000.00	1,187,827.77	
198,645.38	4,838,207.31	5,500,000.00	661,792.69	
7,821.80	3,206,986.11	1,000,000.00	2,206,986.11	
23,296,866.66	26,032,543.39	25,455,200.00	577,343.39	
202,139.33	6,829,499.02	5,000,000.00	1,829,499.02	
44,161.69	1,178,606.42	1,000,000.00	178,606.42	
11,384.76	311,775.28	1,967,100.00	1,655,324.72	
326,923.43	4,816,089.18	2,000,000.00	2,816,089.18	
298,137.91	1,563,034.55	500,000.00	1,063,034.55	
66,929.31	3,534,620.54	5,000,000.00	1,465,379.46	
1,278,580.96	11,758,002.30	2,500,000.00	9,258,002.30	
3,372,518.91	4,650,508.22	1,000,000.00	3,650,508.22	
49,835.90	4,661,155.17	1,000,000.00	3,661,155.17	
2,396,607.81	16,937,116.02	10,000,000.00	6,937,116.02	
48,919,899.34	12,859,930.87	7,000,000.00	5,859,930.87	
1,013,373.13	2,904,914.78	3,500,000.00	595,085.22	
11,653.38	2,944,505.01	3,500,000.00	555,494.99	
$91,506,250.35				
....	$147,216,264.65			
....	....	$98,452,966.66		
....	....	....	$48,763,297.99	
....	....	....	....	$147,216,264.65
....	....	....	....	4,135.25
				$147,220,399.90

NUMBER 62 (See page 267)

AMENDED CERTIFICATE OF INCORPORATION OF THE STANDARD OIL COMPANY OF NEW JERSEY

Resolved, That it is advisable to alter the charter of this company to read as below stated, and that a meeting of the stockholders be called to meet at the principal office of the company in Bayonne, N. J., on the fourteenth day of June, 1899, at 11 A.M., to take action hereon, notice of such meeting to be signed by the president and secretary and given to each stockholder in person or mailed to his proper post-office address at least ten days previous to the time of meeting as provided by the by-law.

First.—The name of the corporation is STANDARD OIL COMPANY.

Second.—The location of the principal office in the State of New Jersey is at the company's refinery, in the City of Bayonne, County of Hudson. The name of the agent therein and in charge thereof, and upon whom process against this company may be served, is J. H. Alexander.

Third.—The objects for which this company is formed are: To do all kinds of mining, manufacturing, and trading business; transporting goods and merchandise by land or water in any manner; to buy, sell, lease, and improve lands; build houses, structures, vessels, cars, wharves, docks, and piers; to lay and operate pipe-lines; to erect and operate telegraph and telephone lines and lines for conducting electricity; to enter into and carry out contracts of every kind pertaining to its business; to acquire, use, sell, and grant licenses under patent rights; to purchase or otherwise acquire, hold, sell, assign and transfer shares of capital stock and bonds or other evidences of indebtedness of corporations, and to exercise all the privileges of ownership including voting upon the stocks so held; to carry on its business and have offices and agencies therefor in all parts of the world, and to hold, purchase, mortgage, and convey real estate and personal property outside the State of New Jersey.

Fourth.—The total authorised stock of the corporation is One Hundred and Ten Million Dollars, divided into One Million and One Hundred Thousand shares of the par value of One Hundred Dollars each. Of said stock the One Hundred Thousand shares now issued and existing shall be preferred stock, and the increase of One Million shares shall be common stock. Said preferred stock shall entitle the holder thereof to receive out of the net earnings a dividend of and not exceeding one and one-half per cent. quarterly before any dividend shall be paid on the common stock. Common stock may at the discretion of the company be issued in exchange for preferred stock, and all preferred stock so received by the company shall be cancelled. Common stock may also be issued in payment for such property as the company has authority to purchase. Holders of preferred and of common stocks shall have like voting power.

Fifth.—The names and post-office addresses of the incorporators and the number of shares subscribed for by each shall remain as set forth in the original certificate of incorporation.

Sixth.—The duration of the corporation shall be unlimited.

Seventh.—The corporation may use and apply its surplus earnings, or accumulated profits authorised by law to be reserved, to the purchase or acquisition of property, and to the purchase or acquisition of its own capital stock from time to time, to such extent and in such manner and upon such terms as its Board of Directors shall determine; and neither the property nor the capital stock so purchased or acquired, nor any of its capital stock taken in payment or satisfaction of any debt due to the corporation, shall be regarded as profits for the purpose of declaration or payment of dividends, unless otherwise determined by a majority of the Board of Directors, or a majority of the stockholders.

The corporation, in its by-laws, may prescribe the number necessary to constitute a quorum of the Board of Directors which may be less than a majority of the whole number.

The number of directors at any time may be increased or diminished by vote of the Board of Directors, and in case of any such increase the Board of Directors shall have power to elect such additional directors, to hold office until the next meeting of stockholders, or until their successors shall be elected.

The Board of Directors shall have power to make, alter, amend, and rescind the by-laws of the corporation, to fix the amount to be reserved as working capital, to authorise and to cause to be executed mortgages and liens upon the real and personal property of the corporation, and from time to time to sell, assign, transfer or otherwise dispose of any or all of the property of the corporation; but no such sale of all of the property shall be made except pursuant to the votes of at least two-thirds of the Board of Directors.

The Board of Directors, by resolution passed by a majority of the whole Board, may designate three or more directors to constitute an executive committee, which committee, to the extent provided in said resolution or in the by-laws of the corporation, shall have, and may exercise, the power of the Board of Directors in the management of the business and affairs of the corporation, and shall have power to authorise the seal of the corporation to be affixed to all papers which may require it.

The Board of Directors from time to time shall determine whether and to what extent, and at what times and places, and under what conditions and regulations, the accounts and books of the corporation, or any of them, shall be open to the inspection of the stockholders; and no stockholder shall have any right of inspecting any account or book or document of the corporation, except as conferred by statute or authorised by the Board of Directors, or by a resolution of the stockholders.

The Board of Directors shall have power to hold its meetings, to have one or more offices, and to keep the books of the corporation (except the stock and transfer books) outside of the state, at such places as may be from time to time designated by them.

I CERTIFY that the above resolution was adopted by the Board of Directors of the STANDARD OIL COMPANY, at a meeting held on the twenty-sixth day of May, A.D. 1899, a majority of directors being present and voting in favour thereof. Witness the seal of said corporation.

L. D. CLARKE,
Secretary.

NUMBER 63 (See page 270)

PRODUCTION OF PENNSYLVANIA AND LIMA CRUDE OIL BY STANDARD OIL COMPANY 1890–1898

(Expressed in barrels of forty-two gallons.)

[Report of Industrial Commission, 1900. Volume I, page 561.]

YEAR	PENNSYLVANIA OIL			LIMA OIL			GRAND TOTAL		
	Total production	Standard Oil Co. production	Standard Oil per cent. of total	Total production	Standard Oil Co. production	Standard Oil per cent. of total	Pennsylvania and Lima production	Standard Oil Co. production	Standard Oil per cent. of total
1890...	30,065,867	2,618,637	8.71	15,014,882	8,400,568	55.95	45,080,749	11,019,205	24.44
1891...	35,742,127	4,913,775	13.74	17,381,923	9,319,156	53.61	53,124,050	14,232,931	26.79
1892...	33,332,306	4,338,822	13.02	16,685,193	7,843,324	47.01	50,017,499	12,182,146	24.36
1893...	31,256,283	6,705,276	21.45	17,823,255	7,260,899	40.74	49,079,538	13,966,175	28.46
1894...	30,696,716	7,210,345	23.49	18,575,603	6,690,951	36.02	49,272,319	13,901,296	28.21
1895...	30,891,868	9,119,920	29.52	21,719,250	6,808,876	31.35	52,611,118	15,928,796	30.28
1896...	33,908,041	9,380,654	27.66	25,222,091	8,031,793	31.84	59,130,132	17,412,447	29.45
1897...	35,170,367	9,787,353	27.83	22,793,033	7,497,349	32.89	57,963,400	17,284,702	29.82
1898...	31,645,151	11,248,443	35.55	20,266,328	7,220,606	35.63	51,911,479	18,469,049	35.58
Total..	292,708,726	65,323,225	22.32	175,481,558	69,073,522	39.36	468,190,284	134,396,747	28.70

NUMBER 64 (See page 270)

BUSINESS OF STANDARD OIL COMPANY AND OTHER REFINERS
1894–1898

(Barrels of fifty gallons. All products, domestic trade.)

[Report of Industrial Commission, 1900. Volume I, page 560.]

YEAR	STANDARD OIL COMPANY		OTHERS		TOTAL
	Barrels	Per cent. of total	Barrels	Per cent. of total	Barrels
1894....	18,118,933	81.4	4,145,232	18.6	22,264,165
1895....	18,348,051	81.8	4,084,720	18.2	22,432,771
1896....	16,341,161	82.1	3,569,719	17.9	19,910,880
1897....	18,141,479	82.4	3,876,706	17.6	22,018,185
1898....	19,999,939	83.7	3,914,999	16.3	23,914,938
Total...	90,949,563	82.3	19,591,376	17.7	110,540,939

INDEX

INDEX

INDEX

M

N

O

S

W

COSIMO

LaVergne, TN USA
17 September 2010
197394LV00002BB/1/P